SENSOR FUSION - FOUNDATION AND APPLICATIONS

Edited by **Ciza Thomas**

Sensor Fusion - Foundation and Applications
Edited by Ciza Thomas

Published by InTech
Janeza Trdine 9, 51000 Rijeka, Croatia

Publishing Process Manager Silvia Vlase

Technical Editor Teodora Smiljanic

Cover Designer InTech Design Team

Additional hard copies can be obtained from orders@intechopen.com

Sensor Fusion - Foundation and Applications, Edited by Ciza Thomas
p. cm.
ISBN 978-953-307-446-7

For EU product safety concerns:
IN TECH d.o.o., Prolaz Marije Krucifikse Kozulić 3, 51000 Rijeka, Croatia,
info@intechopen.com or visit our website at intechopen.com.

Meet the editor

Prof. Ciza Thomas is currently working as Professor and Head, Electronics and Communication Department of College of Engineering, Trivandrum, India. She has publications in more than 40 International Journals and International Conference Proceedings. She has edited four books in the field of Sensor Fusion and Complex Systems and published six book chapters in the field of network security and pattern recognition. She is a reviewer of more than ten reputed International journals. She is a guest editor of the IEEE Security and Privacy Magazine. She is a recipient of achievement award in 2010 and the e-learning IT award in 2014 from Government of Kerala.

Contents

Preface

This book as its name suggests deals with the principles and applications of sensor fusion. Sensor fusion is an important technology, with a very fast growth due to its tremendous application potential in many areas. It is a method of integrating information from several different sources into a unified interpretation that extracts intelligible and more meaningful information. In many cases the source of information are sensors that allow for perception or measurement of changing environment.

Variety of techniques, architectures, levels, etc. of sensor fusion enables to bring solutions in various areas of diverse disciplines. Sensor fusion techniques can be applied to various applications mainly on the data, feature and the decision levels. The function at data level can be spectral data mining using the digital signal processing techniques, or the data adaptation using the coordinate transforms/ unit adjustments or the estimation of parameters using the Kalman filtering/ batch estimation. The function at the feature level is mainly classification using Pattern Recognition/ Fuzzy Logic/ Neural Networks. The function at the decision level is the decide action using Expert Systems/ Artificial Intelligence. This book contains chapters with different methods of sensor fusion for different engineering as well as non-engineering applications. Sufficient evidences and analyses have been provided in the chapters to show the effectiveness of sensor fusion in various applications.

This book provides some novel ideas, theories, and solutions related to the latest practices and research works in the field of sensor fusion. Advanced applications of sensor fusion in the areas of mobile robots, automatic vehicles, airborne threats, agriculture, medical field and intrusion detection are covered in this book.

This book will be of interest to researchers, who need to process and interpret the sensor data in most of the scientific and engineering field. The book provides some projections for the future of sensor fusion are provided along with an assessment of the state-of-the-art and state-of-practice. Hence, this book is intended to serve as a reference guide in the field of sensor fusion applications. This book will be useful to system architects, engineers, scientists, managers, designers, military operations personnel, and other users of sensor fusion for target detection, classification, identification, and tracking.

The chapters in this book provide the foundation on sensor fusion, introducing a particular sensor fusion application, process models, and identification of applicable techniques. The materials presented concentrate upon conceptual issues, problem formulation, computerized problem solution, and results interpretation in various applications of sensor fusion. Solution algorithms will be treated only to the extent necessary to interpret solutions and overview events that may occur during the solution process. A general background in electrical/ electronic engineering, mathematics, or statistics is necessary for a better understanding of the concepts presented in the individual chapters. The readers will benefit by enhancing their understanding of the sensor fusion principles, algorithms, and architectures along with the practical application of modern sensors and sensor fusion.

Acknowledgements

I worked as an undergraduate student in the area of network security under the supervision of Professor N. Balakrishnan, Associate Director, Indian Institute of Science, Bangalore, India. I acknowledge him for introducing me to the applications of sensor fusion.

Several people have made contributions to this book. Special thanks to all authors of the chapters for applying their knowledge in the field of sensor fusion in the real-world problems and also for their co-operation in the timely completion of this book. Ms. Silvia Vlase and all other InTech staff, took keen interest and ensured the publication of the book in good time. I thank them for their persistence and encouragement.

Ciza Thomas
Electronics and Communication Department,
College of Engineering, Trivandrum,
India

1

A Dynamic Context Reasoning based on Evidential Fusion Networks in Home-based Care

Hyun Lee[1], Jae Sung Choi[2] and Ramez Elmasri[3]
[1]*Daegu Gyeongbuk Institute of Science & Technology*
[2,3]*University of Texas at Arlington*
[1]*South Korea*
[2,3]*USA*

1. Introduction

During emergency situations of the patient in home-based care, a Pervasive Healthcare Monitoring System (PHMS) (Lee et al., 2008) is significantly overloaded with pieces of information of different known reliability or unknown reliability. The pieces of the information should be processed, interpreted, and combined to recognize the situation of the patient as accurate as possible. In such a context, the information obtained from different sources such as multi-sensors and Radio Frequency Identification (RFID) devices can be imperfect due to the imperfection of the information itself or unreliability of the sources. In order to deal with different aspects of the imperfection of contextual information, we proposed an evidential fusion network based on Dezert-Smarandache Theory (DSmT) (Dezert & Smarandache, 2009) as a mathematical tool in (Lee et al., 2009). However, context reasoning over time is a difficult in an emergency context, because unpredictable temporal changes in sensory information may happen (Rogova & Nimier, 2004). The (Lee et al., 2009) did not consider dynamic metrics of the context. In addition, some types of contextual information are more important than others. A high respiratory rate may be a strong indication of the emergency of the patient others may not be so important to estimate that specific situation (Padovitz et al., 2005; Wu et al., 2003). The weight of this information may change, due to the aggregation of the evidence and the variation of the value of the evidence over time. For instance, a respiratory rate (e.g., 50 Hz) at current time-indexed state (S_t) should have more weight compared to a respiratory rate (e.g., 21 Hz) at previous time-indexed state (S_{t-1}), because 50 Hz indicates the emergency situation of the patient strongly (Campos et al., 2009; Danninger & Stierelhagen, 2008).

Thus, we propose a Dynamic Evidential Network (DEN) as a context reasoning method to estimate or infer future contextual information autonomously. The DEN deals with the relations between two consecutive time-indexed states of the information by considering dynamic metrics: temporal consistency and relation-dependency of the information using the Temporal Belief Filtering (TBF) algorithm. In particular, we deal with both relative and individual importance of evidence to obtain optimal weights of evidence. By using the proposed dynamic normalized weighting technique (Valiris et al., 2005), we fuse both intrinsic and optional context attributes. We then apply dynamic weights into the DEN in order to infer the situation of the patient based on temporal and relation dependency. Finally, we compare the proposed fusion process with a fusion process based on Dempster-Shafer Theory (DST) (Wu et al., 2003) and Dynamic Bayesian Networks (DBNs) (Murphy, 2002) that has the same assumption of the environments, so as to show the improvement of our proposed method in an emergency situation of the patient. The main contributions of the proposed context reasoning method under uncertainty based on evidential fusion networks are: 1) Reducing the conflicting mass in uncertainty level and improving the confidence level by adapting the DSmT, 2) Distinguishing the sensor reading error from new sensor activations or deactivations by considering the TBF algorithm, and 3) Representing optimal weights of the evidence by applying the normalized weighting technique into related context attributes. These advantages help to make correct decisions about the situation of the patient in home-based care.

The rest of the chapter is organized as follows. Basics of context reasoning are introduced in section 2. In section 3, we introduce a dynamic context reasoning method based on evidential fusion network. In section 4, we perform a case study so as to distinguish the proposed fusion process with traditional fusion processes. We compare and analyze the results of our approach with those of DST and DBNs to show the improvement of our approach in section 5. We introduce some related works in section 6. We then conclude this work in section 7.

2. Basics of context reasoning

2.1 Characteristics of the evidence

Multi-sensors such as medical body sensors, Radio Frequency Identification (RFID) devices, environmental sensors and actuators, location sensors, and time stamps are utilized in a PHMS (Lee et al., 2008). These sensors are operated by pre-defined rules or learning processes of the expert systems. They often have thresholds to represent the emergency status of the patient or to operate actuators. Each sensor can be represented by an evidential form such as 1 (active) and 0 (inactive) based on the threshold. Whenever the state of a certain context associated with a sensor is changed, the value of a sensor can change from 0 to 1 or from 1 to 0. For instance, a medical body sensor activates the emergency signal if the sensor value is over the pre-defined threshold. An environmental sensor operates the actuator based on the fuzzy systems. A location detecting sensor operates if a patient is within the range of the detection area. Thus, we can simply express the status of each sensor as a frame: $\Theta = \{Threshold_{over}, Threshold_{not-over}\} = \{1,0\}$.

Sensor data are inherently unreliable or uncertain due to technical factors and environmental noise. Different types of a sensor may have various discounting factors (D) ($0 \le D \le 1$). Hence we can express the degree of reliability, which is related in an inverse way to the discounting factor. The smaller reliability (R) corresponds to a larger discounting factor (D):

$$R = 1 - D \tag{1}$$

For inferring the activity of the patient based on evidential theory, reliability discounting methods that transform beliefs of each source are used to reflect the sensor's credibility, in terms of discount factor (D) ($0 \le D \le 1$). The discount mass function is defined as:

$$m^D(X) = \begin{cases} (1-D)m(X) & X \subset \Theta \\ D + (1-D)m(\Theta) & X = \Theta \end{cases} \tag{2}$$

where the source is absolutely reliable ($D = 0$), the source is reliable with a discounting factor (D) ($0 < D < 1$), and the source is completely unreliable ($D = 1$).

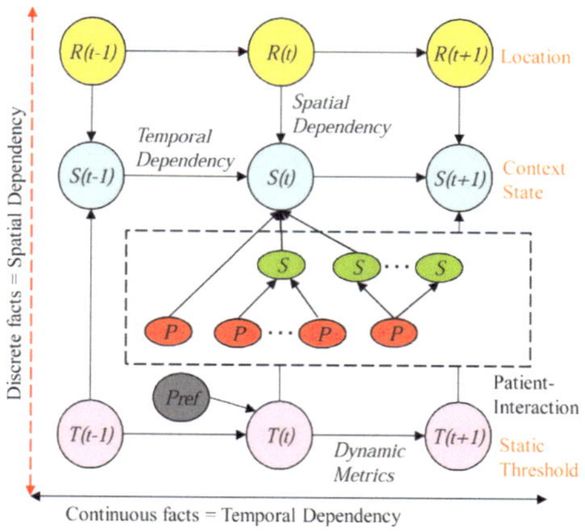

Fig. 1. A Relation-dependency approach

2.2 Context classification

The quality of a given piece of contextual information of a patient should be presented by some generalized forms of context classification (Razzaque et al., 2007) to determine reliable contextual information of a patient. However, it is an impossible task to build a general context classification to capture all aspects of the patient's contextual information in smart spaces. The numbers of ways to describe an event or an object are unlimited and there are no standards or guidelines regarding granularity of contextual information. In particular, the quality of a given piece of contextual information is not guaranteed by uncertainty. Thus, we defined the relation-dependency approach as a context classification based on spatial-temporal limitations which has three categories: 1) discrete environmental facts; 2) continuous environmental facts; and 3) occupant-interaction events as shown in Figure 1. These relation-dependency components consist of "Context state (*S(t)*)", defined as the collection and aggregation of activated or deactivated context attributes (Lee et al., 2009), "Sensor's static threshold (*T(t)*)", "Location of the patient (*R(t)*)", "Primary context (*P*)", "Secondary context (*S*)" and "Preference (*Pref*)".

2.3 Context modeling

We defined a state-space based context modeling with an evidential form as a generalized context modeling to represent the situation of the patient using context concepts that are similarly used in (Padovitz et al., 2005) and to improve the quality of a given piece of contextual information by reducing uncertainty. Within the proposed modeling, all possible values and their ambiguous combinations are considered to improve the quality of data in the given time (t) and location (R). We assign a probability value to each related set to achieve an efficient uncertainty representation. This can transfer a qualitative context information to a quantitative representation. Static weighting factors of the selected data are applied to represent the quality of data initially within the given t and R. This context modeling consists of a hierarchical interrelationship among multi-sensors, related contexts, and relevant activities within a selected region as shown in Figure 2. Each context concept is defined as follow.

A *context attribute*, denoted by α_i, is defined as any type of data that is utilized in the process of inferring situations. It is often associated with sensors, virtual or physical, where the value of a sensor reading denotes the value of a context attribute at a given t, denoted by α_i^t.

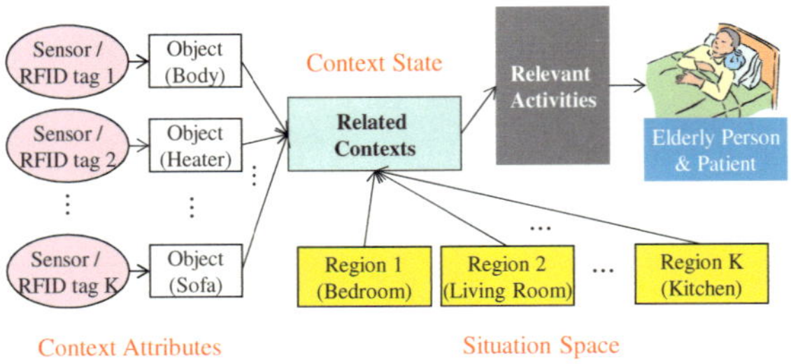

Fig. 2. An inter-relationship based on state-space context modeling

A *context state*, denoted by a vector S_i, describes the current state of the applied application in relation to a chosen context. It is a collection of N context attribute values to represent a specific state of the system at the given t. A context state is denoted as $S_i^t = (\alpha_1^t, \alpha_2^t, \ldots, \alpha_N^t)$, where each value α_i^t corresponds to the value of an attribute α_i at the given t. Whenever contextual information is recognized by certain selected sensors that can be used to make context attributes, a context state changes its current state depending on the aggregation of these context attributes.
A *situation space*, denoted by a vector space $R_i = (\alpha_1^R, \alpha_2^R, \ldots, \alpha_K^R)$, describes a collection of regions corresponding to some pre-defined situations. It consists of K acceptable regions for these attributes. An acceptable region α_i^R is defined as a set of elements V that satisfies a predicate P, (i.e., $\alpha_i^R = V \setminus P(V)$). A particular contextual information can be performed or associated with a certain selected region.
Given a context attribute i, a *quality of data* ψ_i associates weights $\omega_1, \omega_2, \ldots, \omega_M$ with combined attributes of values $\alpha_1^t + \alpha_1^R, \alpha_2^t + \alpha_2^R, \ldots, \alpha_N^t + \alpha_K^R$ of i, respectively, where $\sum_{j=1}^{M} \omega_j = 1$. The weight $\omega_j \in (0, 1]$ represents the relative importance of a context attribute α_j compared to other context attributes in the given t and R. For instance, a higher respiratory rate may be a strong indication of the fainting situation of a patient while other context attributes such as the blood pressure and the body temperature may not be so important to estimate that specific situation of the patient. In addition, a context attribute (α_i^t) within a context state ($S_i^t = (\alpha_1^t, \alpha_2^t, \ldots, \alpha_N^t)$) has various individual weights for α_i^t per different time intervals in the same situation space (α_i^R). For example, a respiratory rate ($50Hz$) at the current time-indexed state (S_t) is a strong indication of the fainting situation of the patient compared to a respiratory rate ($21Hz$) at previous time-indexed state (S_{t-1}). The same context attribute can have different degrees of importance in different contexts. We only consider the quality of data with the pre-defined context attributes, a selected region, and relevant activities initially. We then apply dynamic weights into both relative and individual importance of evidence to obtain an optimal weight of evidence.

2.4 Dezert-Smarandache Theory (DSmT)

The basic idea of DSmT (Dezert & Smarandache, 2004; 2006; 2009) is to consider all elements of Θ as not precisely defined and separated. No refinement of Θ into a new finer set Θ^{ref} of disjoint hypotheses is possible in general, unless some integrity constraints are known, and in such case they will be included in the DSm model of the frame. Shafer's model (Shafer, 1976) assumes Θ to be truly exclusive and appears only as a special case of the DSm hybrid model in DSmT. The hyper-power set, denoted by D^Θ, is defined by the rules 1, 2 and 3 without additional assumption on Θ but the exhaustivity of its elements in DSmT.

1. $\emptyset, \theta_1, \cdots, \theta_n \in D^\Theta$
2. If $\theta_1, \theta_2 \in D^\Theta$, then $\theta_1 \cap \theta_2$ and $\theta_1 \cup \theta_2$ belong to D^Θ
3. No other elements belong to D^Θ, except those obtained by rules 1) or 2)

When Shafer's model $M^0(\Theta)$ holds, D^Θ reduces to 2^Θ. Without loss of generality, G^Θ is equal to D^Θ if the DSm model is used, depending on the nature of the problem.

2.5 Combination rules (conjunctive and disjunctive)

As a conjunctive combination rule, the proportional conflict redistribution no. 5 (PCR5) (Smarandache & Dezert, 2005) are defined based on the conjunctive consensus operator for two sources cases by:

$$m_{12}(X) = \sum_{\substack{X_1, X_2 \in G^\Theta \\ X_1 \cap X_2 = X}} m_1(X_1) m_2(X_2) \tag{3}$$

The total conflicting mass drawn from two sources, denoted by k_{12}, is defined as:

$$k_{12} = \sum_{\substack{X_1, X_2 \in G^\Theta \\ X_1 \cap X_2 = \emptyset}} m_1(X_1) m_2(X_2) = \sum_{\substack{X_1, X_2 \in G^\Theta \\ X_1 \cap X_2 = \emptyset}} m(X_1 \cap X_2) \tag{4}$$

The total conflicting mass is the sum of partial conflicting masses based on Equations (3) and (4). If the total conflicting mass k_{12} is close to 1, the two sources are almost in total conflict. Whereas if the total conflicting mass k_{12} is close to 0, the two sources are not in conflict. Within the DSmT framework, the PCR5 combination rule redistributes the partial conflicting mass only to the elements involved in that partial conflict. For this approach, first, the PCR5 combination rule calculates the conjunctive rule of the belief masses of sources. Second, it calculates the total or partial conflicting masses. And last, it proportionally redistributes the conflicting masses to nonempty sets involved in the model according to all integrity constraints. The PCR5 combination rule is defined for two sources (Dezert & Smarandache, 2009): $m_{PCR5}(\emptyset) = 0$ and $\forall (X \neq \emptyset) \in G^\Theta$,

$$m_{PCR5}(X) = m_{12}(X) + \sum_{\substack{Y \in G^\Theta \setminus \{X\} \\ X \cap Y = \emptyset}} [\frac{m_1(X)^2 m_2(Y)}{m_1(X) + m_2(Y)} + \frac{m_2(X)^2 m_1(Y)}{m_2(X) + m_1(Y)}] \tag{5}$$

where m_{12} and all denominators such as $m_1(X) + m_2(Y)$ and $m_2(X) + m_1(Y)$ differ from zero(0). If a denominator is zero, that fraction is discarded. All sets in formulas are in canonical forms. For example, the canonical form of $X = (A \cap B) \cap (A \cup B \cup C)$ is $A \cap B$.
In addition, a disjunctive combination rule is used for Temporal Belief Filtering (TBF) (Ramasso et al., 2006). For instance, the TBF, which reflects that only one hypothesis concerning activity is true at each time-indexed state, ensures a temporal consistency with an exclusivity. Within a TBF, the disjunctive rule of combination ($m_{\cup}(.)$) is used so as to compute *prediction* from previous mass distributions and model of evolution. $m_{\cup}(.)$ is defined for two sources: $m_{\cup}(\emptyset) = 0$ and $\forall (C) \subset \Theta$,

$$m_{\cup}(C) = \sum_{\substack{i,j \\ C = X_i \cup Y_j}} m_1(X_i) m_2(Y_j), \qquad \forall (C \neq \emptyset) \in \Theta \tag{6}$$

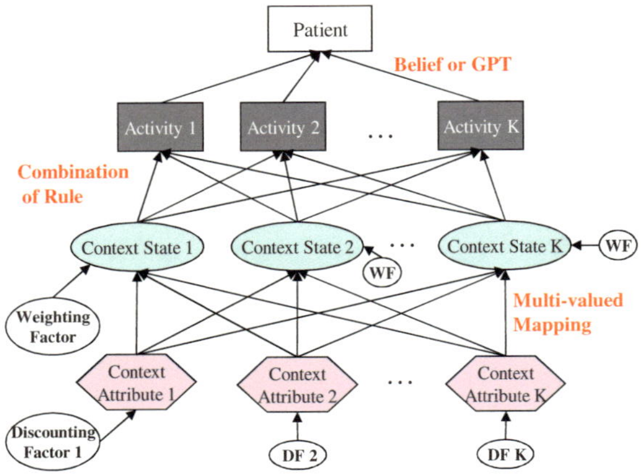

Fig. 3. An Evidential Fusion Network (EFN)

The core of a belief function given by $m_{\cup}(C)$ equals the union of the cores of $Bel(X)$ and $Bel(Y)$. This rule reflects the disjunctive consensus and is usually preferred when one knows that one of the source X or Y is mistaken but without knowing which one between X and Y.

2.6 Pignistic transformations (CPT and GPT)

When a decision must be taken, the expected utility theory, which requires a classical pignistic transformation (CPT) from a basic belief assignment $m(.)$ to a probability function $P\{.\}$, is defined in (Dezert et al., 2004) as follows:

$$P\{A\} = \sum_{X \in 2^{\Theta}} \frac{|X \cap A|}{|X|} m(X) \tag{7}$$

where $|A|$ denotes the number of worlds in the set A (with convention $|0|/|0| = 1$, to define $P\{0\}$). $P\{A\}$ corresponds to $BetP(A)$ in Smets' notation (Smets, 2000). Decisions are achieved by computing the expected utilities. In particular, the maximum of the pignistic probability is used as a decision criterion.

Within the DSmT framework, it is necessary to generalize the CPT to take a rational decision. This generalized pignistic transformation (GPT) is defined by (Dezert et al., 2004): $\forall (A) \in D^{\Theta}$,

$$P\{A\} = \sum_{X \in D^{\Theta}} \frac{C_M(X \cap A)}{C_M(X)} m(X) \tag{8}$$

where $C_M(X)$ denotes the DSm cardinal of a proposition X for the DSm model M of the problem under consideration. In this case, if we adopt Shafer's model $M^0(\Theta)$, Equation (8) reduces to Equation (7) when D^{Θ} reduces to 2^{Θ}. For instance, we gets a basic belief assignment with non null masses only on X_1, X_2 and $X_1 \cup X_2$. After applying GPT, we get:

$$\begin{aligned} P\{\emptyset\} = 0, \qquad P\{X_1 \cap X_2\} = 0 \\ P\{X_1\} = m(X_1) + \tfrac{1}{2} m(X_1 \cup X_2) \\ P\{X_2\} = m(X_2) + \tfrac{1}{2} m(X_1 \cup X_2) \\ P\{X_1 \cup X_2\} = m(X_1) + m(X_2) + m(X_1 \cup X_2) = 1 \end{aligned}$$

2.7 Evidential Fusion Network (EFN)

Based on the proposed state-space context modeling, the Evidential Fusion Network (EFN) is constructed as shown in Figure 3. Within a EFN, context reasoning is performed to make a high confidence level of the situation of the patient. The fusion process is performed to infer the activity of the patient along the EFN as follows.

1. (Define the Frame of Discernment): the evidential form represents all possible values of the sensors and their combination values.
2. (Sensor's Credibility): reliability discounting mass functions defined as Equations (1) and (2) transform beliefs of individual evidence to reflect the credibility of the sensor. A discounting factor (D) is applied to each context attribute within an EFN.
3. (Multi-valued Mapping): a multi-valued mapping represents the evidence to the same problem with different views. In particular, it can be applied to the context attributes so as to represent the relationships between sensors and associated objects by translating mass functions. A multi-valued mapping also can be applied to the related context state so as to represent the relationships among context attributes. Each context state consists of different pre-defined static weight of the evidence (*Relative importance*).
4. (Consensus): several independent sources of the evidence combine the belief mass distributions on the same frame to achieve the conjunctive consensus with the conflict mass. The PCR5 combination rule (Smarandache & Dezert, 2005) is applied to context states to obtain a consensus that helps to recognize the activity of the patient.
5. (Degree of Belief): Lower (*Belief (Bel)*) and upper bounds (*Plausibility (Pl)*) on probability is calculated to represent the degree of belief. Then the uncertainty levels (*Pl - Bel*) of the evidence in evidential framework is measured by using belief functions such as *Belief (Bel)* and *Plausibility (Pl)* after applying the PCR5 combination rule.
6. (Decision Making): The expected utility and the maximum of the pignistic probability such as Generalized Pignistic Transformations (GPT) is used as a decision criterion. The situation of the patient is inferred by calculating the belief, uncertainty, and confidence (i.e., GPT) levels of contextual information within an EFN.

3. Dynamic context reasoning

As shown in Figure 4, contextual information of a patient has the association or correlation between two consecutive time-indexed states. The EFN should include a temporal dimension for dealing with this context reasoning over time. Therefore, we introduce a dynamic context reasoning method in this section.

3.1 Temporal Belief Filtering (TBF) for relation-dependency

Depending on temporal changes, the values of the sensor at the current time-indexed state (S_t) are evolved by the measured values at the previous time-indexed state (S_{t-1}), because the belief mass distribution can not vary abruptly between two consecutive time-indexed states. In order to deal with this evolution, we utilize the Autonomous Learning Process (ALP) principle that has three states: 1) Initial State, 2) Reward State, and 3) Final Decision State as shown in Figure 5. This ALP principle is performed based on the Q-learning technology represented by (Roy et al., 2005). In Equation (9), X_t is the current state, $m(.)$ is the belief mass distribution, D is the discounting factor, and Re is the reward state to help decision making in

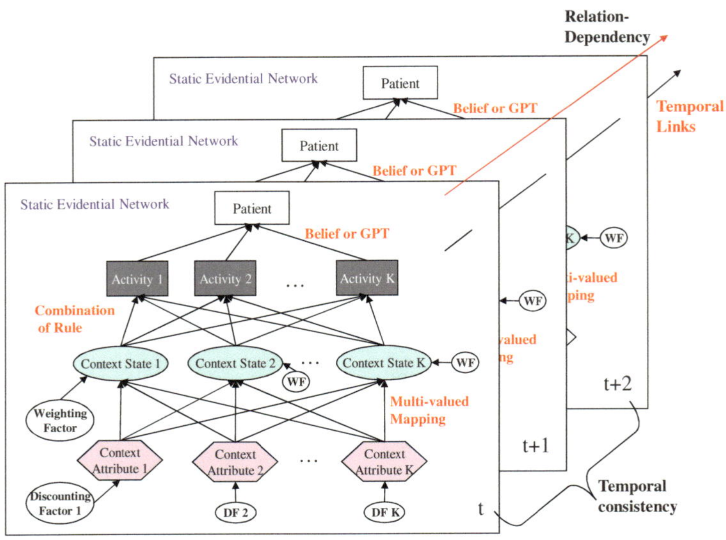

Fig. 4. EFN with a temporal dimension

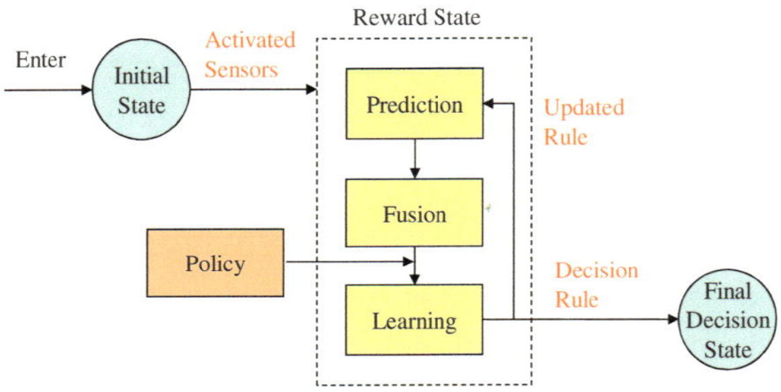

Fig. 5. Autonomous Learning Process (ALP) Principle

final decision state. We can support dynamic metrics (e.g., the evolution of the upper bounds or lower bounds of the pre-defined criteria).

$$Q(X_t, m_t(.)) \leftarrow (1 - m_t(.))Q(X_t, m_t(.)) + \\ m_t(.)(Re + D \max m_{t-1}(.)Q(X_{t-1}, m_{t-1}(.)) \quad (9)$$

In particular, TBF operations: prediction, fusion, learning and update are performed in reward state of the ALP principle to obtain the relation-dependency. The TBF ensures temporal consistency with the exclusivity between two consecutive time-indexed states when only one hypothesis concerning activity is true at each time. The TBF assumes that the general basic belief assignment (GBBA) at the current time stamp t is close to the GBBA at the previous time stamp $t-1$. Based on this assumption, the evolution process predicts a current GBBA taking the GBBA at $t-1$ into account. The TBF that operates at each time stamp t consists in four steps: 1) Prediction, 2) Fusion, 3) Learning and 4) Updated rule if required. For instance, if the activity of the patient was fainting (F) at $t-1$ then it would be partially fainting (F) at t. This is an implication rule for fainting (F) which can be weighted by a confidence value of $m_F\{.\} \in [0, 1]$. In this case, the vector notation of a GBBA defined on the frame of discernment (Θ) is used:

$$m^{\Theta} = [\ m^{\Theta}(\emptyset) \quad m^{\Theta}(\neg F) \quad m^{\Theta}(F) \quad m^{\Theta}(\neg F \cup F)\ ]$$

The evolution process can be interpreted as a GBBA defined as:

$$m_F^{\Theta} = [\begin{array}{cccc} 0 & 1 - Pl_F & Bel_F & Pl_F - Bel_F \end{array}]^T \tag{10}$$

3.1.1 Prediction

Depending on current model M with only two focal sets, the disjunctive rule of combination is used to compute *prediction* from the previous GBBA at $t-1$ and model of evolution using Equation (6). The disjunctive rule of combination does not allow to assign more belief to a hypothesis than does the previous GBBA. It is well suited for the autonomous evolution process under uncertainty:

$$\hat{m}_{t,M}^{\Theta} = \quad m_{t-1}^{\Theta} \quad (M_{\cup}) \quad m_M^{\Theta} \tag{11}$$

where m_{t-1}^{Θ} is the previous GBBA and m_M^{Θ} is model of evolution.
For instance, the prediction for fainting (F) situation of the patient at time stamp t is defined as:

$$\hat{\mathbf{m}}_{t,F}^{\mathbf{\Theta}} = \begin{bmatrix} 0 \\ (1 - Pl_F) \times m_{t-1}^{\Theta}(\neg F) \\ Bel_F \times m_{t-1}^{\Theta}(F) \\ 1 - [((1 - Pl_F) \times m_{t-1}^{\Theta}(\neg F)) + Bel_F \times m_{t-1}^{\Theta}(F)] \end{bmatrix} \tag{12}$$

when $m_F = 1$ or when $m_F = 0$, the prediction reflects a total confidence or a total ignorance with the current time-indexed state, respectively.

3.1.2 Fusion, learning and updated rule

Prediction ($\hat{m}_{t,M}^{\Theta}$) and *measurement* (m_t^{Θ}) represent two distinct pieces of the information. *Fusion* of the two distinct pieces of the information leads to a new GBBA whose conflict value (C_F) is relevant for belief learning and update requirement. In this case, conflict value (C_F), which is similar to k_{12} of Equation (4), is calculated by the conjunctive rule of combination of $\hat{m}_{t,M}^{\Theta}$ and m_t^{Θ}:

$$C_F = \quad \hat{m}_{t,M}^{\Theta} \quad (M_{\cap}) \quad m_t^{\Theta} \quad (\emptyset) \tag{13}$$

In addition, *policy* is required so as to analyze whether the current model M is valid or not. If C_F is not greater than the pre-defined threshold (T), the model at $t-1$ is kept as valid at t. However, if C_F exceeds the T, the model is evolved based on the result of the conjunctive rule of combination of $\hat{m}_{t,M}^{\Theta}$ and m_t^{Θ}. Depending on the applied policy, the evolution process ($m_{t,M}^{\Theta}$) (i.e., *learning*) is performed as below:

$$m_{t,M}^{\Theta} = \begin{cases} \hat{m}_{t,M}^{\Theta} \quad (M_{\cap}) \quad m_t^{\Theta} & ,if \quad C_F \geq T \\ m_{t-1,M}^{\Theta} & ,if \quad C_F < T \end{cases} \tag{14}$$

After a learning, a fading memory process (F_a) has been embedded so as to reduce the relation-dependency of the pieces of long past information even though the cumulative sum of conflict value (C_F) between $\hat{m}_{t,M}^{\Theta}$ and m_t^{Θ} is lower than the pre-defined threshold (T) during long time intervals. A fading memory process (F_a) resets the cumulative sum of C_F as a zero (0) and $\hat{m}_{t+w,M}^{\Theta}$ is equal to m_{t+w}^{Θ} based on time window size (W), which is chosen as a constant value (C). Then, *updated rule* is applied to the model of evolution repeatedly after F_a is applied to $m_{t,M}^{\Theta}$.

$$m_{t+w,M}^{\Theta} = \begin{cases} (1) \quad F_a \leftarrow \begin{cases} \sum C_F = 0 & ,if \quad W = C \\ \hat{m}_{t+w,M}^{\Theta} = m_{t+w}^{\Theta} \end{cases} \\ (2) \quad m_{t,M}^{\Theta} \times (F_a) \end{cases} \tag{15}$$

3.1.3 Decision rule

A decision is taken by the maximum of GPT (i.e., Equation (8)) within the DSmT framework after the evolution process is performed. We adopt Shafer's model (Shafer, 1976) in order to compare our approach with DBNs, which can get a BBA with non null masses only on θ_1 and θ_2 (i.e., $P\{\theta_1 \cup \theta_2\} = m(\theta_1) + m(\theta_2) = 1$) where θ_1 and θ_2 are hypotheses of the frame of discernment (Θ) (i.e., focal elements of the state within the frame of the set).
It is required to assess the recognition performance of a time-indexed state to decide whether a temporal sequence of the state has a false alarm or a new sensor activation/deactivation within the defined time window size (W). It is necessary to find a quality criterion without references to assess this performance. We defined D_F as the differentiation of GPTs of two consecutive time-indexed states. The $\bar{D}_F$ is defined as the mean of D_F (i.e., $\frac{\sum D_F}{W}$) within the defined W as the chosen criterion (i.e., Equation (16)) in order to distinguish a sensor reading error from new sensor activations or deactivations). As shown in Equation (17), if $\bar{D}_F$ is less than δ, there is no error within W. If $\bar{D}_F$ is located between δ and γ, a false alarm happens. And if $\bar{D}_F$ is greater than γ, the emergency situation of the patient progress.

$$\bar{D}_F \triangleq \frac{1}{W} \sum_{i=1,W} D_F^i \tag{16}$$

$$Decision(De) = \begin{cases} \text{No errors within the } W & ,if \quad \bar{D}_F < \delta \\ \text{False alarm} & ,if \quad \delta \leq \bar{D}_F < \gamma \\ \text{Emergency Progress} & ,if \quad \gamma \leq \bar{D}_F \end{cases} \tag{17}$$

where δ is the defined false alarm threshold and γ is the defined emergency progress threshold for the chosen criterion. In this case, the value of δ is always lower than that of γ, because we assume that the false alarm does not often happen when the new sensor activation or deactivation is detected by the expert system in emergency situation of the patient. Based on the defined threshold (T) for conflict value (C_F) and time window size (W), we can distinguish a sensor reading error from new sensor operations. Then, we perform evolution operations with dynamic evidential network (DEN) in order to improve the confidence (i.e., GPT) level of contextual information.

3.2 Evolution operations with DEN

The DEN is constructed based on the EFN with a temporal dimension as shown in Figure 6. Within a DEN, context reasoning is performed to find a false alarm in captured contexts and to make a high confidence level of the situation of the patient. First, we define the threshold (Te) of the GPT level for the emergency situation of the patient. Second, we calculate the GPT level at each time-indexed state using a TBF with defined T and W. And last, if the GPT level is over the defined Te for four continuous time-indexed states, we make a decision about the situation of the patient as an emergency. We assume that the initial prediction is equal to the 1st measurement at 1st time-indexed state (S_1). The consecutive processing of two combination rules (i.e, disjunctive rule and conjunctive rule) is well adapted to EFN to update the belief mass distribution of EFN at time-indexed states. In Figure 6, we define n time intervals and time window sizes W to reflect a fading memory process (F_a) to the pervasive healthcare system. The F_a reduces long past contextual information of the patient. Depending on D_F and $\bar{D}_F$, we trace the emergency progress which can check a false alarm. We then make an optimal time window size (W) that is applied to the evolution process.

Context Type	Sensor Type	Non-Activated			Activated
		Regular (1)	Warning (2)	Warning (3)	Emergency (4)
Intrinsic	Respiratory Rate	15∼20 Hz	13∼14 or 21∼30 Hz	11∼12 or 31∼40 Hz	10 *below* or 41 *over* Hz
	Blood Pressure	120∼90 mmHg	121∼130 or 81∼89 mmHg	131∼140 or 71∼80 mmHg	141 *over* or 61∼70 mmHg
	Body Temperature	36.6∼37 °C	36.1∼36.5 or 37.1∼38 °C	35.6∼36 or 38.1∼39 °C	35.5 °C below or 39.1 °C over
Optional	Location	The motion detector installed in the ceiling catches the RF signal attached on the patient			
	Motion	The motion detector installed in the door catches the RF signal attached on the patient			
	Pressure	The pressure sensor attached on the sofa catches the weight of the patient			

Table 1. Pre-defined Rules of a Context Attribute

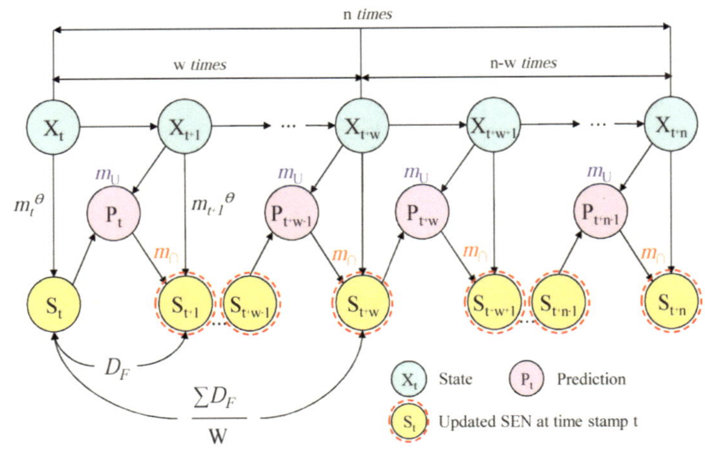

Fig. 6. The Proposed DEN for n time intervals

3.3 An optimal weight for evidence

3.3.1 Pre-defined rule of a context attribute

We define rules of a context attribute to represent dynamic weights of a context attribute as shown in Table 1. We assume that the ratio of total weights of optional context attributes $O(\sum \omega_i)$ is equal to that of intrinsic context attributes $I(\sum \omega_i)$ in order to apply the rule of combination. Within a EFN, each context state has the same weight (e.g., the weight is equal to 0.5). We apply more $C(a_t^k)$, which reflects the increase or decrease degree of a particular context attribute, to the activated case (i.e., Emergency (4)) compared to the non-activated case (i.e., Warning (2 and 3) and Regular (1)), because the activated case is more important than the non-activated case in an emergency situation of the patient. In addition, we apply more $C(a_t^k)$ to the level increased case (i.e., $L(a_{t+1}^k) > L(a_{t+1}^k)$) compared to the level decreased case (i.e., $L(a_{t+1}^k) < L(a_{t+1}^k)$), where $L(a_{t+1}^k)$ reflects the level of a particular context attribute. The level increased case is more important than the level decreased case in an emergency situation of the patient. Thus, we calculate the weight of an intrinsic context attribute as below.

1. initial $O(\sum \omega_i) = I(\sum \omega_i) = 0.5$
2. if all $L(a_t^k) = L(a_{t+1}^k)$, then $C(a_{t+1}^k) = 0$
3. else if $L(a_{t+1}^k) > L(a_t^k)$ and $L(a_{t+1}^k) \neq 4$, then $C(a_{t+1}^k) = 2\alpha$
4. else if $L(a_{t+1}^k) < L(a_t^k)$ and $L(a_t^k) \neq 4$, then $C(a_{t+1}^k) = -\alpha$
5. else if $L(a_{t+1}^k) > L(a_t^k)$ and $L(a_{t+1}^k) = 4$, then $C(a_{t+1}^k) = 3\beta$
6. else if $L(a_{t+1}^k) < L(a_t^k)$ and $L(a_t^k) = 4$, then $C(a_{t+1}^k) = -2\beta$

with two % values α and β (i.e., $\beta \geq \alpha$).

3.3.2 A normalized weighting technique

We calculate the relative weight of a context attribute based on Multi-Attribute Utility Theory (MAUT) (Valiris et al., 2005; Winterfeld & Edwards, 1986) to setup the initial weight of a context attribute within a given context state. The weights are determined by their importance in regarding to a specific situation of the patient. In particular, we construct a scale

Sensor Type	Regular	Emergency	Relative Weight $\tilde{\omega_u}$
Respiratory Rate	Scale-R (5)	Scale-E (55)	0.6
Blood Pressure	Scale-R (5)	Scale-E (15)	0.2
Body Temperature	Scale-R (5)	Scale-E (15)	0.2
Location	Scale-R (5)	Scale-E (10)	0.25
Motion	Scale-R (5)	Scale-E (10)	0.25
Pressure	Scale-R (5)	Scale-E (25)	0.50

Table 2. An example of Relative Weight of a Context Attribute

representing the properties of the levels of a context attribute to evaluate context attributes. For instance, we assume that the scale from 0 (e.g., the least affection) to 55 (e.g., the most affection) for the situation serves as measure of the evaluation as shown in Table 2. We pre-define the scale of a context attribute then calculate the relative importance of a context attribute using Equation (18).

$$\tilde{\omega_u} = \omega_v / \sum_{w=1}^{N} (\omega_w) \tag{18}$$

where $\tilde{\omega_u}$ defines the relative weight of a context attribute, ω_v is the sum of the value of Scale-R and Scale-E for one sensor type, and $\sum_{w=1}^{N}(\omega_w)$ is the total sum of the value of Scale-R and Scale-E. After calculating the relative weight of a context attribute, we redistribute the weight of a context attribute over time based on the pre-defined rule of a context attribute. Let $\omega_1, \omega_2, \cdots, \omega_k, \cdots, \omega_{k+m}, \cdots, \omega_N$ denote an initial relative weight associated with a given context state S_t^i for fusion process. Within the same location, a normalized weighting technique for individual difference between two time-indexed states is applied to each context attribute as below.

1. Repeat for each optional context attribute k:
   $\omega_k = \omega_i$, where i defines an initial weight
2. Repeat for each intrinsic context attribute k:
   if all $L(a_t^k) = L(a_{t+1}^k)$ or all $C(a_{t+1}^k)$ are equal,
   then all $\omega_k = \omega_i$
   else if any $L(a_t^k) \neq L(a_{t+1}^k)$ or any $C(a_{t+1}^k)$ is different,
   then $\hat{\omega_k} = \omega_i / \sum_{j=1}^{N}(\omega_j \pm C(a_{t+1}^j))$,
   where $\hat{\omega_k}$ defines a new weight for a context attribute

3.4 Dynamic context reasoning

Based on the proposed DEN, the dynamic weighting is applied to each evidence to make a high GPT level of the situation of the patient compared to the others. First, we calculate the GBBA of SEN initially using evidential operations at 1st time-indexed state. Second, we apply the updated weight into each context attribute from 2nd time-indexed state using the proposed normalized weighting technique. Finally, we calculate the confidence level (i.e., GPT) of contextual information. The procedures of dynamic context reasoning consist of seven steps.

1. (Measure a GBBA of SEN): Initially, we measure a GBBA of SEN using evidential operations at time stamp t. The first prediction ($\hat{m}_{t_1,M}^{\Theta}$) is equal to measurement ($m_{t_1}^{\Theta}$) at time-indexed state S_1.

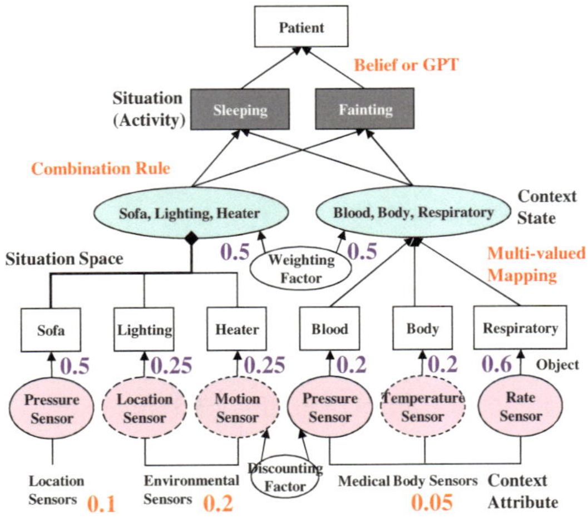

Fig. 7. An example of a patient's situation based on the EFN

2. (Update the Weight of a Context Attribute): First, we calculate the relative importance of a context attribute. Second, we redistribute the weight of a context attribute over time based on the pre-defined rule of a context attribute. Third, we calculate individual difference between two time-indexed states using the proposed normalized weighting technique (i.e., Equation (18)). And last, we apply the updated weight into each context attribute from 2nd time-indexed state so as to obtain the GPT of contextual information.

3. (Prediction and Evolution): We calculate prediction from the previous GBBA and model of evolution using the disjunctive rule of combination (i.e., Equation (6)). The disjunctive rule of combination is well suited for the model evolution under uncertainty because it does not allow to assign more belief to an hypothesis than does the previous GBBA. The GBBA of SEN at time stamp $t+1$ will be affected by prediction ($\hat{m}^{\Theta}_{t+1,M}$).

4. (Learning): We fuse $\hat{m}^{\Theta}_{t+1,M}$ and m^{Θ}_{t+1} using the conjunctive rule of combination so as to make a new GBBA. As a learning, if a conflict value (C_F) is greater than the pre-defined threshold (T), a new GBBA is adapted. Whereas, the previous learned GBBA is adapted as a new GBBA (i.e., Equation (14)).

5. (Fading Memory Process): We apply a fading memory process (F_a) with the defined time window size (W) so as to reduce the affection of long past information. After F_a is performed, the GBBA of $\hat{m}^{\Theta}_{t+w,M}$ is equal to the GBBA of m^{Θ}_{t+w} (i.e., Equation (15)). The previous GBBA of $\hat{m}^{\Theta}_{t+w-1,M}$ is ignored at time stamp $t+w$.

6. (Update and Decision Making): We calculate each GPT of the frame of discernment per time-indexed state (i.e., Equation (8)) by applying the updated rule then calculate differentiation (D_F) of two consecutive time-indexed states. Based on the mean of D_F (i.e., $\bar{D}_F$) and the pre-defined value for δ and γ, we can make a decision: No errors, False alarm, or Emergency progress (i.e., Equation (17)).

7. (Comparison the GPT level): Finally, we compare the GPT level of consecutive time-indexed states. If the GPT level is over the pre-defined threshold (Te), which represents the emergency situation, for four continuous time-indexed states, we make a decision about the situation of the patient as an emergency.

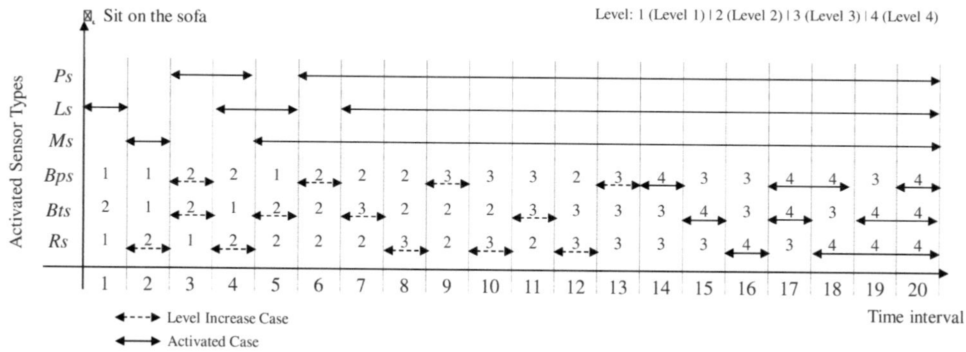

Fig. 8. An example of emergency level changes based on time intervals

% value	Case 1	Case 2	Case 3	Case 4	Case 5	Case 6
β	5 %	10 %	15 %	20 %	25 %	30 %
α	5 %	5 %	10 %	10 %	15 %	30 %

Table 3. An example of the % values of α and β

4. A case study

As shown in Figure 7, many ambiguous situations of the patient can happen in home-based care. We suppose that the situation (i.e., "sleeping" (*Sl*) or "fainting" (*F*)) of the patient can happen in smart home applications. In order to check dynamic emergency level changes based on time intervals, six types of a sensor are randomly activated during 20 time intervals as shown in Figure 8. Among six types of a sensor, three types of a sensor: blood pressure, body temperature and respiratory rate are involved in an intrinsic context attribute type. Whereas three types of a sensor: pressure, location and motion are involved in an optional context attribute type. Within Figure 8, we apply the level increased case and the activated case based on the data of Table 1. Initially, a discounting factor and a relative weight of each sensor are fixed so as to calculate the initial GBBA of EFN. In particular, we assume that a discounting factor of the environmental sensors, the location sensor, and the medical body sensors are 20%, 10% and 5%, respectively. We can obtain an initial relative weight of each sensor using a scale representing method as shown in Table 2. We apply different % values of α and β (i.e., $\beta \geq \alpha$) as shown in Table 3 to check the variations of the weight depending on the selected degree of a level change ($C(a_{t+1}^k)$). For making a simulation, we perform an evidential fusion process with a 95% confidence interval for 500 iterations. Moreover, we use paired observations (Jain, 1991) that construct a confidence interval for the difference in order to compare our method with other methods such as DST and DBNs. The analysis of paired observations deals with two processes as one process of *n* pairs. For each pair, the difference in performance can be computed. Then, if the confidence interval includes zero, two fusion processes are not significantly different.

5. Comparison and analysis

We compare the uncertainty levels of two cases: 1) DST and 2) DSmT and the GPT levels of two cases: 1) DSmT and 2) DBNs. For calculating the "fainting (*F*)" situation of the patient within the applied scenario, we apply three methods: 1) defined static weighting factors, 2) different weighting factors and 3) different discounting factors into the two fusion processes,

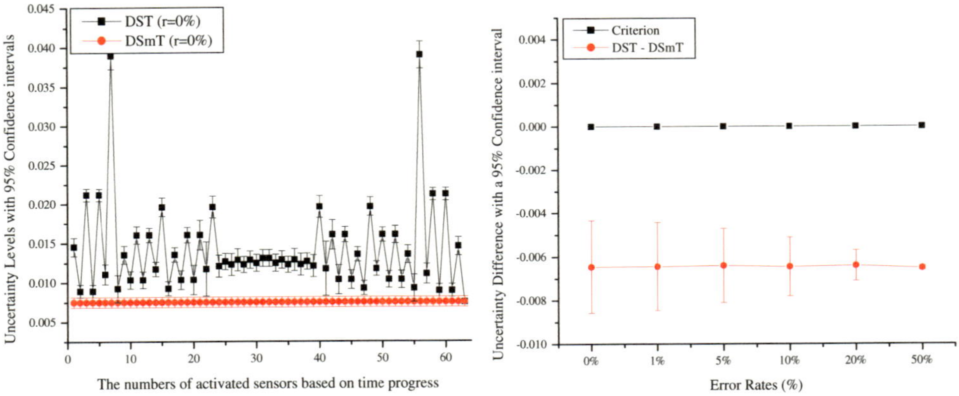

(a) Uncertainty levels of two cases with r=0% (b) Comparison Uncertainty levels of two cases

Fig. 9. Comparison Uncertainty Levels of DSmT and DST with static weighting factors

respectively. In particular, we utilize the paired observation method with different error rates (r) (i.e., 0%, 1%, 5%, 10%, 20% and 50%) so as to compare the two fusion processes.

5.1 Uncertainty levels of DSmT and DST

5.1.1 Comparison with static weighting factors

After we apply a static weights into each context attribute, the evidential fusion process based on DST has more various conflicting mass in the uncertainty level compared to the DSmT approach as shown in Figure 9(a). The reason is that the PCR5 combination rule of DSmT redistributes the total conflicting mass as equal to zero within the DSmT framework. However, Dempster's combination rule of DST takes the total conflicting mass and redistributes it to all non-empty sets within the DST framework, even those not involved in the conflict. In addition, the uncertainty level of DST is higher than that of DSmT when we use the paired observation method as shown in Figure 9(b). Thus, the DSmT approach with static weights reduces the degree of uncertainty (i.e., conflicting mass in uncertainty level) compared to the DST approach.

5.1.2 Comparison with different weighting factors

We apply different static weights into each context attribute based on Table 4 so as to compare the uncertainty levels of the two cases based on different weighting factors. We compare four situations: a) "*Bts*", and "*Rs*" are not activated, b) "*Ls*" and "*Bps*" are not activated, c) only "*Bts*" is not activated, and d) all sensors are activated to see the variation of the uncertainty level of contextual information. We apply 0% and 50% error rates into the fusion process with a 95% confidence interval.

As shown in Figure 10, the uncertainty levels of DSmT have the same degrees for all cases even though those of DST have different degrees depending on the four situations and the used error rates (r) (i.e., 0% and 50%). The degrees of uncertainty of DSmT are lower than those of DST. Only when all sensors are activated will the degrees of uncertainty of DSmT be equal to those of DST. The evidential fusion based on DSmT shows a constant uncertainty level, whether a sensor reading error may happen or whether an emergency situation may progress, by redistributing the total conflicting mass only into the sets involved in the conflict and proportionally to their masses. In this case, the DSmT approach shows the better

No.	*Ps*	*Ls*	*Ms*	*Bps*	*Bts*	*Rs*
Case 1	0.9	0.05	0.05	0.05	0.05	0.9
Case 2	0.8	0.1	0.1	0.1	0.1	0.8
Case 3	0.7	0.1	0.2	0.1	0.2	0.7
Case 4	0.6	0.2	0.2	0.2	0.2	0.6
Case 5	0.5	0.2	0.3	0.2	0.3	0.5
Case 6	0.4	0.3	0.3	0.3	0.3	0.4
Case 7	0.3	0.4	0.3	0.3	0.4	0.3
Case 8	0.2	0.4	0.4	0.4	0.4	0.2
Case 9	0.1	0.45	0.45	0.45	0.45	0.1

Table 4. An example of different static weighting factors

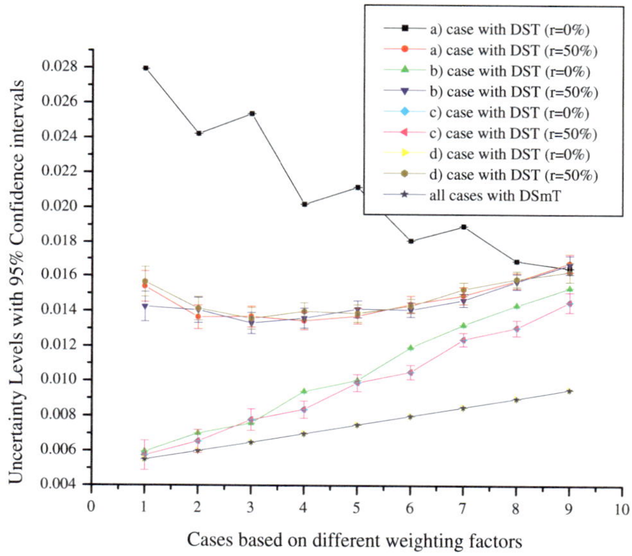

Fig. 10. Uncertainty levels of DSmT and DST with different weights

performance than the DST approach in order to reduce the conflicting mass in uncertainty level of contextual information.

5.1.3 Comparison with different discounting factors (*D*)

We apply different discounting factors (*D*), which are related to sensor's credibility, into *"Ps"* and *"Rs"* to calculate the uncertainty levels of the two cases based on Table 5. In this case, we calculate four situations: a) "*Bts*", and "*Bps*" are not activated, b) "*Ps*" and "*Bts*" are not activated, c) only "*Bps*" is not activated, and d) all sensors are activated to see the variation of the uncertainty level of contextual information. Depending on different *D* on *"Ps"* and *"Rs"*, the two cases show different degrees of uncertainty as shown in Figure 11. The degrees of uncertainty of the two cases are increased based on the increase of the *D* as expected. The uncertainty levels of DSmT have the same degrees for all cases even though those of DST have different degrees for the four situations. The degrees of uncertainty of DSmT are lower than those of DST. This result shows that the DSmT approach is better than the DST approach in order to reduce the conflicting mass in uncertainty level of contextual information.

No.	Ps	Ls	Ms	Bps	Bts	Rs
Case 1	0%	20%	20%	5%	5%	0%
Case 2	1%	20%	20%	5%	5%	1%
Case 3	2%	20%	20%	5%	5%	2%
Case 4	5%	20%	20%	5%	5%	5%
Case 5	10%	20%	20%	5%	5%	10%
Case 6	20%	20%	20%	5%	5%	20%
Case 7	50%	20%	20%	5%	5%	50%

Table 5. An example of different discounting factors

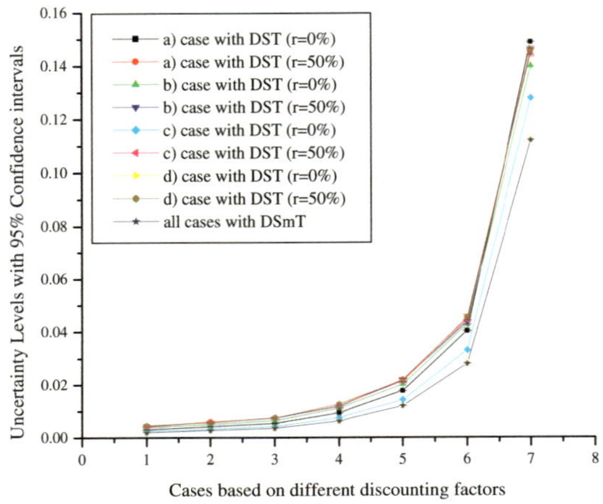

Fig. 11. Uncertainty levels of DSmT and DST with different discounting factors D

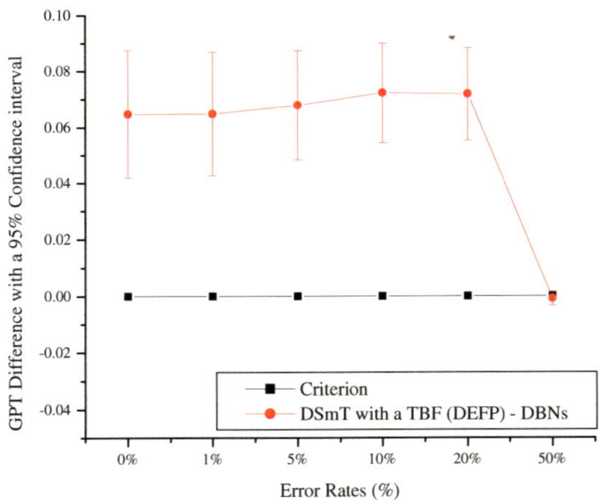

Fig. 12. Comparison GPT levels of DSmT and DBNs with static weighting factors

5.2 GPT levels of DSmT and DBNs

5.2.1 Comparison with static weighting factors

We compare the GPT level of DSmT with that of DBNs by calculating the GPT difference with a 95% confidence interval. We consider the same static weighting factors with $T = 0$ and $W = 5$. We use paired observations depending on the GPT level of DSmT when the degree of

Case	*Ps*	*Ls*	*Ms*	*Bps*	*Bts*	*Rs*
1 (DEN)	0.9	0.05	0.05	0.05	0.05	0.9
2 (DEN)	0.5	0.25	0.25	0.2	0.2	0.6
3 (DEN)	0.3	0.4	0.3	0.3	0.3	0.4
4 (DEN)	0.1	0.4	0.5	0.4	0.4	0.2
5 (DBN)	0.9	0.05	0.05	0.05	0.05	0.9
6 (DBN)	0.5	0.25	0.25	0.2	0.2	0.6
7 (DBN)	0.3	0.4	0.3	0.3	0.3	0.4
8 (DBN)	0.1	0.4	0.5	0.4	0.4	0.2

Table 6. An example of different weights for DSmT and DBNs

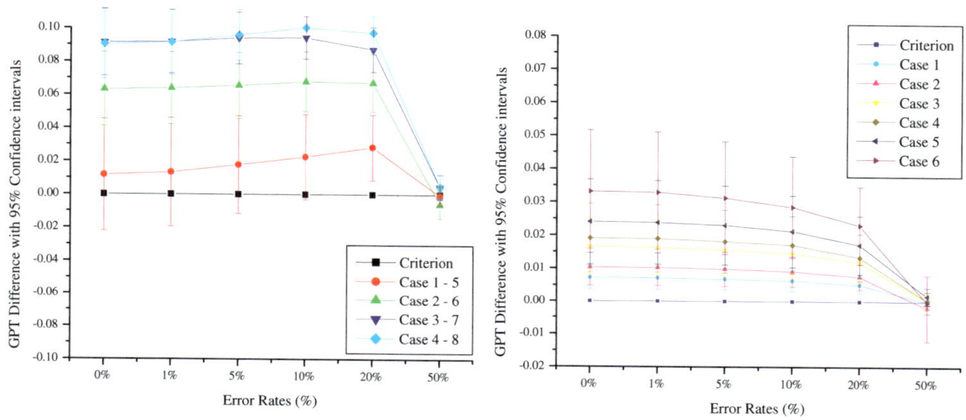

(a) Comparison DSmT with DBNs (b) Comparison DWEFP with DEFP

Fig. 13. Comparison GPT levels of the two cases with different weighting factors using paired observations

GPT level is over 0.5 case, because the aggregation of the degree of GPT is not over 0.5 reduces the total GPT level. The GPT level of DSmT with static weighting factor is higher than that of DBNs as shown in Figure 12. This result shows that the GPT level of the DSmT is higher than that of DBNs when the degree of GPT is over 0.5.

5.2.2 Comparison with different weighting factors

In order to compare the GPT level of DSmT with that of DBNs with different weighting factors, first, we apply different static weights to each context attribute based on "*Ps*" and "*Rs*" as shown in Table 6. As shown in Figure 13(a), the GPT levels of eight cases have different paired observation results. When we compare the case 1 and case 5, the confidence interval includes zero so it is impossible to distinguish which one is better than the other. The reason is that the degree of GPT is lower than 0.5 sometimes. Whereas the confidence intervals of the case 2 and 4, the case 3 and 7, and the case 4 and 8 do not have zero so we can prove that the GPT levels of DSmT with static weights are better than those of DBNs.
Second, we apply dynamic weights to each context attribute based on different % values of α and β (i.e., from Case 1 to Case 6 in Table 3) in order to compare the GPT levels of the two cases: 1) DSmT with dynamic weights (DWEFP) and 2) DSmT with static weights (DEFP). When we utilize the paired observation method, the confidence intervals do not include zero

No.	Ps	Ls	Ms	Bps	Bts	Rs
Case 1 - error rate 0%	0%	0%	0%	0%	0%	0%
Case 2 - error rate 0%	5%	5%	5%	5%	5%	5%
Case 3 - error rate 5%	5%	5%	5%	5%	5%	5%
Case 4 - error rate 5%	10%	10%	10%	10%	10%	10%
Case 5 - error rate 10%	10%	10%	10%	10%	10%	10%
Case 6 - error rate 10%	20%	20%	20%	20%	20%	20%
Case 7 - error rate 20%	20%	20%	20%	20%	20%	20%
Case 8 - error rate 20%	50%	50%	50%	50%	50%	50%
Case 9 - error rate 50%	50%	50%	50%	50%	50%	50%

Table 7. Different discounting factors (D) with selected error rates (r)

except for the error rate is 50% case as shown in Figure 13(b). With a 50% error rate, it is impossible to prove anything, because an error rate make the wrong simulation operation then it is nothing. The GPT level of DSmT with dynamic weights is higher than that of DSmT with static weights. We can improve the GPT level of DSmT using dynamic weights compared to the DEFP approach that applies static weights into each context attribute.

5.2.3 Comparison with different discounting factors (D)

In order to compare the GPT level of DSmT with that of DBNs with different discounting factors, first, we apply different discounting factors into each context attribute. Depending on different D on "Ps" and "Rs", the two cases show different degrees of GPT levels. In addition, the GPT levels of DBNs are lower than those of the DSmT except for the 50% error rate case when we compare the two fusion processes using the paired observation method for all cases in Table 5. Based on the result of Figure 14(a), we know that the DSmT approach with different discounting factors gets the better performance than the DBNs for improving the confidence level of contextual information.

Second, we apply different discounting factors (D) with selected error rates (r) (i.e., 0%, 5%, 10%, 20% or 50%) into context attributes as shown in Table 7 in order to compare the GPT levels of DSmT with dynamic weights with those of DSmT with static weights. We apply updated weights into each sensor by calculating the % values of α and β as shown in Case 1 and Case 6 of Table 3, because the % value of α and β is the smallest and the biggest in Table 3, respectively. According to Figure 14(b), the confidence intervals do not include zero except for the error rate is 50% case. Thus, the GPT level of DSmT with dynamic weights (DWEFP) is higher than that of DSmT with static weights (DEFP) in this scenario. As a result, we can improve the degree of GPT using DSmT with dynamic weights compared to DSmT without dynamic weights.

Finally, we can infer the situation of the patient by using the mean of the D_F (i.e., $\bar{D}_F$) and pre-defined rule of a decision. For example, we assume that the pre-defined threshold (Te) for an emergency situation is equal to 0.7. If the degree of GPT is over 0.7 for four continuous time-indexed states, we estimate that the patient is an emergency. For instance, we catch a false alarm between 10th and 12th time intervals in Figure 8. Then, we estimate that the emergency situation starts from 8th time interval. This is helpful to make a decision about the situation of a patient in home-based care.

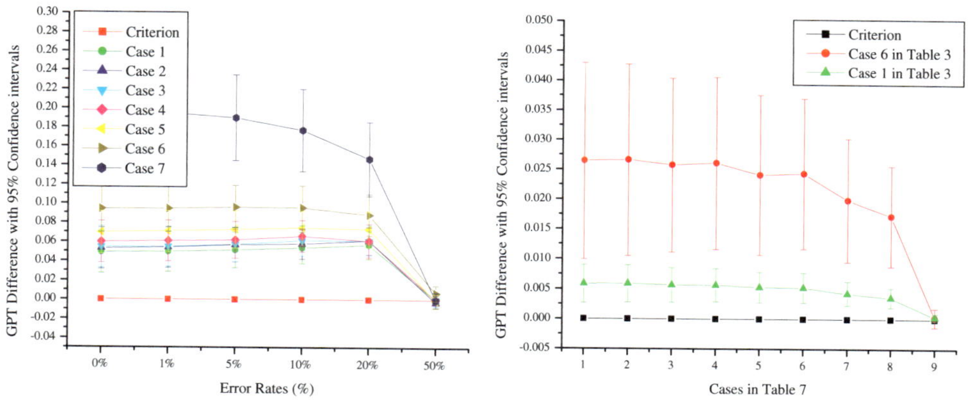

(a) Comparison DSmT with DBNs (b) Comparison DWEFP with DEFP

Fig. 14. Comparison GPT levels of the two cases with different discounting factors (D) using paird observations

6. Related work

In context-aware applications, situations (Dey, 2001; Gellersen et al., 2002) are external semantic interpretations of low-level sensor data by permitting a higher-level specification of human behavior and the corresponding system services and the way of changing situation is called context reasoning and interpretation (Loke, 2006). It means that we need reasoning context models that can adapt the situation definitions based on discovered changes with changing environments and changing user needs (Jayaraman et al., 2009). However, both the physical world itself and our measurements of it are prone to uncertainty. Thus, different types of entities in the pervasive environment must be able to reason about uncertainty. In order to solve this problem, a number of mechanisms have been proposed in the literature for reasoning on uncertainty and there are two main purposes for reasoning on uncertainty: 1) improving the quality of contextual information and 2) inferring new kinds of contextual information. Reasoning to improve the quality of contextual information typically takes the form of multi-sensor fusion where data from different sensors are used so as to increase confidence, resolution or any other context quality metrics. Reasoning to infer new contextual information typically takes the form of deducing higher-level contexts (e.g., activity of a user) or situations from lower-level contexts (e.g., location information of a user), because we can not directly sense the higher-level contexts. These contexts may be associated with a certain level of uncertainty depending on both the accuracy of the sensed information and precision of the deduction process (Bettini et al., 2010; Lee et al., 2010a). Therefore, we introduce some context reasoning approaches such as Fuzzy logic, Probabilistic logic, Bayesian Networks (BNs), Hidden Markov Models (HMMs), Kalman Filtering Models (KFMs), Dynamic Bayesian Networks (DBNs) and Dempster-Shafer Theory (DST) of the evidence in order to compare them with our context reasoning approach.

6.1 Puzzy logic, probabilistic logic and BNs

In fuzzy logic, a degree of membership represented by a pair (A:m) where A is a set and m is a possibility distribution in real unit interval [0,1] is used to show an imprecise notion such as confidence values (Lemmi & Betti, 2006; Zadeh, 1999). The elements of two or more fuzzy sets can be combined in order to create a new fuzzy set with its own membership function then it is used for reasoning models which need more than the probabilistic theory with uncertainty. For instance, the fuzzy logic is used so as to capture a clinical uncertainty in medical data of pervasive computing applications in (Agarwal et al., 2010). In addition, fuzzy logic is well suited for describing subject contexts by resolving conflicts between different contexts (e.g., Actuator's operation in (Lee et al., 2008a)). In this work, we assume that the environmental sensors are operated based on the fuzzy logic of the selected sensors.

Probabilistic logic and Bayesian networks (BNs) can be used for improving the quality of contextual information through multi-sensor fusion as well as for deriving the higher-level probabilistic contexts. They also can be used for resolving conflicts between contextual information obtained from different sources. According to (Ranganathan et al., 2004), the probabilistic logic is used for encoding access control policies and the BNs is used for combining uncertain information from a large number of sources and deducing higher-level contexts. However, these rules can not represent the ignorance (Maskell, 2008), which manages the degree of uncertainty, caused by the lack of information.

6.2 HMMs, KFMs and DBNs

In order to deal with unpredictable temporal changes in sensory information, Hidden Markov Models (HMMs) (Dargie, 2007; Soyer et al., 2003), Kalman Filtering Models (KFMs) (Welch & Bishop, 2006) or Dynamic Bayesian Networks (DBNs) (Dezert et al., 2004; Murphy, 2002; Zhang & Ji, 2006) are utilized as fusion techniques. In terms of probabilistic networks, HMMs represent stochastic sequences as Markov chains; the states are not directly observed, but are associated with observable evidences, and their occurrence probabilities depend on the hidden states. This model can be used for location prediction by using a hierarchical Markov model that can learn and infer a user's daily movements (Liao et al., 2007). KFMs represent the state of the system refers to a set of variables that describe the inherent properties of the system at a specific instant of time. This is a useful technique for estimating, or updating the previous estimate of, a system's state by using indirect measurements of the state variables and using the covariance information of both state variables and indirect measurements (Olfati-Saber, 2007). However, DBNs, which were proposed as a generalization of HMMs and KFMs, have some distinct features. DBNs allow much more general graph structures compared with HMMs or KFMs. DBNs represent the hidden state in terms of a set of random variable compared with HMMs, which represent the state space with a single random variable. DBNs allow general hybrid and nonlinear conditional probability densities (CPDs) compared with KFMs, which require all CPDs to be linear-Gaussian. This is a useful feature to manage the causality between random variables as well as time series data. For instance, a high level user behavior is inferred from low level sensor data by adding knowledge of real-world constraints to user location data in (Patterson et al., 2003). A variant of DBNs is used in an unsupervised way in order to predict transport routes based on GPS data. By adding constraints on the routes that could be learned by the training algorithm, the prediction accuracy was significantly improved.

DBNs are made up of the interconnected two time-indexed states of a static Bayesian Network (BN) and the transition of a static BN between two consecutive time t and $t+1$ satisfies the Markov property (Padhraic, 1997) as shown in Figure 15. DBNs can be implemented by keeping in memory two states at any one time-indexed state, representing a previous time-indexed state and current time-indexed state, respectively. In Figure 15, the two time-indexed states, which have an associated conditional probability, are such rotated that old states are dropped and new states are used as time progress. The arcs between two

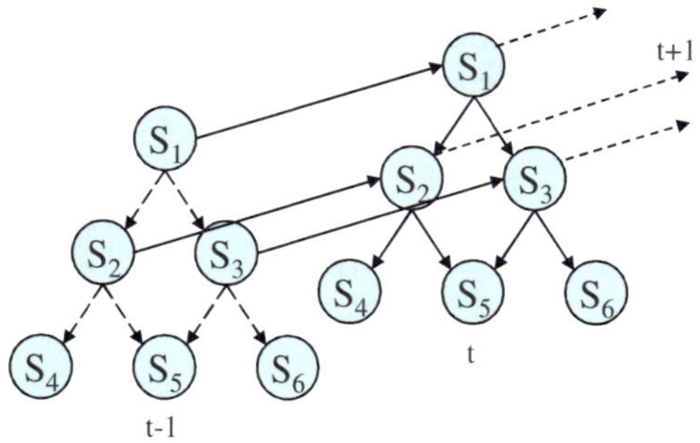

Fig. 15. An example of Dynamic Bayesian Networks (DBNs)

time-indexed states reflect temporal causality and they are parameterized by transitional probabilities. The joint distribution from the initial moment of time (t = 1) until the time boundary (t = T) is then given by

$$P(S_{1:T}) = \prod_{t=1}^{T} \prod_{i=1}^{n} P(S_i^t | k(S_i^t)) \tag{19}$$

where S_i^t is the ith node at time t and $k(S_i^t)$ stands for the parents of a node S_i^t at time t. They can either be in the same time-indexed state or in the previous time-indexed state. In this work, we use the Markov property, which is similar to DBNs, in order to represent temporal and state links between two consecutive time-indexed states of a Static Evidential Network (SEN) (i.e., Dynamic Evidential Network (DEN)) then compare it with the original process of DBNs.

6.3 Dempster-Shafer Theory (DST)

DST is a mathematical theory of the evidence based on belief and plausible reasoning, which is used to combine separate pieces of information in order to calculate the probability of the event. It is often used method of sensor fusion to deal with uncertainty associated with context reasoning by combining the independent observations of multiple sensors (e.g., the user's activity monitoring in smart home) (Hong et al., 2009; Wu et al., 2003). However, the DST has limitations and weaknesses. In particular, the Dempster's combination rule has limitations. The results of the combination has low confidences when a conflict becomes important between sources. Thus, we use the Dezert-Smarandache Theory (DSmT), which is an extended DST, as a context reasoning method. No one applies the DSmT into the ubiquitous or pervasive computing area.

7. Conclusion

Until now, we proposed context reasoning under uncertainty based on evidential fusion networks in home-based care in order to support both consistency verification of the model and context reasoning techniques. The proposed reasoning technique improved the quality of contextual information and inferred new kinds of contextual information.

Based on the defined pragmatic context classification, generalized context modeling, and proposed evidential fusion network (EFN), we proposed a dynamic context reasoning method. A dynamic context reasoning method deals with dynamic metrics such as preference, temporal consistency and relation-dependency of the context using the autonomous learning

process (ALP) and the temporal belief filtering (TBF). In addition, A dynamic context reasoning method improve the confidence level of contextual information using the proposed normalized weighting technique compared to previous fusion networks such as DST and DBNs.
To show the improvement of our approach, we compared the uncertainty levels of two fusion processes such as DSmT and DST and the confidence (i.e., GPT) levels of two fusion processes such as DSmT and DBNs using paired observations. Finally, we got the better performance compared to DST and DBNs.
In the future, we will continuous work on user experience in order to adapt the user's feelings stemming both from pragmatic and hedonic aspects of the system into the pervasive healthcare monitoring system (PHMS).

8. Acknowledgement

This work was 100% supported by the General R/D Program of the Daegu Gyeongbuk Institute of Science and Technology (DGIST).

9. References

Agarwal S.; Joshi A.; Finin T.; Yesha Y.; Ganous T. (2007). A pervasive computing system for the operating room of the future, *Mobile Networks and Applications*, Vol. 12, No. 2, March, 2007, pp. 215-228, ISSN:1383-469X

Bettini C.; Brdiczka O.; Henricksen K.; Indulska J.; Nicklas D.; Ranganathan A.; Riboni D. (2010). A survey of context modeling and reasoning techniques, *Pervasive and Mobile Computing*, Vol. 6, No. 2, April, 2010, pp. 161-180, ISSN:1574-1192

Campos M.; Juarez J. M.; Salort J.; Palma J.; Marin R. (2009). Reasoning in dynamic systems: From raw data to temporal abstract information, *Neurocomputing*, Vol. 72, No. 4 – 6, January, 2009, pp. 871-878, ISSN:0925-2312

Danninger M.; Stierelhagen R. (2008). A Context-Aware Virtual Secretary in a Smart Office Environment, *Proc. of the 16th ACM Intl. Conf. on Multimedia*, pp. 529-538, New York, NY, USA, 2008

Dargie W. (2007). The Role of Probabilistic Schemes in Multisensor Context-Awareness, *Proc. of the 5th IEEE Intl. Conf. on Pervasive Computing and Communications Workshops*, pp. 27-32, Washington, DC, USA, 2007

Dey A. K. (2001). Understandings and using context, *Personal and Ubiquitous Computing*, Vol. 5, No. 1, January, 2001, pp. 4-7, ISSN: 1617-4909

Dezert J.; Smarandache F.; Daniel M. (2004). The Generalized Pignistic Transformation, *Proc. of the 7th Intl. Conf. on Information Fusion*, pp. 384-391, Stockholm, Sweden, 2004

Dezert J.; Smarandache F. (2004). *Advances and Applications of DSmT for Information Fusion, Vol. 1*, American Research Press, ISBN: 1931233829, Rehoboth

Dezert J.; Smarandache F. (2006). *Advances and Applications of DSmT for Information Fusion, Vol. 2*, American Research Press, ISBN: 1599730006, Rehoboth

Dezert J.; Smarandache F. (2009). *Advances and Applications of DSmT for Information Fusion, Vol. 3*, American Research Press, ISBN: 1599730731, Rehoboth

Gellersen H. W.; Schmidt A.; Beigl M. (2002). Multi-sensor context-awareness in mobile devices and smart artifacts, *Mobile Networks and Applications*, Vol. 7, No. 5, October, 2002, pp. 341-351, ISSN:1383-469X

Hong X.; Nugent C.; Mulvenna M.; McClean S.; Scotney B.; Devlin S. (2009). Evidential fusion of sensor data for activity recognition in smart homes, *Pervasive and Mobile Computing*, Vol. 5, No. 3, June, 2009, pp. 236-252, ISSN:1574-1192

Jain R. (1991). *The Art of Computer Systems Performance Analysis Techniques for Experimental Design, Measurement, Simulation, and Modeling*, John Wiley & Sons, Inc., ISBN: 0471503363, Digital Equipment Corporation, Littleton, Massachusetts

Jayaraman P. P.; Zaslavsky A.; Delsing J. (2009). Dynamic Situation Modeling and Reasoning under Uncertaint, *Proc. of the Intl. Conf. on Pervasive Services*, pp. 113-122, London, UK, 2009

Lee H.; Choi J.; Elmasri R. (2008a). A conflict resolution architecture for the comfort of occupants in intelligent office, *Proc. of the 4th IET Intl. Conf. on Intelligent Environments*, pp. 1-8, University of Washington, Seatle, USA, 2008

Lee H.; Choi J.; Elmasri R. (2009). Sensor Data Fusion using DSm Theory for Activity Recognition under Uncertainty in Home-based Care, *Proc. of the 23th IEEE Intl. Conf. on Advanced Information Networking and Applications*, pp. 517-524, Bradford, UK, 2009

Lee H.; Lee B.; Park K.; Elmasri R. (2010). Fusion Techniques for Reliable Information: A Survey, *International Journal of Digital Content Technology and its Applications*, Vol. 4, No. 2, June, 2010, pp. 74-88, doi: 10.4156/jdcta

Lee H.; Park K.; Lee B.; Choi J.; Elmasri R. (2008). Issues in data fusion for healthcare monitoring, *Proc. of the 1st ACM Intl. Conf. on Pervasive Technologies Related to Assistive Environments*, pp. 1-8, Athens, Greece, 2008

Lemmi A.; Betti G. (2006). Capability approach and fuzzy set theory: description, aggregation and inference issues, *Fuzzy set approach to multidimensional proverty measurement*, Vol. 3, No. 1, pp. 93-113

Liao L.; Patterson D. J.; Fox D.; Kautz H. (2007). Learning and inferring transportation routines, *Artificial Intelligence*, Vol. 171, No. 5 – 6, April, 2007, pp. 311-331, ISSN:0004-3702

Loke S. (2006). On representing situations for context-aware pervasive computing: Six ways to tell if you are in a meeting, *Proc. of the 4th IEEE Intl. Workshop on Pervasive Computing and Communications*, pp. 35-39, Pisa, Italy, 2006

Maskell S. (2008). A Bayesian approach to fusing uncertain, imprecise and conflicting information, *Information Fusion*, Vol. 9, No. 2, April, 2004, pp. 259-277, ISSN:1566-2535

Murphy K. (2002). *Dynamic Bayesian Networks: Representation, Inference and Learning*, Ph. D. Disseration, Univ. of California, Berkeley, 2002

Olfati-Saber R. (2007). Distributed Kalman Filtering for Sensor Networks, *Proc. of the 46th IEEE Intl. Conf. on Decision and Control*, pp. 5492-5498, New Orleans, Louisiana, USA, 2007

Padhraic S. (1997). Belief Networks, Hidden Markov Model, and Markov Random Field: A Unifying View, *Pattern Recognition letters*, Vol. 18, No. 11 – 13, November, 1997, pp. 1261-1268, ISSN:0167-8655

Padovitz A.; Loke S. W.; Zaslavsky A.; Bartolini C. (2005). An approach to data fusion for context awareness, *Proc. of the 5th Intl. Conf. on Modeling and Using Context*, pp. 353-367, Paris, France, 2005

Patterson D. J.; Liao L.; Fox D.; Kautz H. (2003). Inferring high-level behavior from low level sensors, *Proc. of the 5th Intl. Conf. on Ubiquitous Computing*, pp. 73-89, Seattle, Washington, USA, 2003

Ramasso E.; Rombaut M.; Pellerin D. (2006). A Temporal Belief Filter Improving Human Action Recognition in Videos, *Proc. of the IEEE Intl. Conf. on Acoustics, Speech, and Signal Processing*, pp. 141-144, Toulouse, France, 2006

Ranganathan A.; Al-Muhtadi J.; Campbell R. H. (2004). Reasoning about uncertain contexts in pervasive computing environments, *IEEE Pervasive Computing*, Vol. 3, No. 2, April, 2004, pp. 62-70, ISSN:1536-1268

Razzaque M. A.; Dobson S.; Nixon P. (2007). Classification and Modeling of the quality of Contextual Information, *IEEE TCAAS Letters*, 2007

Rogova G. L.; Nimier V. (2004). Reliability in information fusion: literature survey, *Proc. of the 7th Intl. Conf. on Information fusion*, pp. 1158-1165, Stockholm, Sweden, 2004

Roy N.; Roy A.; Das S.; Basu K. (2005). A cooperative learning framework for mobility-aware resource management in multi-inhabitant smart homes, *Proc. of the 2nd IEEE Intl. Conf. on Mobile and Ubiquitous Systems: Networking and Services*, pp. 393-403, San Diego, CA, USA, 2005

Shafer G. (1976). *A Mathematical Theory of Evidence*, Princeton Univ. Press, ISBN: 0691081751, Princeton, New Jersey

Smarandache F.; Dezert J. (2005). Information fusion based on new Proportional Conflict Redistribution rules, *Proc. of the 8th Intl. Conf. on Information Fusion*, pp. 907-914, Philadelphia, PA, USA, 2005

Smets P. (2000). Data Fusion in the transferable Belief Model, *Proc. of the 3rd Intl. Conf. on Information Fusion*, pp. 21-33, Paris, France, 2000

Soyer C.; Bozma H. I.; Isterfanopulos Y. (2003). Attentional Sequence-based Recognition: Markovian and Evidential Reasoning, *IEEE Transactions on Systems, Mans, and Cybernetics, Part B: Cybernetics*, Vol. 33, No. 6, November, 2003, pp. 937-950, ISSN:1083-4419

Valiris G.; Chytas P.; Glykas M. (2005). Making decisions using the balanced scorecard and the simple multi-attribute raitng technique, *Performance Measurement and Metrics*, Vol. 6, No. 3, pp. 159-171, ISSN: 1467-8047

Welch G.; Bishop G. (2006). *An Introduction to the Kalman Filter*, UNC-Chapel Hill, TR 95-041, Chapel Hill, NC, USA

Winterfeld D.; Edwards W. (1986). *Decision Analysis and Behavioral Research*, Cambridge, ISBN: 0521273048, University Press, Cambridge, UK

Wu H.; Siegel M.; Ablay S. (2003). Sensor fusion using dempster-shafer theory II: static weighting and kalman filter-like dynamic weighting, *Proc. of the 20th IEEE Instrumentation and Measurement Technology Conference*, pp. 907-912, Vail, CO, USA, 2003

Zadeh L. A. (1999). Fuzzy sets as a basis for a theory of possibility, *Fuzzy Sets and Systems*, Vol. 100, No. 1, April, 1999, pp. 9-34, ISSN:0165-0114

Zhang Y.; Ji Q. (2006). Active and Dynamic Information Fusion for Multisensor Systems with Dynamic Bayesian Networks, *IEEE Transactions on Systems, Mans, and Cybernetics, Part B: Cybernetics*, Vol. 36, No. 2, March, 2006, pp. 467-472, ISSN:1083-4419

2

Sensor Fusion for Precision Agriculture

Viacheslav I. Adamchuk[1], Raphael A. Viscarra Rossel[2],
Kenneth A. Sudduth[3] and Peter Schulze Lammers[4]
[1]McGill University,
[2]CSIRO Land & Water,
[3]USDA-ARS
[4]University of Bonn
[1]Canada
[2]Australia
[3]USA
[4]Germany

1. Introduction

With the rapid rise in demand for both agricultural crop quantity and quality and with the growing concern of non-point pollution caused by modern farming practices, the efficiency and environmental safety of agricultural production systems have been questioned (Gebbers and Adamchuk, 2010). While implementing best management practices around the world, it was observed that the most efficient quantities of agricultural inputs vary across the landscape due to various naturally occurring, as well as man-induced, differences in key productivity factors such as water and nutrient supply. Identifying and understanding these differences allow for varying crop management practices according to locally defined needs (Pierce and Nowak, 1999). Such spatially-variable management practices have become the central part of precision agriculture (PA) management strategies being adapted by many practitioners around the world (Sonka et al., 1997). PA is an excellent example of a system approach where the use of the sensor fusion concept is essential.

Among the different parameters that describe landscape variability, topography and soils are key factors that control variability in crop growing environments (Robert, 1993). Variations in crop vegetation growth typically respond to differences in these microenvironments together with the effects of management practice. Our ability to accurately recognize and account for any such differences can make production systems more efficient. Traditionally differences in physical, chemical and biological soil attributes have been detected through soil sampling and laboratory analysis (Wollenhaupt et al., 1997; de Gruijter et al., 2006). The cost of sampling and analysis are such that it is difficult to obtain enough samples to accurately characterize the landscape variability. This economic consideration resulting in low sampling density has been recognized as a major limiting factor.

Both proximal and remote sensing technologies have been implemented to provide high-resolution data relevant to the soil attributes of interest. Remote sensing involves the deployment of sensor systems using airborne or satellite platforms. Proximal sensing requires the operation of the sensor at close range, or even in contact, with the soil being measured, allowing *in situ* determination of soil characteristics at, or below, the soil surface at specific locations (Viscarra Rossel et al., 2011).

Alternatively, the crop itself can be viewed as a bioindicator of variable growing conditions. The most frequently used crop-related data source is a yield map, particularly in locations where grain cropping is practiced in large fields. Yield maps summarize the overall impact of management activities and of natural conditions, such as weather and soils. However, yield data provide only a retrospective analysis and does not allow the user to address any spatial and temporal inconsistencies in crop growth during the corresponding growing season. Therefore, different in-season sensing scenarios have been implemented to provide feedback on crop performance in time to alter management decisions according to local needs. One example of this is online crop canopy sensing for in-season fertilizer management. Crop canopy reflectance in visible and near-infrared wavelengths is normally used to calculate vegetation indexes, which can be related to plant biomass, chlorophyll content, and/or nitrate stress (Shanahan et al., 2008). It has been demonstrated that detection and identification of weeds using machine vision systems is feasible as well; other crop status sensing techniques such as laser florescence, thermal imaging and ultrasonic proximity sensing are the subject of ongoing research.

One of the main limitations of any sensor-based management is that virtually every layer of information can respond to more than one soil, landscape, or crop property used to describe growing conditions and process. This makes a corresponding decision-making strategy uncertain and/or complex when attempting to deploy it over different production settings (McBratney et al. 2005). Using a combination of conceptually different sensing techniques and integrating the subsequent data holds promise for providing more accurate property estimates, leading to more robust management and increased adoptability of sensor-based crop management. The goal of this publication is to discuss the concept of sensor fusion relevant to precision agriculture and to provide the framework for future research in this area.

2. Proximal sensing sechnologies

Some proximal sensor systems can be operated in a stationary field position and can be used to: 1) make a single site measurement; 2) produce a set of measurements related to different depths at a given site; or 3) monitor changes in soil properties when installed at a site for a period of time. Although single site measurements can be beneficial for a variety of applications, high-resolution thematic soil maps are typically obtained when measurements are conducted while the sensor systems are moved across the landscape. These on-the-go proximal soil sensing technologies have become an interdisciplinary field of research and their development provides essential tools for precision agriculture and other areas of natural resources management (Hummel et al., 1996; Sudduth et al., 1997; Adamchuk et al., 2004; Shibusawa, 2006; Viscarra Rossel et al., 2011). Proximal crop sensors have been used to determine such physiological parameters as biomass, chlorophyll content, height, etc. that indicate a spatially non-consistent status of agricultural crops, such as nitrogen deficiency or water stress (Solari et al., 2008; Samborski et al., 2009).

Sensors have been used to supplement either predictive or reactive approaches to differentiated crop management. As shown in Figure 1, the reactive (real-time) method of sensor deployment means that the application rate changes in response to local conditions assessed by a sensor at the time of application. In contrast, for a predictive (map-based) strategy, many soil sensors are used to generate soil property maps that can be processed and interpreted off site prior to making decisions about the optimized distribution of agricultural inputs. Unfortunately, real-time sensing is not feasible due to the time delay or is not optimal if the spatial distribution of sensed soil properties (e.g., soil electrical conductivity) does not change during the growing season. On the other hand, more dynamic parameters (e.g. crop performance indices) need to be defined in real-time so that differentiation of an agricultural input can be accomplished on time to address the cause of variable crop performance. Therefore, some research groups have focused their recent studies on the most promising integrated method (Figure 1c).

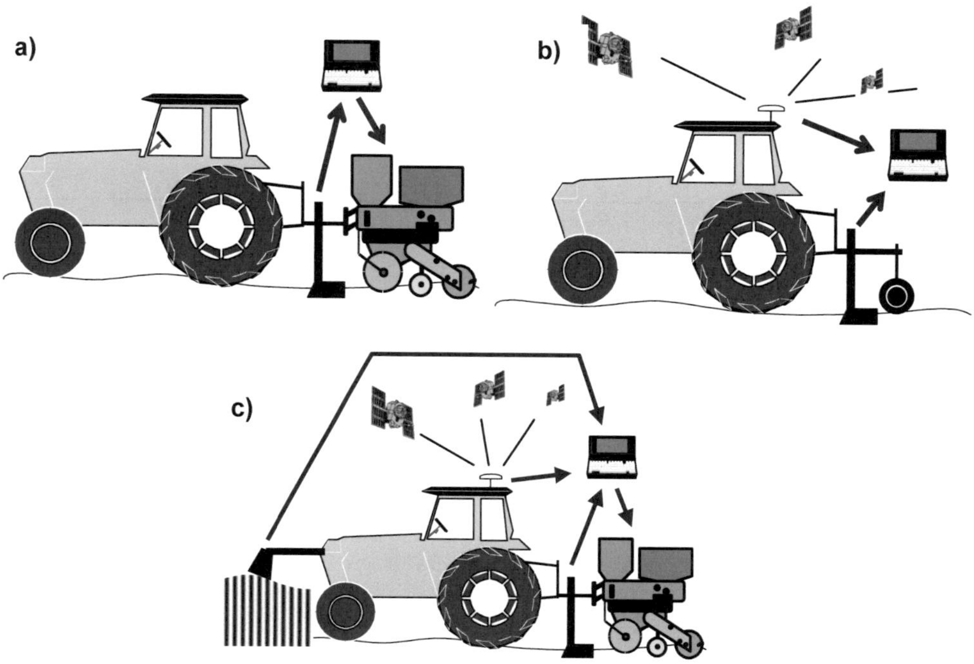

Fig. 1. Proximal sensing deployment strategies that are based on: real time (a), map-based (b), and integrated (c) approaches.

A great variety of design concepts exist, but most proximal soil sensors being developed rely on measuring the soil's ability to reflect or emit electromagnetic energy. In addition, some sensors have been used to quantify the amount of electrical charge that soil media can conduct and/or accumulate. Also, electrochemical sensors directly detect the activity of specific ions, while mechanistic sensors provide signals relevant to the physical interaction between soil and a measuring tool. Table 1 summarizes different proximal soil sensing systems and classifies them according to the source of energy (active versus passive) and principle of operation (invasive versus non-invasive or stationary versus mobile). The physical and chemical properties of soil have been reported to have a direct (D) or indirect (I) relationship with the signal obtained using different types of sensors listed in Table 2.

In addition to locating sensor measurements, the availability of accurate global navigation satellite system (GNSS) receivers permit the collection of low cost digital elevation data. This data can then be used to provide information on surface geometry (e.g. slope, aspect,

Sensor Type (wavelength, m)	Method[1]	Energy[2]	Interaction[3]	Operation[4]
Gamma-ray (10^{-12})	INS	A	N	S/M
	TNM	A	I	S
	Spectroscopy	A/P	N	S/M
X-ray (10^{-10})	XRF	A	N	S
	XRD	A	N	S
Optical (10^{-8}-10^{-4})	UV	A	N	S
	Visible	A/P	N/I	S/M
	NIR	A/P	N/I	S/M
	MIR	A	N	S
	LIBS	A	N	S
Microwave (10^{-2})	Microwave	A	N	S
Radio wave (10^{1}-10^{6})	TDR	A	I	S
	FDR/Capacitance	A	I	S/M
	GPR	A	N	S/M
	NMR	A	N	S
	EMI	A	N	S/M
Electrical	EC/ER	A	I	M
	Soil matric potential	P	I	S
Electrochemical	ISE/ISFET	P	N	S/M
Mechanistic	Implement draft	P	I	M
	Mechanical impedance	P	I	S/M
	Fluid permeability	A	I	S/M
	Acoustic	P	I	S/M

1 - inelastic neutron scattering (INS), (TNM), x-ray fluorescence (XRF), x-ray difraction (XRD), ultraviolet (UV), near-infrared (NIR), mid-infrared (MIR), laser induced breakdown spectroscopy (LIBS), time domain reflectrometry (TDR), frequency domain reflectrometry (TDR), ground penetrating radar (GPR), nuclear magnetic resonance (NMR), electromagnetic induction (EMI), electrical conductivity (EC), electrical resistivity (ER), ion-selective electrode (ISE), ion-selective field effect transistor (ISFET)

2 - active sensors (A) provide their own source of energy, passive sensors (P) rely on ambient or emitted energy

3 - invasive sensors (I) rely on a direct contact with soil, non-invasive sensors (N) are operated without any soil distortion

4 - stationary operation (S) requires placing the sensor in a specific geographic location at a fixed or variable depth, mobile operation (M) allows on-the-go soil sensing.

Table 1. Classification of Proximal Soil Sensors (adapted from Viscarra Rossel et al., 2011)

landscape position) as an indirect descriptor of soil. Local variations in terrain control the movement of sediments, water and solutes in the landscape. Soil formation is strongly influenced by these processes and terrain-related attributes can be used to help characterize the spatial distribution of soil properties (Moore et al., 1993). Elevation data also provides the landscape framework for interpreting results from other sensors.

3. Sensor fusion

As every soil-sensing technology has strengths and weaknesses and no single sensor can measure all soil properties, the selection of a complementary set of sensors to measure the

Soil property[1]	Sensor type							
	Gamma-ray	X-ray	Optical	Microwave	Radio wave	Electrical	Electrochemical	Mechanistic
Chemical								
Total carbon	D	D	D					
Organic carbon	I		D					
Inorganic carbon	I		D					
Total nitrogen	D	D	D					
Nitrate-nitrogen			I		I	I	D	
Total Phosphorus	D	D	I					
Extractable phosphorus								
Total Potassium	D	D	D					
Extractable potassium			I				I	
Other major nutrients	D	D	D					
Micronutrients, elements	D	D	D					
Total Iron	D	D	D		I			
Iron oxides	I		D		I			
Heavy metals	D	D	I					
CEC	I		I			I		
Soil pH	I		I		D		D	
Buffering capacity and LR			I				I	
Salinity and sodicity					D	D	D	
Physical								
Color			D					
Water content	D		D	D	D	D		I
Soil matric potential	I					D		I
Particle size distribution	I		I		I	I		I
Clay minerals	I	D	D			I		I
Soil strength								D
Bulk density	I		I		D			I
Porosity								D
Rooting depth					I			D

[1] – soil properties directly (D) or indirectly (I) predictable using different types of proximal soil sensors

Table 2. Predictability of Main Soil Properties Using Different Soil Sensing Concepts (adapted from Viscarra Rossel et al., 2011)

required suite of soil properties is important. Integrating multiple proximal soil sensors in a single multisensor platform can provide a number of operational benefits over single-sensor systems, such as: robust operational performance, increased confidence as independent measurements are made on the same soil, extended attribute coverage, and increased dimensionality of the measurement space (e.g., conceptually different sensors allow for an emphasis on different soil properties).

There are few reports of multisensor systems directed at PSS in the literature. For example, Lund et al. (2005) and later Jonjak et al. (2010) reported on a mobile sensor platform that simultaneously measures soil pH and apparent electrical conductivity (Figure 2). This system has been used to develop lime prescription maps, as electrical conductivity helps differentiate liming needs for soils with different texture at the same level of acidity. Adding a real time kinematic (RTK) level GNSS receiver allowed for the development of accurate elevation maps that together with the map of apparent electrical conductivity helped delineate field areas with different water holding capacity (Pan et al., 2008). Adamchuk et al. (2005) used the same apparatus to measure soil nitrate, soluble potassium and sodium at the same time as pH. An NIR sensor has also been suited for a later version of this multisensor platform (Christy, 2008).

Fig. 2. Sensor system integrating soil electrical conductivity and pH mapping along with a centimeter-level GNSS receiver (Jonjak et al., 2010).

In other research, Adamchuk and Christensen (2005) described a system that simultaneously measured soil mechanical resistance, optical reflectance and capacitance (Figure 3). Integrating the three types of sensors addressed spatial variability in soil organic matter, water content and compaction. Taylor et al. (2006) reported on the development of a multisensor platform consisting of two EMI instruments, ER and pH sensors, a gamma-radiometer and a high-resolution DGPS (Figure 4). Such a system can be used to investigate the entire array of physical and chemical soil characteristics and represents an ultimate solution that can be simplified when adopted for a given application.

In addition to mapping spatial soil variability, there is a need to explore the way in which soil properties change with depth and time. For that reason a variety of penetrometers

Fig. 3. Prototype sensor integrating optical, mechanical and capacitance components (Adamchuk et al., 2005).

Fig. 4. A multisensor platform integrating several EC/ER sensors with gamma radiometry and soil pH sensing capabilities (Taylor et al., 2006).

integrating different sensors has been developed. For example, Sun et al. (2008) reported on the development of a multisensor technique for measuring the physical properties of soil, including soil water, mechanical strength and electrical conductivity (Figure 5).

Wireless sensor networking allows sensor fusion to be employed in mobile or stationary sensor applications. A stationary soil probe application provides the instrumentation for the long term monitoring of soil conditions. For example, a network of soil water content monitoring sites (Figure 6) can be used to blend temporal data obtained from different locations across the landscape to alter irrigation scheduling to optimize water use efficiency (Pan et al., 2010). In addition, the wireless transfer of data and signals from mobile sensors extends multiple sensor integration to various positions on agricultural machinery. By minimizing the physical connections between sensors, smart sensor operations can be designed.

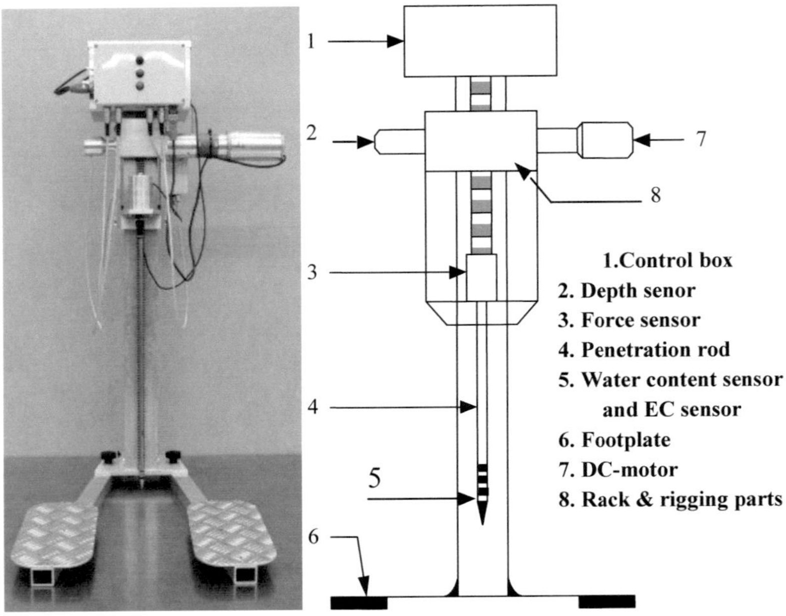

Fig. 5. Vertical cone penetrometer with sensors for soil water content and apparent electrical conductivity (Sun et al., 2008).

Fig. 6. An example of wireless sensor network (Pan et al., 2010).

With regards to proximal crop sensing, our on-going research (Shiratsuchi et al., 2009) employs a system integrating active crop canopy reflectance sensing with crop height assessment using ultrasonic sensors along with crop canopy temperature sensing (Figure 7). The need for such integration can be explained by the difference in crop physiology when either nitrogen or water stress conditions are observed.

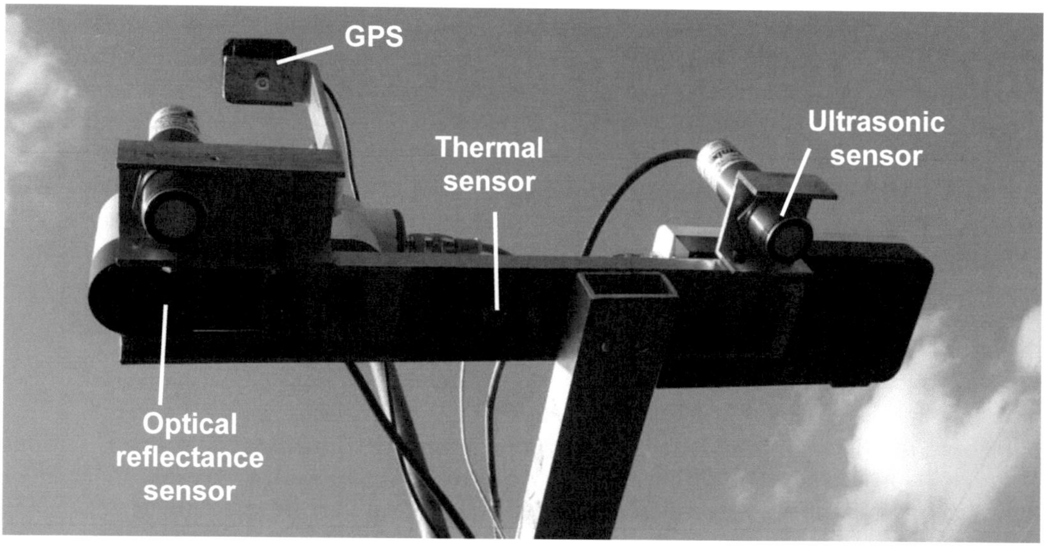

Fig. 7. An integrated crop sensing system (upgraded prototype from Shiratsuchi et al., 2009).

4. Data integration

Producers prefer sensors that provide direct inputs for existing prescription algorithms. Instead, commercially available sensors provide measurements such as apparent electrical conductivity that cannot be used directly since the absolute value depends on a number of physical and chemical soil properties such as texture, organic matter, salinity, moisture content, temperature, etc. In contrast, these sensors give valuable information about soil differences and similarities which make it possible to divide the field into smaller and relatively homogeneous areas referred to as finite management elements (FME) or management zones. For example, such FME could be defined according to the various soil types found within a field. In fact, electrical conductivity maps usually reveal boundaries of certain soil types better than conventional soil survey maps. Various anomalies such as eroded hillsides or ponding can also be easily identified on an EC map. Different levels of productivity observed in yield maps also frequently correspond to different levels of electrical conductivity.

Therefore, it seems reasonable to use electrical conductivity maps along with other data layers (e.g., yield maps, aerial imagery, terrain, management history, etc.) to discover the heterogeneity (variability) of crop growing conditions within a field. When based on multiple data layers, FMEs with a similar EC and relatively stable yield may receive a uniform treatment that can be prescribed based on a reduced number of soil samples located within each FME. In addition, soil sensors may be useful in identifying areas within fields which are less profitable or environmentally risky to farm. Work by Corwin and Lesch (2003), and by Heiniger et al. (2003), can serve as examples of site-specific data management that includes processing of electrical conductivity maps.

With regards to proximal crop sensing, optimization of application rates of crop inputs may require combining data from both crop and soil sensors. One type of crop sensor has been used to detect parameters related to the physical crop size using mechanical, ultrasonic or other proximal sensing. Recently, optical reflectance sensors that detect the ability of the crop canopy to reflect light in the visible and near-infrared parts of the electromagnetic spectrum have become popular (Shanahan et al., 2008). Sensor-based information on physical crop size has been used to vary the application rate of agricultural chemicals according to the predicted demand, while crop reflectance sensing has been used to alter the in-season supply of fertilizer and/or water to supplement what is locally available from the soil. However, in both cases information on soil variability may need to be combined with plant information to optimize in-season fertilization to account for a spatially different crop response (Roberts et al., 2010). Discussed earlier field terrain and soil electrical conductivity maps can be used to account for spatial differences in soil conditions.

Figure 8 illustrates the process of combining different sources of precision agriculture data that can be applied to assist with crop management decisions. Data can be obtained both from mobile, real-time sensing and from georeferenced maps of parameters such as crop yield and topography. The integration process may lead to management zone delineations and interpolated high-resolution maps that can be used to prescribe the spatially-variable management of agricultural inputs such as fertilizer. Alternatively, data integrated temporally could be used to manage an in-season farming operation, such as irrigation.

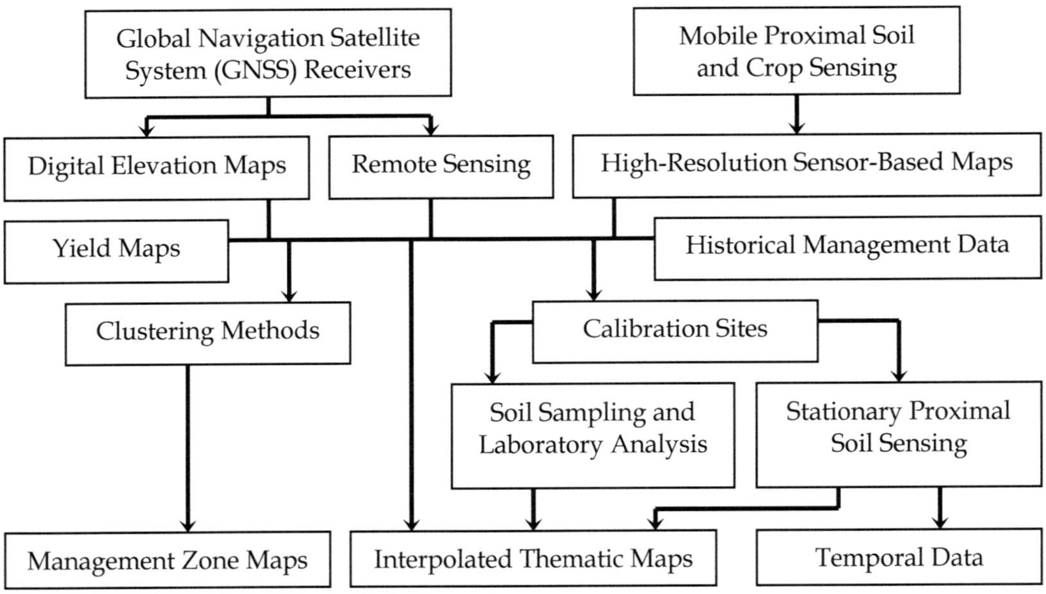

Fig. 8. Data integration in precision agriculture.

As illustrated in the flowchart, discrete management of finite field areas (zones) can be conducted based on maps produced using clustering methods that can integrate multiple layers of crop performance (e.g., yield) and/or remote/proximal sensing data (Fridgen et al, 2004. In many cases, each zone will need additional investigation or data collection to determine the most appropriate treatment plan. Another approach is to algorithmically convert one or multiple layers of high-resolution sensor data into a thematic map. In this case, additional point-based measurements or calibration sampling may be needed to relate the sensor signal to the parameter of interest. Finally, high-density data can be used to locate temporal monitoring sites that will provide information on how different field areas behave during the growing season.

As many precision agricultural practices are specific to a given geographical area and to particular cropping systems, the set of most informative sensors and data layers may also vary from location to location and from practice to practice. On one hand, adding new data always requires additional costs, but does not always bring new information as many spatial data layers are highly correlated. Nevertheless, when different sensors are assigned different functions in the development of a multisensor system, more robust solutions can be found and deployed over a wider range of farm operations. New research in the area of sensor fusion for precision agriculture is expected to provide a variety of such examples.

6. Summary

Precision agriculture encompasses identifying, understanding and utilizing information that quantifies variations in soil and crop within agricultural fields. The information needed is generally spatially and/or temporally intensive, which has led to the development of various sensing technologies that assess the soil or crop. These sensing systems are based on diverse measurement concepts, including electrical and electromagnetic, optical and radiometric, mechanistic, and electrochemical.

Robustness of single-sensor measurements is often less than ideal because virtually all currently used sensor technologies can respond to more than one basic parameter of interest. For example, crop canopy reflectance sensors can be affected by multiple stressors such as water or nitrogen deficiencies, the reflectance of the underlying soil, and the size of the crop plants. A sensor fusion approach that integrates canopy reflectance sensing with other sensors measuring plant size and soil parameters has the potential to improve the measurement accuracy of agronomically important stresses in the crop. Accurate measurements are important to determine the best management treatment because the economic and/or environmental risk associated with applying the wrong treatment to the crop can be large.

Some examples of integrated soil and crop sensing systems that combine multiple sensors already exist, and others are in various stages of development. However, multisensor platforms are difficult to implement in an agricultural setting due to constraints such as cost and durability. Typically low profit margins mean that agricultural producers are not willing to adopt technology with a high added cost. Reliability of sensor systems in field conditions including dust, moisture and vibration is difficult to attain, particularly given the cost constraints. The need to keep multiple sensors functioning simultaneously magnifies this problem. Nevertheless, researchers and developers have recognized the benefits of integrating multiple sensor datasets for agricultural decision-making. Finding an appropriate set of sensors and spatial data layers for a given application is a research topic of current interest around the world. We expect that many more examples of sensor fusion for precision agriculture will appear in the near future.

7. References

Adamchuk, V.I.; Lund, E.; Sethuramasamyraja, B. ; Morgan, M.T. ; Dobermann, A. & Marx, D.B. (2005) Direct measurement of soil chemical properties on-the-go using ion-selective electrodes. *Computers and Electronics in Agriculture* 48: 272-294.

Adamchuk, V.I. & Christenson, P.T. (2005) An integrated system for mapping soil physical properties on-the-go: the mechanical sensing component. In: *Precision Agriculture: Papers from the Fifth European Conference on Precision Agriculture*, Uppsala, Sweden, 9-12 June 2005, ed. J. Stafford, 449-456. Wageningen Academic Publishers, Wageningen, The Netherlands.

Adamchuk, V.I. ; Hummel, J.W. ; Morgan, M.T. & Upadhyaya, S.K. (2004) On-the-go soil sensors for precision agriculture. *Computers and Electronics in Agriculture* 44: 71-91.

Christy, C.D. (2008) Real-time measurement of soil attributes using on-the-go near infrared reflectance spectroscopy. *Computers and Electronics in Agriculture* 61: 10–19.

Corwin, D.L. & Lesch, S.M. (2003) Application of soil electrical conductivity toprecision agriculture: theory, principles, and guidelines. *Agronomy Journal* 95: 455–471.

de Gruijter, J., D.J. Brus, M.F.P. Bierkens, and M. Knotters. 2006. *Sampling for Natural Resource Monitoring*. Springer, New York, New York, USA.

Fridgen, J.J.; Kitchen, N.R.; Sudduth, K.A.; Drummond, S.T.; Wiebold, W.J. & Fraissed, C.W. (2004) Management zone analyst (MZA): software for subfield management zone delineation. *Agronomy Journal* 96: 100-108.

Gebbers, R. & Adamchuk, V.I. (2010) Precision agriculture and food security. *Science* 327: 828-831.

Heiniger, R.W.; McBride, R.G. & Clay, D.E. (2003) Using soil electrical conductivity to improve nutrient management. *Agronomy Journal* 95: 508–519.

Hummel, J.W., Gaultney, L.D. & Sudduth, K.A. (1996) Soil property sensing for site-specific crop management. *Computers and Electronics in Agriculture* 14: 121–136.

Jonjak, A.K.; Adamchuk, V.I.; Wortmann, C.S.; Ferguson, R.B. & Shapiro, C.A. (2010) A comparison of conventional and sensor-based lime requirement maps. In: *Proceedings of the Tenth International Conference on Precision Agriculture*, Denver, Colorado, 18-21 July 2010, ed. R. Kholsa. Colorado State University, Fort Collins, Colorado, USA (CD publication, 15 pages).

Lund, E.D.; Adamchuk, V.I.; Collings, K.L.; Drummond, P.E. & Christy, C.D. (2005) Development of soil pH and lime requirement maps using on-the-go soil sensors. In: *Precision Agriculture: Papers from the Fifth European Conferenceon Precision Agriculture*, Uppsala, Sweden, 9-12 June 2005, ed. J.Stafford, 457-464. Wageningen Academic Publishers, Wageningen, The Netherlands.

McBratney, A.; Whelan, B.; Ancev, T. & Bouma, J. (2005) Future directions of precision agriculture. *Precision Agriculture* 6: 7-23.

Moore, I.D.; Gessler, P.E.; Nielsen, G.A. & Peterson, G.A. (1993). Soil attribute prediction using terrain analysis. *Soil Science Society of America Journal* 57: 443–452.

Pan, L.; Adamchuk, V.I.; Martin D.L.; Schroeder M.A. & Ferguson, R.B. (2010) Analysis of water use efficiency using on-the-go soil sensing and a wireless network. In: *Proceedings of the Tenth International Conference on Precision Agriculture*, Denver, Colorado, 18-21 July 2010, ed. R. Kholsa. Colorado State University, Fort Collins, Colorado, USA (CD publication, 13 pages).

Pierce, F.J. & Nowak, P. (1999) Aspects of precision agriculture. *Advances in Agronomy* 67: 1-85.

Robert, P.C. (1993) Characterization of soil conditions at the field level for soil specific management. *Geoderma* 60: 57-72.

Roberts, D.F.; Shanahan J.F.; Ferguson R.B.; Adamchuk V.I. & Kitchen N.R. (2010) A crop and soil strategy for sensor-based variable-rate nitrogen management. In: *Proceedings of the Tenth International Conference on Precision Agriculture*, Denver, Colorado, 18-21 July 2010, ed. R. Kholsa. Colorado State University, Fort Collins, Colorado, USA (CD publication, 15 pages).

Samborski, S.M.; Tremblay, N. & Fallon, E. (2009) Strategies to make use of plant sensors-based diagnostic information for nitrogen recommendations. *Agronomy Journal* 101: 800-816.

Shanahan, J.F.; Kitchen N.R.; Raun W.R. & Schepers J.S. (2008) Responsive in-season nitrogen management for cereals. *Computers and Electronics in Agriculture* 61: 51–62.

Shibusawa, S. (2006) Soil sensors for precision agriculture. In: *Handbook of Precision Agriculture. Principles and Applications*, A. Srinivasan, ed., 57.90. Food Products Press, New York, New York, USA.

Shiratsuchi, L.S.; Ferguson, R.B.; Adamchuk, V.I.; Shanahan, J.F. & Slater, G.P. (2009) Integration of ultrasonic and active canopy sensors to estimate the in-season nitrogen content for corn. In: *Proceedings of the 39th North Central Extension-Industry Soil Fertility Conference*, Des Moines, Iowa, 18-19 November 2009. International Plant Nutrition Institute, Norcross, Georgia, USA.

Simbahan, G.C. & Dobermann, A. (2006) An algorithm for spatially constrained classification of categorical and continuous soil properties. *Geoderma* 136: 504-523.

Solari, F.; Shanahan, J.; Ferguson, R.; Schepers, J. & Gitelson, A. (2008) Active sensor reflectance measurements of corn nitrogen status and yield potential. *Agronomy Journal*, 100: 571–579.

Sonka, S.T.; Bauer, M.E.; Cherry, E.T.; Colburn, J.W.; Heimlich, R.E.; Joseph, D.A.; Leboeuf, J.B.; Lichtenberg, E.; Mortensen, D.A.; Searcy, S.W.; Ustin, S.L. & Ventura, S.J. (1997) *Precision agriculture in the 21st century. Geospatial and information technologies in crop management*. Committee on Assessing Crop Yield: Site-Specific Farming, Information Systems, and Research Opportunities, Board of Agriculture, National Research Council. National Academy Press, Washington, DC, USA.

Sudduth, K.A.; Hummel J.W. & Birrell, S.J. (1997) Sensors for site-specific management. In: *The State of Site-Specific Management for Agriculture*, F.T. Pierce and E.J. Sadler, eds., 183-210. ASA-CSSA-SSSA, Madison, Wisconsin, USA.

Sun, Y.; Schulze Lammers, P.; Ma, D.; Lin, J. & Zeng, Q. (2008) Determining soil physical properties by multi-sensor technique. *Sensors and Actuators* 147: 352-357.

Taylor, J.A.; McBratney, A.B.; Viscarra Rossel, R.A.; Minansy, B.; Taylor, H.; Whelan, B. & Short, M. (2006). Development of a multi-sensor platform for proximal soil sensing. In: *Proceedings of the 18th World Congress of Soil Science,* July 9–15, 2006. World Congress of Soil Science, Philadelphia, Pennsylvania, USA.

Viscarra Rossel, R.A.; Adamchuk, V.A.; Sudduth, K.A.; McKenzie, N.J. & Lobsey, C. (2011) Proximal soil sensing: a spectrum of opportunities. *Advances in Agronomy* (accepted).

Viscarra Rossel, R.A., Adamchuk, V.A., Sudduth, K. McKenzie, N.J., Lobsey, C. 2011. Proximal soil sensing. Advances in Agronomy, (in press).

Wollenhaupt, N.C.; Mulla, D.J. & Gotway Crawford, C.A. (1997) Soil sampling and interpolation techniques for mapping spatial variability of soil properties. In: *The State of Site-Specific Management for Agriculture,* F.T. Pierce and E.J. Sadler, eds., 19-53. ASA-CSSA-SSSA, Madison, Wisconsin, USA.

3

Localization and Tracking Using Camera-Based Wireless Sensor Networks

J.R. Martínez-de Dios, A. Jiménez-González and A. Ollero
Robotics, Vision and Control Research Group, University of Seville
Spain

1. Introduction

This chapter presents various methods for object detection, localization and tracking that use a Wireless Sensor Network (WSN) comprising nodes endowed with low-cost cameras as main sensors. More concretely, it focuses on the integration of WSN nodes with low-cost micro cameras and describes localization and tracking methods based on Maximum Likelihood and Extended Information Filter. Finally, an entropy-based active perception technique that balances perception performance and energy consumption is proposed.

Target localization and tracking attracts significant research and development efforts. Satellite-based positioning has proven to be useful and accurate in outdoor settings. However, in indoor scenarios and in GPS-denied environments localization is still an open challenge. A number of technologies have been applied including inertial navigation (Grewal et al., 2007), ultra-wideband (Gezici et al., 2005) or infrared light signals (Depenthal et al., 2009), among others.

In the last decade, the explosion of ubiquitous systems has motivated intense research in localization and tracking methods using Wireless Sensor Networks (WSN). A good number of methods have been developed based on Radio Signal Strength Intensity (RSSI) (Zanca et al., 2008) and ultrasound time of flight (TOF) (Amundson et al., 2009). Localization based on Radio Frequency Identification (RFID) systems have been used in fields such as logistics and transportation (Nath et al., 2006) but the constraints in terms of range between transmitter and reader limits its potential applications. Note that all the aforementioned approaches require active collaboration from the object to be localized -typically by carrying a receiver- which imposes important limitations in some cases.

Also, recently, multi-camera systems have attracted increasing interest. Camera based localization has high potentialities in a wide range of applications including security and safety in urban settings, search and rescue, and intelligent highways, among many others. In fact, the fusion of the measurements gathered from distributed cameras can reduce the uncertainty of the perception, allowing reliable detection, localization and tracking systems. Many efforts have been devoted to the development of cooperative perception strategies exploiting the complementarities among distributed static cameras at ground locations (Black & Ellis, 2006), among cameras mounted on mobile robotic platforms (Shaferman & Shima, 2008) and among static cameras and cameras onboard mobile robots (Grocholski et al., 2006).

In contrast to other techniques, camera-based Wireless Sensor Networks, comprised of distributed WSN nodes endowed with a camera as main sensor, require no collaboration from the object being tracked. At the same time, they profit from the communication infrastructure, robustness to failures, and re-configurability properties provided by Wireless Sensor Networks.

This chapter describes various sensor fusion approaches for detection, localization and tracking of mobile objects using a camera-based Wireless Sensor Network. The main advantages of using WSN multi-camera localization and tracking are: 1) they exploit the distributed sensing capabilities of the WSN; 2) they benefit from the parallel computing capabilities of the distributed nodes; 3) they employ the communication infrastructure of the WSN to overcome multi-camera network issues. Also, camera-based WSN have easier deployment and higher re-configurability than traditional camera networks making them particularly interesting in applications such as security and search and rescue, where pre-existing infrastructure might be damaged.

This chapter is structured as follows:

- Section 2 includes a brief introduction to Wireless Sensor Networks and describes the basic scheme adopted for the camera-based WSN.
- Section 3 presents a basic data fusion based on Maximum Likelihood approach. The method has bad performance in case of losses of WSN messages, which can be not infrequent in some applications.
- Section 4 proposes a data fusion method based on Extended Information Filter. This method has good performance at moderate computer cost.
- Section 5 summarizes an entropy-based active perception technique that dynamically balances between perception performance and use of resources.
- Section 6, which describes implementation details and presents some experimental results.

Finally, Section 7 is devoted to the final discussions and conclusions.

2. Camera-based WSN

2.1 Brief introduction to wireless sensor networks

A Wireless Sensor Network (WSN) consists of a large number of spatially distributed devices (nodes) with sensing, data storage, computing and wireless communication capabilities. Low size, low cost and particularly low power consumption are three of the key issues of WSN technology. Nodes are designed to operate with minimal hardware and software requirements, see basic scheme of the main modules in Fig. 1Left. They often use 8 or 16-bit microcontrollers at low processing rates and a limited RAM capacity for data storage. Nodes often require a few milliwatts for operation. Most nodes can be set in a standby state, from which they wake up occasionally, for instance when one sensor detects an event. Their radio transceivers are also very energy efficient and their transmission range is typically less than 100 m in the open air. Besides, its bandwidth is often low.

In contrast to the simplicity of each node, the main strengths of WSN rely on the cooperation of a number of nodes for cooperatively performing tasks. In fact, a good number of algorithms have been developed to provide them significant flexibility, scalability, tolerance to failures and self-reconfiguration. WSN are typically organized in tree-like channels between data sources (nodes) and data sinks (WSN base), see Fig. 1Right. Despite this apparent simplicity, algorithms for network formation and information routing have been intensively researched with the objective of optimizing the energy consumption, the communication delays expressed as the number of hops, the network dynamic reconfiguration or its reliability to failures. A survey on WSN routing protocols can be found in (Al-Karaki & Kamal, 2004).

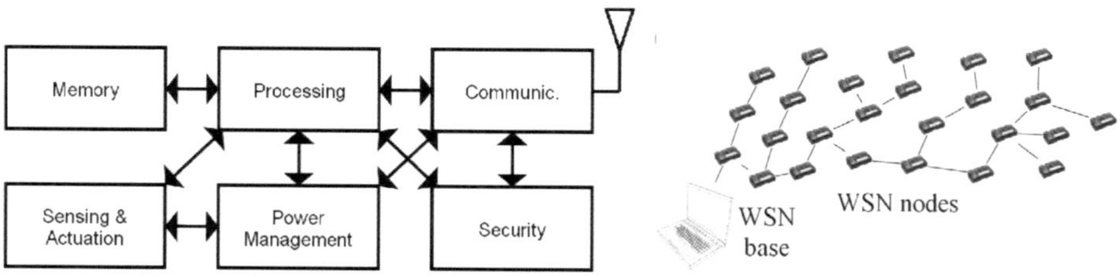

Fig. 1. Left) Basic module scheme of a WSN node. Right) Tree-like channels established in WSN between nodes (data sources) and the WSN base (data sink).

The nodes of the WSN can be equipped with a growing variety of sensors including light intensity sensors, optical barrier, presence sensors, gas sensors and GPS. These features together with battery operation facilitate its deployment with minimal invasion and low installation and maintenance costs. The standardization of communication protocols, such as IEEE 802.15.4, has facilitated the effort to extend its range of possible applications. WSN have already been applied to building control (Sandhu et al., 2004), environmental monitoring (Polastre et al., 2004) and manufacturing automation (Hanssmann et al., 2008), among others (Akyildiz et al. 2002).

2.2 Camera-based WSN

A camera-based WSN uses cameras as main sensors of the distributed WSN nodes. In contrast to the advantage of using robust and reconfigurable WSN communication, camera-based WSN must face the main constraints of WSN technology, i.e. limited computational and data storage capacity and low communication bandwidth. Thus, centralized schemes in which all the images are processed by one node are not suitable for camera-based WSN. One proposed solution is to transmit the images gathered by the cameras through the WSN (Wark et al, 2007). However, this approach has bad scalability in terms of bandwidth, being critical in problems that require images of certain resolution at a certain rate. Also, additional constraints arise in centralized schemes when considering the computational and memory capacity required to process the images from all the cameras and fuse their results in only one WSN node. Lack of reliability to failures in the central node is another important drawback.

In our case a distributed scheme is adopted: the images captured by each camera are processed locally at each node. Camera nodes have sufficient computational capabilities to execute efficient image-processing algorithms in order to extract from the images the required information, for instance the location of an object on the image plane. Hence, only distilled bandwidth-reduced information from each camera node is transmitted through the WSN. Then, a node receiving the measurements from all the camera nodes can perform data fusion algorithms to determine the location of the object in real-world coordinates. This scheme reduces drastically the bandwidth requirements and distributes the overall computational burden among the nodes of the WSN.

In the adopted scheme, each camera node applies image-processing segmentation techniques to identify and locate the object of interest on the image plane. Thus, the possibility of flexibly programming image processing algorithms is a strong requirement. The selected camera board is the *CMUcam3*, an open source programmable board endowed with a color detector with 352x288 pixels. Figure 2 shows a picture of a camera node comprised of an *Xbow* node and one *CMUcam3* board.

Fig. 2. *CMUcam3* integrated with a *Xbow* node.

We implemented a robust algorithm based on a combination of color and motion segmentations capable of being efficiently executed with limited computational and memory resources. The result of the segmentation algorithm is a rectangular region on the image plane characterized by the coordinates of its central pixel, its width and height.
Several data fusion methods are used to merge the results from the segmentation algorithms running in every camera node. Data fusion reduces the influence of errors in measurements and increases the overall system accuracy. On the other hand, it requires having the measurements from all the cameras expressed in the same reference frame. In the methods presented, each camera node obtains the coordinates of the region of interest on the image plane applying image segmentation algorithms and corrects its own optical distortions transforming them to the undistorted normalized pin-hole projection on the image plane. Each camera is internally calibrated and the calibration parameters are known at each camera node. Hence, camera nodes message packets include the distortion-corrected normalized measurements for each image analyzed. These messages are transmitted through the WSN for data fusion. This approach standardizes the measurements from all camera nodes facilitating data fusion method and distributes the computational cost among the camera nodes. For further details of the implementations refer to Section 6.

3. Localization using maximum likelihood

Maximum Likelihood (ML) is one of the basic statistical data fusion methods, (Mohammad-Djafari, 1997). Its objective is to estimate the state of an event that best justifies the observations maximizing a statistical likelihood function that can be expressed as the probability of measurement z conditioned to state x:

$$\hat{x} = \arg\max_{x} \{P(z \mid x)\} \tag{1}$$

Assume that the state is measured synchronously from N different sensors $z_1,\ldots,z_N$, where z_i is the measurement gathered by sensor i. Supposed the measurements of all the sensor $z_1,\ldots,z_N$ can be considered statistically independent, the overall likelihood function can be expressed by:

$$p(z_1,\ldots,z_N \mid x) = \prod_{i=1}^{N} p(z_i \mid x) \tag{2}$$

Assume that each measurement is subject to errors that can be considered to be originated by the influence of a high number of independent effects. By virtue of the Central Limit Theorem it can be considered to have Gaussian distribution, (Rice, 2006):

$$p(z_i \mid x) = \det(2\pi\Sigma_i)^{-1/2} \exp\left\{\frac{-1}{2}(z_i - x)^T \Sigma_i^{-1} (z_i - x)\right\} \tag{3}$$

where Σ_i is the covariance of measurements from sensor i. The ML method estimates the state as the following weighted sum:

$$x = \left(\Sigma_1^{-1} + \cdots + \Sigma_N^{-1}\right)^{-1} \left(\Sigma_1^{-1} z_1 + \cdots + \Sigma_N^{-1} z_N\right) \tag{4}$$

where each measurement is weighted proportionally to the inverse of its covariance: measurements with more noise have lower weigh in (4). The overall estimated covariance follows the expression:

$$\Sigma_x = \left(\Sigma_1^{-1} + \cdots + \Sigma_N^{-1}\right)^{-1} \tag{5}$$

It should be noted that since $\Sigma_x < \Sigma_i$ the estimate is more accurate than any measure.
The following describes the ML method adopted for camera-based WSN. Consider that a point P in the environment is observed by N camera nodes. $p_i=[x_i \;\; y_i]$ are the distortion-corrected pixel coordinates of P viewed from camera node i. Before applying (4) it is necessary to have the measurements from all the cameras in the same reference frame.
Let frame F_i be a reference frame local to camera i. The location and orientation of the N cameras in a global reference frame G are known. Let T_i be the transformation matrix from frame F_i to frame G. Assume that Z_i, the Z coordinate of P in frame F_i, is known. Taking into account the pin-hole model it is possible to project pixel p_i at distance Z_i:

$$P_i = [X_i \;\; Y_i \;\; Z_i]^T = [x_i Z_i \;\; y_i Z_i \;\; Z_i]^T \tag{6}$$

P_i represents the coordinates of P measured by camera i and expressed in frame F_i. Using T_i P_i can be transformed to frame G applying:

$$\begin{bmatrix} P_i^G \\ 1 \end{bmatrix} = T_i^{-1} \begin{bmatrix} P_i \\ 1 \end{bmatrix} \tag{7}$$

P_i^G represents the coordinates of P in frame G as measured by camera i. With the measurements from all the cameras in the same frame G, the ML method can be applied. Assume that P_i^G contains Gaussian errors with covariance matrix Σ_i. Supposed the measurements from different cameras statistically independent, the ML method estimates P with measurements P_i^G using (4).

Σ_i can be decomposed in an eigenvector matrix and an eigenvalue matrix, $\Sigma_i = L\Lambda L^{-1}$. The eigenvectors of Σ_i form the columns of L. The eigenvectors are orthonormal vectors that represent the axes of frame F_i in the global frame G. Λ is a diagonal matrix. The elements of the diagonal are the eigenvalues of Σ_i, which are the variance associated to P_i^G at each axis of frame F_i. L and Λ, and thus Σ_i can be easily constructed knowing the orientation of camera i and estimating the noise in the measurements.

Figure 3 shows an illustration of the method with two cameras. The probability distribution of the measurements from *Camera1* and *Camera2* are represented in cyan color. The probability distribution of the fused estimate is in blue. The remarkable reduction in the covariance denotes an increment in the fused estimate.

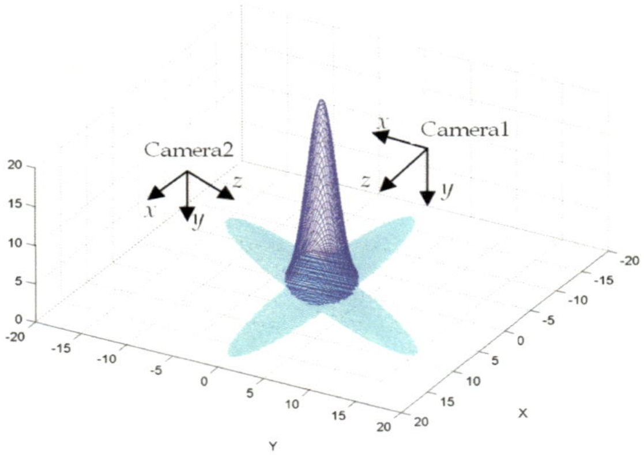

Fig. 3. Example illustrating ML data fusion method.

The described ML method can be executed in a WSN node in few milliseconds. This high efficiency facilitates schemes where camera nodes observing the same object interchange their observations and apply data fusion.

This method can be used for object localization but it is not suitable for object tracking, and even when used for localization the ML method has important constraints. Applying the ML data fusion requires having previously determined Z_i, the location of P in frame F_i. A typical approach is to set Z_i with an average value and compensate the error assuming a high value for the variance of the error at the Z axis of F_i. However, this artificial increase of uncertainty decreases the quality of the overall estimate. Another approach is to use, under an iterative scheme, the value of Z_i at time t-1. Nevertheless, this method requires assuming an initial value and, errors in estimation at time t-1 involve errors at subsequent iterations.

Furthermore, ML has high sensitivity to failures in measurements, for instance in cases where the object is out of the field of view of the camera, occluded in the image or in case of losses of WSN messages, not infrequent in some environments. This sensor fusion method relies totally on the measurements and its performance degrades when some of them are lost. Other sensor fusion techniques such as Bayesian Filters rely on observations and on models, which are very useful in case of lack of measurements.

4. Localization and tracking using EIF

Bayesian Filters (RBFs) provide a well-founded mathematical framework for data fusion. RBFs estimate the state of the system assuming that measurements and models are subject to uncertainty. They obtain an updated estimation of the system state as a weighted average using the prediction of its next state according to a system model and also using a new measurement from the sensor to update the prediction. The purpose of the weights is to give more trust to values with better (i.e., smaller) estimated uncertainty. The result is a new state estimate that lies in between the predicted and measured state, and has a better estimated uncertainty than either alone. This process is repeated every step, with the new estimate and measure of uncertainty used as inputs for the following iteration.
The Kalman Filter (KF) is maybe the most commonly used RBF method. The Kalman Filter and its dual, the Information Filter (IF), use a prediction model, that reflects the expected evolution of the state, and a measurement model, that takes into account the process through which the state is observed to respectively predict and update the system state:

$$\begin{aligned} x_t &= g(x_{t-1}) + \varepsilon_t \\ z_t &= h(x_t) + \delta_t \end{aligned} \tag{8}$$

x_t is the current system state vector, x_{t-1} is the previous state vector, z_t is the measurement vector and ε_t and δ_t are White Gaussian Noise (WGN) parameterized by their mean value and a covariance matrix.
In our problem the measurements considered are the location of the object on the image of the distributed camera nodes. Even assuming simple pin-hole cameras, these observation models are non-linear and a first order linearization is required. In this case, having non-linear prediction and measurement models leads to the Extended Information Filter (EIF). After linearizing the IF equations via Taylor Expansion, we can assume that the predicted state probability, written as Gaussian, is as follows:

$$\begin{aligned} p(x_t \mid x_{t-1}) &= \det(2\pi R_t)^{-1/2} \\ &\exp\left\{-\frac{1}{2}\left[\left(x_t - g(\mu_{t-1}) - G_t(x_{t-1} - \mu_{t-1})\right)\right]^T R_t^{-1}\left[\left(x_t - g(\mu_{t-1}) - G_t(x_{t-1} - \mu_{t-1})\right)\right]\right\} \end{aligned} \tag{9}$$

where μ_{t-1} is the mean of the previous state, R_t is the covariance of the prediction model (correspondent to ε_t) and G_t is the Jacobian matrix of g. The next state probability, written as Gaussian, is as follows:

$$\begin{aligned} p(z_t \mid x_t) &= \det(2\pi Q_t)^{-1/2} \\ &\exp\left\{-\frac{1}{2}\left[\left(z_t - h(\overline{\mu}_t) - H_t(x_t - \overline{\mu}_t)\right)\right]^T Q_t^{-1}\left[\left(z_t - h(\overline{\mu}_t) - H_t(x_t - \overline{\mu}_t)\right)\right]\right\} \end{aligned} \tag{10}$$

where $\overline{\mu}_t$ is the mean of the predicted state, Q_t is the covariance of the measurement model (correspondent to δ_t) and H_t is the Jacobian matrix of h.
Information Filters (IF) employ the so-called *canonical representation*, which consists of an information vector $\xi=\Sigma^{-1}\mu$ and matrix $\Omega=\Sigma^{-1}$. Figure 4 shows the full EIF algorithm. In each recursive iteration it computes the current system state (ξ_t , Ω_t) from the previous state and

the new measurement (ξ_{t-1}, Ω_{t-1}, z_t). Each iteration is divided in two steps: prediction (lines 1-4 in Fig. 4) and update (lines 5, 6). For more details, refer to (Thrun et al., 2005).
Since both g and h require the state as an input, it is mandatory to recover the state estimate μ from canonical parameters (see step 1 of the EIF algorithm in Fig. 4) which makes the prediction stage from the algorithm lose efficiency compared to the EKF. Nevertheless, the update stage of EIF is much more efficient than EKF and thus the former is more suitable when there are a large number of observations. In this sense, the efficiency of this algorithm with respect to other implementations is improved when a simple prediction model together with a large measurement vector z_t are used. Besides, Information Filters are also numerically more stable and are more suitable for characterizing and representing information and its absence, $\Omega=0$.

Extended_Information_Filter (ξ_{t-1}, Ω_{t-1}, z_t):

1: $\mu_{t-1} = \Omega_{t-1}^{-1}\xi_{t-1}$

2: $\overline{\Omega}_t = \left(G_t\Omega_{t-1}^{-1}G_t^T + R_t\right)^{-1}$

3: $\overline{\xi}_t = \overline{\Omega}_t g(\mu_{t-1})$

4: $\overline{\mu}_t = g(\mu_{t-1})$

5: $\Omega_t = \overline{\Omega}_t + H_t^T Q_t^{-1} H_t$

6: $\xi_t = \overline{\xi}_t + H_t^T Q_t^{-1}\left[z_t - h(\overline{\mu}_t) + H_t\overline{\mu}_t\right]$

7: return ξ_t, Ω_t

Fig. 4. EIF algorithm.

Therefore, the selection of the state and models has critical impact on the performance and computational burden of the filter. We selected a state vector typical in tracking problems that considers only the current object position and velocity $x_t=[X_t\ Y_t\ Z_t\ Vx_t\ Vy_t\ Vz_t]^T$. In our problem we can have a large number of inexpensive camera nodes. We preferred EIF over EKF due to its better scalability with the number of observations. Also, we assumed a very simple local linear motion model to reduce the burden of the prediction stage in EIF:

$$\begin{cases} X_t = X_{t-1} + Vx_{t-1} \\ Y_t = Y_{t-1} + Vy_{t-1} \\ Z_t = Z_{t-1} + Vz_{t-1} \end{cases} \qquad \begin{cases} Vx_t = Vx_{t-1} \\ Vy_t = Vy_{t-1} \\ Vz_t = Vz_{t-1} \end{cases} \tag{11}$$

Of course, we do not know a priori what kind of movement would the object perform. So we assume local linear motion and we include Gaussian noise in each coordinate to consider errors in the model. This model can efficiently represent local motions and has been extensively applied in RBFs. Also, more complex models increase the computation burden and would require a priori knowledge of the motion, unavailable in tracking of objects with no collaboration, as is the case of security applications.
The EIF uses a different observation model for each camera that is seeing the object. The observation model adopted for camera i uses as measurements the distortion-corrected pin-hole projections from camera i at time t, $p_{i,t}$. To allow the estimation of the object velocity, we

also include in the measurement the projection from camera i at time t-1, $p_{i,t-1}$. The measurement vector including measurements from all the N cameras that are tracking the object can be written as $z_t=[p_{1,t}\ p_{1,t-1}\ p_{2,t}\ p_{2,t-1} \dots p_{N,t}\ p_{N,t-1}]^T$.

The location of the object at time t in the global reference frame G, P_t, can be computed from $p_{i,t}$, its projection in the image plane of camera node i, as described in (6) and (7). Provided T_i is the transformation matrix of F_i, the reference frame of camera i, and $t_{i,j}$ represents the j-th row of T_i, the measurement from each camera node i can be related to the target position as:

$$p_{i,t} = \begin{bmatrix} x_{i,t} \\ y_{i,t} \end{bmatrix} = \begin{bmatrix} t_{i,1}[P_t\ \ 1]^T / t_{i,3}[P_t\ \ 1]^T \\ t_{i,2}[P_t\ \ 1]^T / t_{i,3}[P_t\ \ 1]^T \end{bmatrix} \tag{12}$$

Thus, the overall measurement model h which relates z_t with x_t can be written as:

$$h = \begin{bmatrix} h_{1,t} & h_{1,t-1} & \cdots & h_{N,t} & h_{N,t-1} \end{bmatrix}^T, \quad h_{i,t} = \begin{bmatrix} t_{i,1}[X_t\ \ Y_t\ \ Z_t\ \ 1]^T / t_{i,3}[X_t\ \ Y_t\ \ Z_t\ \ 1]^T \\ t_{i,2}[X_t\ \ Y_t\ \ Z_t\ \ 1]^T / t_{i,3}[X_t\ \ Y_t\ \ Z_t\ \ 1]^T \end{bmatrix} \tag{13}$$

This observation model is, as already stated, non-linear. At the updating stage the EIF requires using the Jacobian matrices of h, H_t.

Each measurement at each camera node i requires only one prediction step and one updating step. Assuming 3 cameras, the execution of an iteration of an EIF for 2D localization and tracking with 3 cameras requires approximately 6,000 floating point operations, roughly 400 ms. in a Xbow TelosB mote, such as those used in the experiments. The Bayesian approach provides high robustness in case of losses of measurements. If at time t there are no measurements, only the prediction stage of the EIF algorithm is executed. In this case, the uncertainty of the state grows more and more until new measurements are available. This behavior naturally increases the robustness in case of failures of the segmentation algorithm or losses of measurement messages. Thus, EIF exhibits higher robustness than ML to noisy measurements and particularly to the lack of measurements. Some experimental results can be found in Section 7.

5. Active perception techniques

In the previous schemes all the cameras that are seeing the object at any time t are used for data fusion regardless of the usefulness of the measurement they provide for the overall estimation. In this section we briefly summarize an entropy-based active perception approach that dynamically activates or deactivates each camera node balancing the information it effectively provides and the cost of the measurement.

The active perception problem can be broadly defined as the procedure to determine the best actions that should be performed. In our problem there are two types of actions, activate or deactivate camera i. Given a certain system state x, each action a involves an impact on the perception, i.e. it obtains a certain reward $r(x,a)$. Also, each action has a certain cost $c(x,a)$. For instance, by activating camera node i, the reward is a perception with lower uncertainty, and the cost is the increase of energy consumption.

In most active perception strategies the selection of the actions is carried out using reward VS cost analyses. In the so-called greedy algorithms the objective is to decide the next best action to be carried out without taking into account long-term goals. POMDPs (Kaelbling et al., 1998), on the other hand, consider the long-term goals providing an elegant way to model the interaction of an agent in an environment, both of them uncertain. Nonetheless, POMDPs require intense computing resources and memory capacity. POMDPs also scale badly with the number of camera nodes. Thus, in our problem we adopted an efficient greedy active perception scheme.

At each time step, the strategy adopted activates or deactivates one camera node taking into account the expected information gain and the cost of the measurement. In our approach the reward is the information gain about the target location due to the new observation. Shannon entropy is used to quantify the information gain.

Consider the prior target location distribution at time t to be $p(x_t)$. If camera node i, currently unused, is activated and its measurement is available at t, then the posterior target location distribution will be $p(x_t | z_i)$. Then, the gain of information from activating camera node i can be expressed by $H(x_t)$-$H(x_t | z_i)$, where $H(x_t)$ and $H(x_t | z_i)$ stand for the Shannon entropy of $p(x_t)$ and $p(x_t | z_i)$. $H(x_t)$-$H(x_t | z_i)$ also denotes the mutual information between x_t and z_i.

Entropy is a measure of the uncertainty associated to a random variable, i.e. the information content missing when one does not know the value of a random variable. The reward for action a=$A(i)$ -activating camera node i- is expressed by:

$$r(x_t, a = A(i)) = H(x_t) - H(x_t|z_i) \tag{14}$$

There are analytical expressions to express the entropy of a Gaussian distribution. Assuming $p(x_t)$ and $p(x_t | z_i)$ are Gaussians the reward of an action can be computed with:

$$r(x_t, a = A(i)) = \frac{1}{2}\log\left(\frac{|\Sigma_1|}{|\Sigma_2|}\right) \tag{15}$$

where Σ_1 and Σ_2 are the covariance matrices of distributions $p(x_t)$ and and $p(x_t | z_i)$.

On the other hand, the cost of activating a camera node is mainly expressed in terms of the energy consumed by camera. However, note that there are other costs, as those associated to the use of the wireless medium for transmitting the new measurements or the increase in computational burden required to consider the measurements from the new camera in the EIF. Also, these costs can vary depending on the camera node and the currently available resources. For instance, the cost of activating a camera with low battery level is higher than activating one with full batteries.

An action a_j is defined as advantageous at certain time t if the reward is higher than the cost, i.e. $r(x_t,a_j)>c(x_t,a_j)$. In a system with a set of potential advantageous actions, $a \in A^+$, the more advantageous action is selected to be carried out:

$$\hat{a} = \arg\max_{a_j \in A^+}\left(r(x_t, a_j) - c(x_t, a_j)\right) \tag{16}$$

This active perception method can be easily incorporated within a Bayesian Recursive Filter. In our case it was integrated in the EIF described in Section 5. To simplify the complexity and computer burden, the number of actions that can be done at each time is limited to one. Thus, in a deployment with N cameras the number of actions analyzed at each time is N: deactivation of each of the currently active camera nodes and activation of each of the currently unused camera nodes. The most advantageous action is selected to be carried out.

The main disadvantage of (14) is that the action to be carried out should be decided without actually having the new measurement. We have to rely on estimations of future information gain. At time t the information matrix of the EIF at t is Ω_t. In the prediction stage the information matrix is predicted, $\overline{\Omega}_{t+1}$, see the EIF algorithm in Fig. 4. In the update stage, it is updated, Ω_{t+1}, using the observation models of the sensors currently used. In case of performing sensory action a, the observation model would change and involve a new updated information matrix Ω^a_{t+1}. The expectation of the information gain can be approximated by $\frac{1}{2} \cdot \log(|\Omega^a_{t+1}| / |\Omega_{t+1}|)$.
This expression assumes that the location distribution of the target is Gaussian, which is not totally exact due to the nonlinearities in the observation pin-hole models. Also, they provide expectation of the information gain instead of the information gain itself. Despite these inaccuracies, it is capable of providing a useful measure of the information gain from a sensory action in an efficient way. In fact, the active perception method for a setting with 3 cameras adopted requires approximately 3,400 floating point operations, roughly 300 ms in Xbow TelosB motes, but can imply remarkable resources saving rates, up to 70% in some experiments shown in Section 6. It should be noted that its computational burden scales well since it is proportional to the number of cameras in the setting.

6. Implementation and some results

This Section provides details of the camera-based WSN implementation and presents some experimental results.

6.1 Implementation of camera-based WSN with COTS equipment

Requirements such as energy consumption, size and weight are very important in these systems. In our experiments we used *TelosB* motes from *Xbow Inc* (http://www.xbow.com). These motes use a *Texas Instruments* MPS430 16-bit microprocessor at 8 MHz, which can be enough to execute algorithms with low computer-burden but is not capable of applying image processing methods with sufficient image resolution and frame rate. The RAM memory of *TelosB* (10 KB) is also insufficient for most image processing techniques. In previous developments we also used *Xbow Mica2* motes, with lower resources.
The micro camera board selected is the *CMUcam3* (http://www.cmucam.org). It is an open source programmable embedded platform connected to an *Omnivision* 1/4'' *CMOS* 352x288 color detector. Its main processor, the *NXP LPC2106*, allows implementing, in Custom C, code-efficient real-time image processing algorithms. Different lenses (of up to 150° FOV_H) were used in the experiments to accommodate the dimensions of the environment. Figure 5 shows a set of camera nodes. They were mounted on small tripods to facilitate deployment and orientation. In preliminary works we used *CMUcam2* boards. The main practical advantages of *CMUcam3* over *CMUcam2* are the possibility of being programmed (*CMUcam2* used fixed pre-programmed algorithms instead) and a high reduction in energy consumption.
Each *CMUcam3* is connected to a single *Xbow* mote though a RS-232 link. The *CMUcam3* board captures the images and executes the algorithms for object segmentation while the *Xbow* mote runs a series of algorithms required for cooperative location and tracking including control of the *CMUcam3*, correction of optical distortions, algorithms for synchronization among the camera nodes and wireless transmission of the measurements. From the *Xbow* side, *CMUcam3* operates transparently as any other sensor.

Fig. 5. Set of camera nodes equipped with *CMUcam2* and *CMUcam3* micro cameras.

6.1.1 Image segmentation

Although *CMUcam3* offers programming facilities, its limited computational and memory resources require efficient algorithm design and coding to achieve near to real-time processing capabilities. In fact, the constraints in their memory capacity prevent from loading the whole image in the RAM memory and block-based processing is required.

We assume that the objects of interest are mobile. First, assuming a static environment, the moving objects are identified through difference with respect to a reference image. A pixel of image k $Im_k(x,y)$ is considered part of a mobile object if $|Im_k(x,y)-Im_{ref}(x,y)|>T$, where T is a color threshold and $Im_{ref}(x,y)$ is the reference image. To reduce computer burden, images are divided in windows and if the number of pixels which color has changed is above NP, the window is considered with motion.

In case the color of the object of interest can be characterized, then a color-based segmentation is applied only to the windows with motion previously identified. For this operation the HSI color field is preferred in order to achieve higher stability of color with lighting changes. Then, an efficient 8-neighbours region-growing algorithm is used. Finally, the characteristics of the region of interest such as coordinates of the central pixel, region width and height are obtained. Figure 6 shows the results of each step over an image from a *CMUcam3* in which a fireman is segmented.

Fig. 6. Left) Object segmentation using motion.Right) Object segmentation using color.

The algorithm has been efficiently programmed so that the complete segmentation (the images are 352x288 pixels) takes 560 ms., 380 ms. of which are devoted to downloading the image from the internal camera buffer to the *CMUcam3* board memory.

6.1.2 Image distortions correction

In the next step, before transmitting the measurements for data fusion, each camera node corrects its own optical distortions, transforming them to the normalized pin-hole projection. Let $P_i=[X_i\ Y_i\ Z_i]^T$ be the coordinates of a point in the environment expressed in reference frame local to camera i, F_i. Assuming an ideal pin-hole model, the normalized projection of P_i on the image plane of camera i is $p_i=[X/Z\ Y/Z]^T=[x_i\ y_i]^T$. After including lens radial and tangential distortions, the distorted point $p_i^d=[x_i^d\ y_i^d]^T$ is defined as follows:

$$p_i^d = d_i^r p_i + d_i^t \tag{17}$$

where d_i^r and d_i^t are simplified radial and tangential distortions terms as defined in the model described in (Heikkilä & Silven, 1997):

$$d_i^r = 1 + ar_i^2 + br_i^4 \qquad d_i^t = \begin{bmatrix} 2cx_iy_i + d(r_i^2 + 2x_i^2) \\ c(r_i^2 + 2y_i^2) + 2dx_iy_i \end{bmatrix} \tag{18}$$

where $r_i^2=x_i^2+y_i^2$. Finally, assuming the skew factor is zero, the pixel coordinates on the image $p_i^p=[x_i^p\ y_i^p]^T$ are determined considering the focal length f and the coordinates of the principal point of the lens cc of camera i by using the following expression:

$$p_i^p = fp_i^d + cc \tag{19}$$

The internal calibration parameters -optical distortion parameters a, b, c and d, the focal distance f and the coordinates of the principal point of the lens cc- are considered known at each camera node. Consider p_i^p is the pixel coordinates of the centre of a region of interest segmented in the images. The correction is applied in two steps: obtain p_i^d using (19) and compute p_i using (17) and (18). For practical purposes (18) is usually approximated using x_i^d and y_i^d instead of x_i and y_i. Thus, $p_i=[x_i\ y_i]^T$ can be computed efficiently involving only two divisions and few products and sums. It is executed in the *Xbow* node itself.

Also, the position and orientation for each camera in a global reference frame are assumed known at each camera node. These 6 parameters -3 for camera position and 3 for orientation- are included in the measurements packets sent for data fusion so that it can cope with static and mobile cameras, for instance on very light-weight UAVs. The time stamps of the measurements are also included in these packets for synchronization.

6.1.3 Interface and synchronization modules

Several software modules were implemented on the *Xbow* mote. One of them implements the command interface with the *CMUcam3* using low-level *TinyOS* routines. The node commands the *CMUcam3* to start capturing images at a certain rate and to execute object segmentation to the captured images. For each image, the *CMUcam3* replies the characteristics of the region segmented (centre, width and height). Then, the distortion correction method described in Section 6.1.2 is applied. Finally, the resulting measurements are sent through the WSN for data fusion. It should be noted that the *Xbow* nodes can disable or enable the operation of the *CMUcam3* board, allowing active perception techniques such as those described in Section 5.

Another software module was devoted to synchronization among camera nodes. The method selected is the so-called *Flooding Time Synchronization Protocol* (*FTSP*) (Maróti et al., 2004). This algorithm establishes hierarchies among the WSN nodes. The leader node periodically sends a synchronization message. Each camera node that receives the message resends it following a broadcast strategy. The local time of each camera node is corrected depending on the time stamp on the message and the sender of the message. The resulting synchronization error is of few milliseconds.

6.2 Some results

Figure 7Left shows a picture of one localization and tracking experiment. The objective is to locate and track mobile robots that follow a known trajectory, taken as ground truth. Figure 7Right depicts a scheme of an environment involving 5 camera nodes at distributed locations and with different orientations. The local reference frames of each of the camera are depicted. The global reference frame is represented in black.

In this Section the three data fusion methods presented are compared in terms of accuracy and energy consumption. Accuracy is measured as the mean error with the ground truth. For the consumption analysis we will assume that the energy dedicated by the *Xbow* node to execute any of the three data fusion algorithms is significantly lower than the energy devoted by a camera node to obtain observations. The latter includes the energy required for image acquisition and segmentation in the *CMUcam3* boards. The energy consumed by a camera node during an experiment is proportional to the number of measurements made.

In all the experiments the commands given to the robot to generate the motion were the same. The object locations are represented with dots in Fig. 7Right. In all the experiments the measurements computed by cameras 2 and 3 from t=10 s. to t=25 s. are considered lost and cannot be used for data fusion. The object locations within this interval are marked with a rectangle in Fig. 7Right.

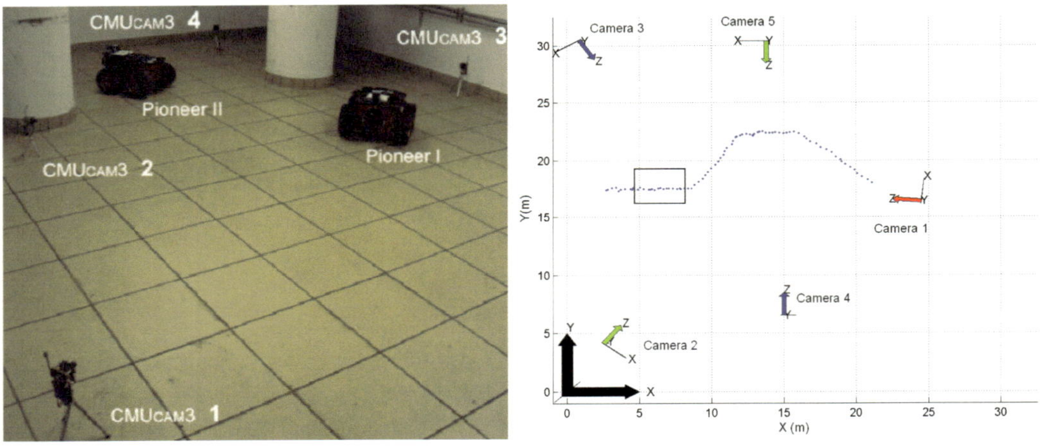

Fig. 7. Left) Object tracking experiment using 3 *CMUcam3* micro cameras. Right) Scheme of the environment involving 5 camera nodes.

Four different cases were analyzed: ML using cameras 1, 2 and 3; EIF using cameras 1, 2 and 3; EIF using the five cameras; and active perception with all the cameras. A set of ten repetitions of each experiment were carried out. Figures 8a-d shows the results obtained for axis X (left) and Y (right) in the four experiments. The ground truth is represented in black color and the estimated object locations are in red. In Figs. 8b-d the estimated 3σ confidence interval is represented in blue color. Table 1 shows the average of the mean error and the number of measurements used by each method.

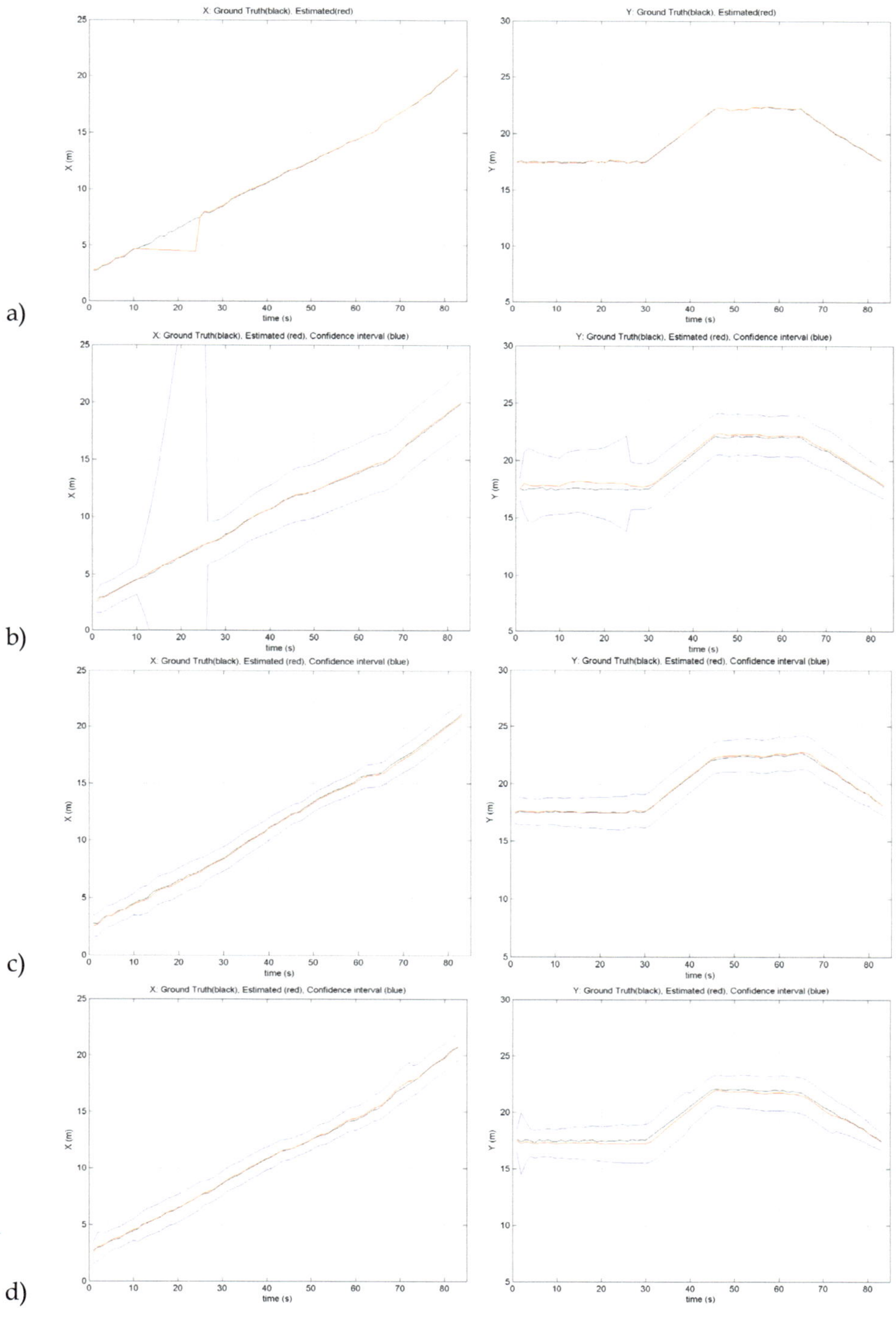

Fig. 8. Estimated X(Right) and Y(Left) location of the object: a) ML using cameras 1, 2 and 3, b) EIF with cameras 1, 2 and 3, c) EIF with the five cameras and, d) active perception.

	ML	EIF 3 cameras	EIF 5 cameras	Active Perception
Mean error (m.)	0.42	0.37	0.18	0.24
Number of measurements	225	225	395	239

Table 1. Average values of the mean error and number of measurements in the experiments.

In Fig. 8a it can be observed that the ML method performs quite well when the level of noise in the measurements is low. On the other hand, losses of measurements from cameras 2 and 3 originate important errors in the X coordinate of the estimated object location while measurements from camera 1 are enough to provide accuracy in the Y coordinate.

The EIF with cameras 1, 2 and 3 exhibits a more robust performance. The loss of measurements from cameras 2 and 3, prevents the EIF from having valid measurements for the X coordinate and thus, it relies on the system prediction model. Note that the covariance of the estimation in X increases gradually until measurements from camera 2 and 3 are again available, see Fig. 8b. Loss of measurements from cameras 2 and 3 have moderate effect in the confidence interval in Y. Globally the EIF achieved higher accuracy than the ML method in the experiments, see Table 1.

When all the cameras are fused, the estimation of the EIF is even more accurate: the 3σ confidence interval becomes narrower, see Fig. 8b,c, and the mean error becomes significantly lower, see Table 1. Loss of measurements from cameras 2 and 3 has negligible effect in the estimation because other cameras provide that information into the filter. On the other hand, using a higher number of cameras requires using often constrained in WSN applications.

The active perception method dynamically activates the camera nodes required to reduce the uncertainty and deactivates the non-informative camera nodes to save resources. The practical effect is that it obtains good object localization and tracking accuracy, see Fig. 8d, with a drastic reduction in the number of measurements used, see Table 1. In the experiments carried out the mean errors achieved by the active perception method were almost as good as those achieved by the EIF with 5 cameras (0.24 versus 0.18) but they needed 39.49% less measurements (239 versus 395).

Figure 9 shows the results in an experiment assuming a cost of $c(x_t,a)=0.2$ for all sensory actions. Figure 9Left shows which camera nodes are active at each time. It can be noted that camera 5 is the most informative one and is active during the whole experiment. In contrast, camera 2 is the less informative. Figure 9Right shows the estimated standard deviation for X (blue) and Y (red). The values of standard deviation are in the range 0.5-0.8 m. during the experiment except for the filter initialization. It can be noted that the standard deviation for X has a slight increase in the interval 10–25 s. originated by the lack of measurements from camera 2 and 3. In this experiment the mean error was 0.17 m. and the number of cameras used was 249.

The performance of the active perception is highly dependant on the values of the cost adopted to decide on the sensory action. The higher the cost, the higher has to be the information gain of an action to become advantageous. Figure 10 shows results obtained in an experiment using $c(x_t,a)=1.0$ for all sensory actions. It can be noted that the number of camera nodes active decreases significantly -only a total of 99 measurements were used in this experiment needing 74.93% less resources than EIF with 5 cameras- without much degradation in the standard deviation, which keeps in the range 0.6-1.3 m. In this experiment the mean error was 0.26 m.

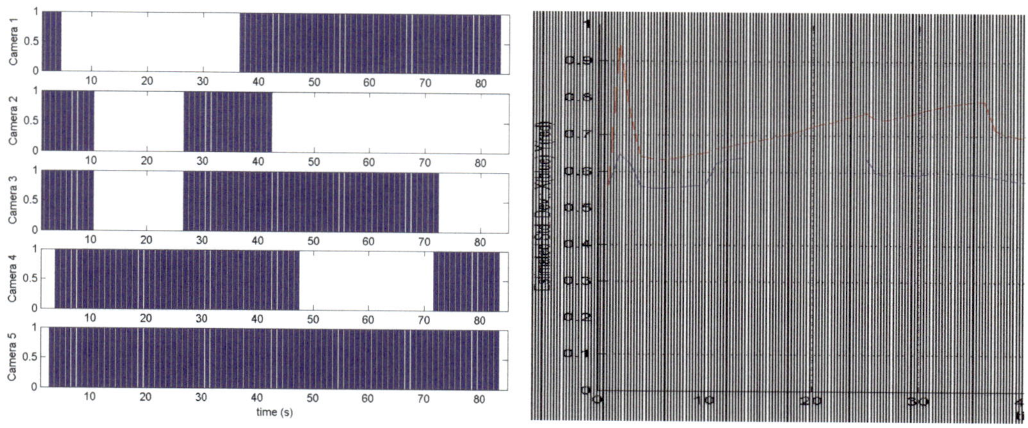

Fig. 9. Results in an experiment using $c(x_t,a)=0.2$ for all sensory actions.

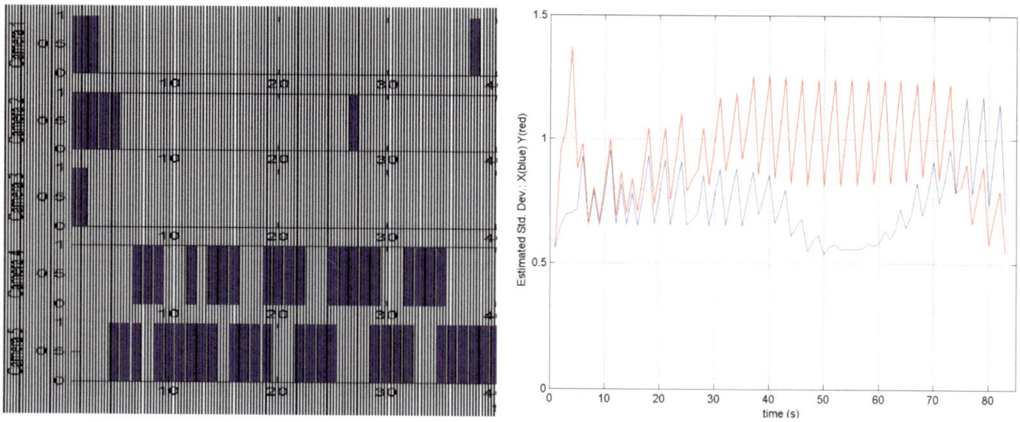

Fig. 10. Results in an experiment using $c(x_t,a)=1.0$ for all sensory actions.

7. Conclusions

This chapter describes three efficient data fusion methods for localization and tracking with WSN comprising nodes endowed with low-cost cameras as main sensors. The approach adopted is a partially decentralized scheme where the images captured by each camera node are processed locally using segmentation algorithms in order to extract the location of the object of interest on the image plane. Only low-bandwidth data is transmitted through the network for data fusion.

First, a Maximum Likelihood technique that fuses camera observations in a very efficient way is described. ML carries out data fusion using only the information contained in the measurements. It has good performance when the level of noise in the measurements is low but degrades with noisy measurements and particularly with lacks of measurements, for instance in cases of losses of WSN messages.

Then, an Extended Information Filter is proposed. Bayesian Filters compute the estimation based on measurements and observation and system models. We preferred EIFs instead of its dual EKF since the update stage of EIF is more efficient than EKF and thus it is more suitable when there are a high number of observations, such as it is our case, where a good number of low-cost camera nodes can be used. The uncertainty of the perception using EIF is reduced by using more camera nodes at the expense of requiring more resources such as energy, bandwidth and computer and memory capacity.

Finally, an Active Perception method based on a greedy algorithm balances between the information that can be obtained from a camera node and the cost of that information. The method dynamically activates the most-informative camera nodes required to reduce the uncertainty and deactivates the least-informative ones to save resources.

Several experiments with WSN comprising *Xbow* nodes and *CMUcam3* boards are used to compare the methods and illustrate their advantages and disadvantages.

The described methods have limited scalability with the number of camera nodes due to the computational and memory constraints of WSN nodes and limitations in the effective WSN bandwidth. The reliability to failures of the node performing the data fusion is also an important drawback. Decentralized data fusion can help to improve these issues. Efficient fully decentralized schemes suitable for camera-based WSN are object of current research.

8. References

Akyildiz, I.F.; Su, W.; Sankarasubramaniam, Y. & Cayirci, E. (2002). A Survey on Sensor Networks. *IEEE Communications Magazine,* Vol.40, No.8, (August 2002), pp. 102-114, ISSN 0163-6804

Al-Karaki, J.N. & Kamal, A.E. (2004). Routing Techniques in Wireless Sensor Networks: a Survey. *IEEE Wireless Communications*, Vol.11, No.6, pp. 6-28, ISSN 1536-1284

Amundson, I. & Koutsoukos, X. (2009). A Survey on Localization for Mobile Wireless Sensor Networks, *Proceedings of the Second International Workshop on Mobile Entity Localization and Tracking in GPS-less Environments*, pp. 235-254, ISBN 3-642-04378-X 978-3-642-04378-9, Orlando FL, USA, September 30, 2009

Black, J. & Ellis, T. (2006). Multi Camera Image Tracking. *Image and Vision Computing*, Vol.24, No.11, (November 2006), pp. 1256-1267, ISSN 0262-8856

Depenthal, C. & Schwendemann, J. (2009). IGPS - A New System for Static and Kinematic Measurements, *Proceedings of the 9th Conference on Optical 3D Measurement Techniques*, pp. 131-140, ISBN 978-3-9501492-5-8, Viena, Autria, July 1-3, 2009

Gezici, S.; Zhi Tian; Giannakis, G.B.; Kobayashi, H.; Molisch, A.F.; Poor, H.V. & Sahinoglu, Z. (2005). Localization via Ultra-Wideband Radios. *IEEE Signal Processing Magazine*, Vol.25, No.4, (July 2005), pp. 70-84, ISSN 1053-5888

Grewal, S. M.; Weill, L.R. & Andrews, A.P. (January 2007). *Positioning Systems, Inertial Navigation, and Integration*, Wiley-Interscience, ISBN 0470041900

Grocholsky, B.; Keller, J.; Kumar, V. & Pappas, G. (2006). Cooperative Air and Ground Surveillance, *IEEE Robotics & Automation Magazine*, Vol.13, No.3, pp. 16-25, ISSN 1070-9932

Hanssmann, M.; Rhee, S. & Liu, S. (2008). The Applicability of Wireless Technologies for Industrial Manufacturing Applications Including Cement Manufacturing, *Proceedings of the IEEE Cement Industry Technical Conference 2008*, pp.155-160, ISBN 978-1-4244-2080-3, Miami FL, USA, May 18-22, 2008

Heikkila, J. & Silven, O. (1997). A Four-step Camera Calibration Procedure with Implicit Image Correction, *Proceedings of the 1997 Conference on Computer Vision and Pattern Recognition*, pp. 1106, ISBN 0-8186-7822-4, San Juan, Puerto Rico, June 17-19, 1997.

Kaelbling, L.P.; Littman, M.L. & Cassandra A.R. (1998). Planning and acting in partially observable stochastic domains. *Artificial Intelligence*, Vol.101, No.1-2, pp. 99–134, ISSN 0004-3702

Maróti, M.; Kusy, B.; Simon G. & Lédeczi, A. (2004). The Flooding Time Synchronization Protocol, *Proceedings of the ACM Second International Conference on Embedded Networked Sensor Systems*, pp. 39-49, ISBN 1-58113-879-2, Baltimore MD, USA, November 3-5, 2004

Mohammad-Djafari, A. (1997). Probabilistic Methods for Data Fusion, *Proceedings of the 17th International Maximum on Entropy and Bayesian Methods*, pp. 57-69, ISBN 978-0-7923-5047-7, Boise Idaho, USA, August 4-8, 1997

Nath, B.; Reynolds, F. & Want, R. (2006). RFID Technology and Applications. *IEEE Pervasive Computing*, Vol.5, No.1, pp 22-24, ISSN 1536-1268

Polastre, J.; Szewczyk, R.; Mainwaring, A.; Culler D. & Anderson, J. (2004). Analysis of Wireless Sensor Networks for Habitat Monitoring, In: *Wireless Sensor Networks*, C. S. Raghavendra, K.M. Sivalingam, T. Znati, (Eds.), pp. 399-423, Kluwer Academic Publishers, ISBN 1-4020-7883-8, Norwell, MA, USA

Rice, J. (April 2006). *Mathematical Statistics and Data Analysis*. (3rd ed.), Brooks/Cole, ISBN 0534399428

Sandhu, J. S.; Agogino, A.M. & Agogino A.K. (2004). Wireless Sensor Networks for Commercial Lighting Control: Decision Making with Multi-agent Systems, *Proceedings of the AAAI Workshop on Sensor Networks*, pp. 88-89, ISBN 978-0-262-51183-4, San Jose CA, USA, July 25-26, 2004

Shaferman, V; & Shima, T. (2008). Cooperative UAV Tracking Under Urban Occlusions and Airspace Limitations, *Proceedings of the AIAA Conf. on Guidance, Navigation and Control*, , ISBN 1-56347-945-1, Honolulu, Hawaii, USA, Aug 18-21, 2008

Thrun, S.; Burgard, W. & Fox, D. (September 2005). *Probabilistic Robotics*. The MIT Press, ISBN 0262201623, Cambridge, Massachusetts, USA

Wark, T.; Corke, P.; Karlsson, J.; Sikka, P. & Valencia, P. (2007). Real-time Image Streaming over a Low-Bandwith Wireless Camera Network, *Proceedings of the Intl. Conf. on Intelligent Sensors*, pp. 113-118, ISBN 978-1-4244-1501-4, Melbourne, Australia, December 3-6, 2007

Zanca, G.; Zorzi, F.; Zanella, A. & Zorzi, M. (2008). Experimental Comparison of RSSI-based Localization Algorithms for Indoor Wireless Sensor Networks, *Proceedings of the Workshop on Real-World Wireless Sensor Networks*, pp. 1-5, ISBN 978-1-60558-123-1, Glasgow, UK, April 1, 2008

4

Sensor Fusion for Enhancement in Intrusion Detection

Ciza Thomas[1] and Balakrishnan Narayanaswamy[2]
[1]*Professor, College of Engineering, Trivandrum*
[2]*Associate Director, Indian Institute of Science, Bangalore*
India

1. Introduction

An Intrusion Detection System (IDS) gathers information from a computer or a network, and analyzes this information to identify possible security breaches against the system or the network. An observation of various IDSs available in literature shows distinct preferences for detecting a certain class of attack with improved accuracy, while performing moderately on the other classes. The availability of enormous computing power has made it possible for developing and implementing IDSs of different types on the same network. The integration of the decisions coming from different IDSs has emerged as a technique that could strengthen the final decision. Sensor fusion can be defined as the process of collecting information from multiple and possibly heterogeneous sources and combining them to obtain a more descriptive, intuitive and meaningful result (1).

An analysis of the poorly detected attacks reveals the fact that the attacks are characterized by features that do not discriminate them much. In this chapter, we prove the distinct advantages of sensor fusion over individual IDSs. All the related work in the field of sensor fusion has been carried out mainly with one of the methods like probability theory, evidence theory, voting fusion theory, fuzzy logic theory or neural network in order to aggregate information. The Bayesian theory is the classical method for statistical inference problems. The fusion rule is expressed for a system of independent learners, with the distribution of hypotheses known *a priori*. The Dempster-Shafer evidence theory is considered a generalized Bayesian theory. It does not require *a priori* knowledge or probability distribution on the possible system states like the Bayesian approach and it is mostly useful when modeling of the system is difficult or impossible (2). The improved performance of multiple IDSs using rule-based fusion and deta-dependent decision fusion has been demonstrated in the work of Thomas and Balakrishnan (3).

An attempt to prove the distinct advantages of sensor fusion over individual IDSs is done in this chapter using the Chebyshev inequality. Fusion threshold bounds were derived using the principle of Chebyshev inequality at the fusion center using the false positive rates and detection rates of the IDSs. The goal was to achieve best fusion performance with the least amount of model knowledge, in a computationally inexpensive way. The anomaly-based IDSs detect anomalies beyond a set threshold level in the features it detects. Threshold bounds instead of a single threshold give more freedom in steering system properties. Any threshold

within the bounds can be chosen depending on the preferred level of trade-off between detection and false alarms.
The remaining part of the chapter is organized as follows. Section 2 discusses the related work of sensor fusion in IDS. In section 3, the modeling of the Intrusion Detection System is presented. Section 4 includes the modeling of the fusion of Intrusion Detection Systems. Section 5 contains the experimental results along with the discussions regarding the higher performance of the proposed fused IDS. Finally, the concluding comments are presented in section 6.

2. Related work

Tim Bass (4) presents a framework to improve the performance of intrusion detection systems based on data fusion. A few first steps towards developing the engineering requirements using the art and science of multi-sensor data fusion as an underlying model is provided in (4). Giacinto et al. (5) propose an approach to intrusion detection based on fusion of multiple classifiers. Didaci et al. (6) attempt the formulation of the intrusion detection problem as a pattern recognition task using data fusion approach based on multiple classifiers. Wang et al. (7) present the superiority of data fusion technology applied to intrusion detection systems. The use of data fusion in the field of DoS anomaly detection is presented by Siaterlis and Maglaris (1). The detection engine is evaluated using the real network traffic. Another work incorporating the Dempster-Shafer theory of evidence is by Hu et al. (8).
Siraj et al. (9) discuss a Decision Engine for an Intelligent Intrusion Detection System (IIDS) that fuses information from different intrusion detection sensors using an artificial intelligence technique. Thomopolous in one of his work (10), concludes that with the individual sensors being independent, the optimal decision scheme that maximizes the probability of detection at the fusion for fixed false alarm probability consists of a Neyman-Pearson test at the fusion unit and the likelihood ratio test at the sensors. The threshold based fusion of combining multiple IDSs by fixing a certain number of false alarms is discussed in the work of Thomas and Balakrishnan (11). This is a case of combining the top ranking outputs of each IDS after removing the duplicate alerts and setting the maximum acceptable false alarm rate.
The other somewhat related works albeit distantly are the alarm clustering method by Perdisci et al. (12), aggregation of alerts by Valdes et al. (13), combination of alerts into scenarios by Dain et al. (14), the alert correlation by Cuppens et al. (15), the correlation of Intrusion Symptoms with an application of chronicles by Morin et al. (16), and aggregation and correlation of intrusion-detection alerts by Debar et al. (17). In the work of Thomas and Balakrishnan (3), a sensor fusion architecture, which is data-dependent and different from the conventional fusion architecture is demonstrated. The focus of the present work is modeling the fusion of IDSs using threshold bounds in an attempt to optimize both the fusion rule as well as the sensor rules.

3. Modeling the Intrusion Detection Systems

Consider an IDS that either monitors the network traffic connection on the network or the audit trails on the host. The network traffic connection or the audit trails monitored are given as $x \in X$, where X is the entire domain of network traffic features or the audit trails respectively. The model is based on the hypothesis that the security violations can be detected by monitoring the network for traffic connections of malicious intent in the case of network-based IDS and a system's audit records for abnormal patterns of system usage in

the case of host-based IDS. The model is independent of any particular operating system, application, system vulnerability or type of intrusion, thereby providing a framework for a general-purpose IDS.
When making an attack detection, a connection pattern is given by $x_j \in \Re^{jk}$ where j is the number of features from k consecutive samples used as input to an IDS. As seen in the DARPA dataset, for many of the features the distributions are difficult to describe parametrically as they may be multi-modal or very heavy-tailed. These highly non-Gaussian distributions has led to investigate non-parametric statistical tests as a method of intrusion detection in the initial phase of IDS development. The detection of an attack in the event x is observed as an alert. In the case of network-based IDS, the elements of x can be the fields of the network traffic like the raw IP packets or the pre-processed basic attributes like the duration of a connection, the protocol type, service etc. or specific attributes selected with domain knowledge such as the number of failed logins or whether a superuser command was attempted. In host-based IDS, x can be the sequence of system calls, sequence of user commands, connection attempts to local host, proportion of accesses in terms of TCP or UDP packets to a given port of a machine over a fixed period of time etc. Thus IDS can be defined as a function that maps the data input into a normal or an attack event either by means of absence of an alert (0) or by the presence of an alert (1) respectively and is given by:

$$IDS : X \to \{0, 1\}.$$

To detect attacks in the incoming traffic, the IDSs are typically parameterized by a threshold T. The IDS uses a theoretical basis for deciding the thresholds for analyzing the network traffic to detect intrusions. Changing this threshold allows the change in performance of the IDS. If the threshold is very low, then the IDS tends to be very aggressive in detecting the traffic for intrusions. However, there is a potentially greater chance for the detections to be irrelevant which result in large false alarms. A large value of threshold on the other hand will have an opposite effect; being a bit conservative in detecting attacks. However, some potential attacks may get missed this way. Using a 3σ based statistical analysis, the higher threshold (T_h) is set at $+3\sigma$ and the lower threshold (T_l) is set at -3σ. This is with the assumption that the traffic signals are normally distributed. In general the traffic detection with s being the sensor output is given by:

$$Sensor\ \ Detection = \begin{cases} attack, & T_l < s < T_h \\ normal, & s \le T_l,\ \ s \ge T_h \end{cases}$$

The signature-based IDS functions by looking at the event feature x and checking whether it matches with any of the records in the signature database D.

$$\begin{aligned} Signature - based\ IDS &: X \to \{1\} \qquad \forall x \in D, \\ &: X \to \{0\} \qquad \forall x \notin D. \end{aligned}$$

Anomaly-based IDS generates alarm when the input traffic deviates from the established models or profiles P.

$$\begin{aligned} Anomaly - based\ IDS &: X \to \{1\} \qquad \forall x \notin P, \\ &: X \to \{0\} \qquad \forall x \in P. \end{aligned}$$

4. Modeling the fusion of Intrusion Detection Systems

Consider the case where n IDSs monitor a network for attack detection and each IDS makes a local decision s_i and these decisions are aggregated in the fusion unit f. This architecture is often referred to as the parallel decision fusion network and is shown in Figure 1. The fusion unit makes a global decision, y, about the true state of the hypothesis based on the collection of the local decisions gathered from all the sensors. The problem is casted as a binary detection

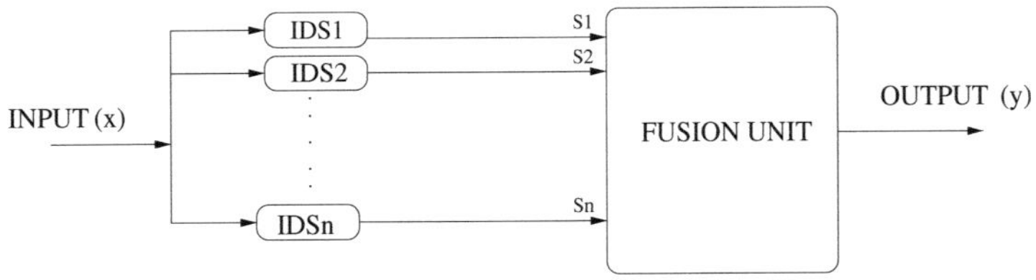

Fig. 1. Parallel Decision Fusion Network

problem with the hypothesis "*Attack*" or "*Normal*". Every IDS participating in the fusion has its own detection rate D_i, and false positive rate F_i, due to the preferred heterogeneity of the sensors in the fusion process. Each IDS indexed i gives an alert or no-alert indicated by s_i taking a value one or zero respectively, depending on the observation x.

$$s_i = \begin{cases} 0, & \textit{normal is declared to have been detected} \\ 1, & \textit{attack is declared to have been detected} \end{cases}$$

The fusion center collects these local decisions s_i and forms a binomial distribution y as given by y=$s = \sum_{i=1}^{n} s_i$, where n is the total number of IDSs taking part in fusion.

4.1 The effect of setting threshold

To detect the attack in the incoming traffic, the IDSs are typically parameterized with a threshold, T. Changing this threshold allows the change in performance of the IDS. If the threshold is very large, some potentially dangerous attacks get missed. A small threshold on the other hand results in more detections, with a potentially greater chance that they are not relevant.

The final step in the approach towards solving of the fusion problem is taken by noticing that the decision function $f_i(.)$ is characterized by the threshold T_i and the likelihood ratio (if independence is assumed). Thus the necessary condition for optimal fusion decision occurs if the thresholds $(T_1, T_2, ..., T_n)$ are chosen optimally. However, this does not satisfy the sufficient condition. These refer to the many local minima, each need to be checked to assure the global minimum.

The counterintuitive results at the individual sensors with the proper choice of thresholds will be advantageous in getting an optimum value for the fusion result. They are excellent paradigms for studying distributed decision architectures, to understand the impact of the limitations, and even suggest empirical experiments for IDS decisions.

The structure of the fusion rule plays a crucial role regarding the overall performance of the IDS since the fusion unit makes the final decision about the state of the environment. While a few inferior IDSs might not greatly impact the overall performance, a badly designed fusion rule can lead to a poor performance even if the local IDSs are well designed. The fusion IDS can be optimized by searching the space of fusion rules and optimizing the local thresholds for each candidate rule. Other than for some simple cases, the complexity of such an approach is prohibitive due to exponential growth of the set of possible fusion rules with respect to the number of IDSs. Searching for the fusion rule that leads to the minimum probability of error is the main bottleneck due to discrete nature of this optimization process and the exponentially large number of fusion rules. The computation of thresholds couples the choice of the local decision rules so that the system-wide performance is optimized, rather than the performance of the individual detector.

4.2 Threshold optimization

Tenney and Sandell in their work (21) establish the optimum strategy that minimizes a global cost in the case where the *a priori* probabilities of the hypotheses, the distribution functions of the local observations, the cost functions, and the fusion rule are given. They concluded that each local detector is optimally a likelihood ratio detector but that the computation of the optimum thresholds for these local detectors is complicated due to cross coupling.
The global optimization criterion for a distributed detection system would encompass local decision statistics, local decision thresholds, the fusion center decision statistic, and the fusion center decision threshold. For each input traffic observation x, the set of n local thresholds should be optimized with respect to the probability of error. With a fusion rule given by a function f, the average probability of error at the fusion unit is given by the weighted sum of false positive and false negative errors.

$$P_e(T, f) = p * P(s = 1|Normal) + q * P(s = 0|Attack) \tag{1}$$

where p and q are the respective weights of false positive and false negative errors.
Assuming independence between the local detectors, the likelihood ratio is given by:

$$\frac{P(s|Attack)}{P(s|Normal)} = \frac{P(s_1, s_2, ..., s_N|Attack)}{P(s_1, s_2, ..., s_N|Normal)} = \prod_{i=1}^{n} \frac{P(s_i|Attack)}{P(s_i|Normal)}.$$

The optimum decision rule for the fusion unit follows:

$$f(s) = log \frac{P(s|Attack)}{P(s|Normal)}$$

Depending on the value of $f(s)$ being greater than or equal to the decision threshold, T, or less than the decision threshold, T, the decision is made for the hypothesis as "*Attack*" or "*Normal*" respectively. Thus the decisions from the n detectors are coupled through a cost function. It is shown that the optimal decision is characterized by thresholds as in the decoupled case. As far as the optimum criterion is concerned, the first step is to minimize the average probability of error in equation 1. This leads to sets of simultaneous inequalities in terms of the generalized likelihood ratios at each detector, the solutions of which determine the regions of optimum detection.

4.3 Dependence on the data and the individual IDSs

Often, the data in the databases is only an approximation of the true data. When the information about the goodness of the approximation is recorded, the results obtained from the database can be interpreted more reliably. Any database is associated with a degree of accuracy, which is denoted with a probability density function, whose mean is the value itself. In order to maximize the detection rate it is necessary to fix the false alarm rate to an acceptable value, taking into account the trade-off between the detection rate and the false alarm rate. The threshold (T) that maximizes the TP_{rate} and thus minimizes the FN_{rate} is given as:

$$FP_{rate} = P[alert|normal] = P\left[\sum_{i=1}^{n} w_i s_i \geq T\,|normal\right] = \alpha_0 \tag{2}$$

$$TP_{rate} = P[alert|attack] = P\left[\sum_{i=1}^{n} w_i s_i \geq T\,|attack\right] \tag{3}$$

The fusion of IDSs becomes meaningful only when $FP \leq FP_i \quad \forall i$ and $TP \geq TP_i \quad \forall i$; where FP and TP correspond to the false positives and the true positives of the fused IDS and FP_i and TP_i correspond to the false positives and the true positives of the individual IDS indexed i. It is required to provide low value of weight to any individual sensor that is unreliable, hence meeting the constraint on false alarm as given in equation 2. Similarly, the fusion improves the TP_{rate} as the detectors get weighted according to their performance.

4.4 Modeling the fusion IDS by defining proper threshold bounds

Every IDS participating in the fusion has its own detection rate D_i, and false positive rate F_i, due to the preferred heterogeneity of the sensors in the fusion process. Each IDS indexed i gave an alert or no-alert indicated by s_i taking a value of one or zero respectively. The fusion center collected these local decisions and formed a binomial distribution s as given by $s=\sum_{i=1}^{n} s_i$, where n is the total number of IDSs taking part in the fusion.

Let D and F denote the unanimous detection rate and the false positive rate respectively. The mean and variance of s in case of attack and no-attack, are given by the following equations:

$$E[s|alert] = \sum_{i=1}^{n} D_i, \quad Var[s|alert] = \sum_{i=1}^{n} D_i(1 - D_i)$$

; in case of attack

$$E[s|alert] = \sum_{i=1}^{n} F_i, \quad Var[s|alert] = \sum_{i=1}^{n} F_i(1 - F_i)$$

; in case of no-attack

The fusion IDS is required to give a high detection rate and a low false positive rate. Hence the threshold T has to be chosen well above the mean of the false alerts and well below the mean of the true alerts. The figure 2 shows a typical case where the threshold T is chosen at the point of overlap of the two parametric curves for normal and attack traffics. Consequently, the threshold bounds are given as:

$$\sum_{i=1}^{n} F_i < T < \sum_{i=1}^{n} D_i$$

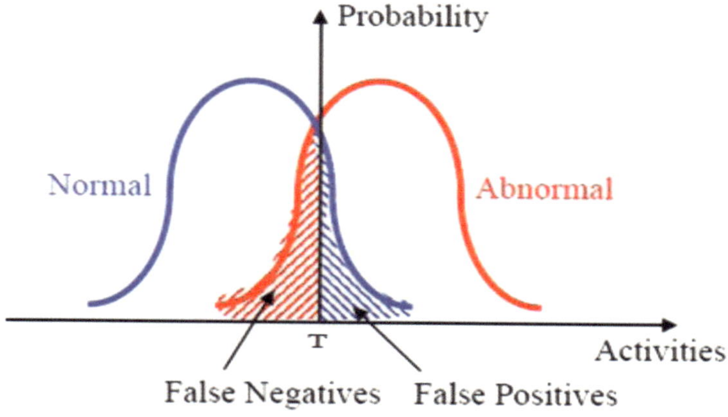

Fig. 2. Parametric curve showing the choice of threshold T

The detection rate and the false positive rate of the fusion IDS is desired to surpass the corresponding weighted averages and hence:

$$D > \frac{\sum_{i=1}^{n} D_i^2}{\sum_{i=1}^{n} D_i} \tag{4}$$

and

$$F < \frac{\sum_{i=1}^{n} (1-F_i)F_i}{\sum_{i=1}^{n} (1-F_i)} \tag{5}$$

Now, using simple range comparison,

$$D = Pr\{s \geq T|attack\} = Pr\{|s - \sum_{i=1}^{n} D_i \leq (\sum_{i=1}^{n} D_i - T)|attack\}.$$

Using Chebyshev inequality on the random variable s, with $Mean = E[s] = \sum_{i=1}^{n} D_i$ and $Variance = Var[s] = \sum_{i=1}^{n} D_i(1-D_i)$,

$Pr\{|s - E(s)| \geq k\} \leq \frac{Var(s)}{k^2}$

With the assumption that the threshold T is greater than the mean of normal activity,

$$Pr\{|s - \sum_{i=1}^{n} D_i \leq (\sum_{i=1}^{n} D_i - T)|attack\} \geq 1 - \frac{\sum_{i=1}^{n} D_i(1-D_i)}{(\sum_{i=1}^{n} D_i - T)^2}$$

From equation 4 it follows that $1 - \frac{\sum_{i=1}^{n} D_i(1-D_i)}{(\sum_{i=1}^{n} D_i - T)^2} \geq \frac{\sum_{i=1}^{n} D_i^2}{\sum_{i=1}^{n} D_i}$

The upper bound of T is derived from the above equation as:

$$T \leq \sum_{i=1}^{n} D_i - \sqrt{\sum_{i=1}^{n} D_i}$$

Similarly, for the false positive rate, $F = Pr\{S \geq T \mid no\text{-}attack\}$, in order to derive the lower bound of T,

From equation 5 it follows that $\frac{\sum_{i=1}^{n} F_i(1-F_i)}{(T - \sum_{i=1}^{n} F_i)^2} \leq \frac{\sum_{i=1}^{n} F_i(1-F_i)}{\sum_{i=1}^{n} (1-F_i)}$

The lower bound of T is derived from the above equation as:

$$T \geq \sum_{i=1}^{n} F_i + \sqrt{\sum_{i=1}^{n} (1-F_i)}$$

The threshold bounds for the fusion IDS is:

$$\left[\sum_{i=1}^{N} F_i + \sqrt{\sum_{i=1}^{n} (1-F_i)}, \quad \sum_{i=1}^{n} D_i - \sqrt{\sum_{i=1}^{n} D_i} \right]$$

.

Since the threshold T is assumed to be greater than the mean of normal activity, the upper bound of false positive rate F can be obtained from the Chebyshev inequality as:

$$F \leq \frac{Var[s]}{(T - E[s])^2} \tag{6}$$

In a statistical intrusion detection system, a false positive is caused due to the variance of network traffic during normal operations. Hence, to reduce the false positive rate, it is important to reduce the variance of the normal traffic. In the ideal case, with normal traffic the variance is zero. The equation 6 shows that as the variance of the normal traffic approaches zero, the false positive rate should also approach zero. Also, since the threshold T is assumed to be less than the mean of the intrusive activity, the lower bound of the detection rate D can be obtained from the Chebyshev inequality as:

$$D \geq 1 - \frac{Var[s]}{(E[s] - T)^2} \tag{7}$$

For an intrusive traffic, the factor $D_i(1 - D_i)$ remains almost steady and hence the variance given as:

$Variance = \sum_{i=1}^{n} D_i(1 - D_i)$, is an appreciable value. Since the variance of the attack traffic is

above a certain detectable minimum, from equation 7, it is seen that the correct detection rate can approach an appreciably high value. Similarly the true negatives will also approach a high value since the false positive rate is reduced with IDS fusion.
It has been proved above that with IDS fusion, the variance of the normal traffic is clearly dropping down to zero and the variance of the intrusive traffic stays above a detectable minimum. This additionally supports the proof that the fusion IDS gives better detection rate and a tremendously low false positive rate.

5. Results and discussion

5.1 Test set up

The test set up for the experimental evaluation consisted of a combination of shallow and deep sensors. Hence, for the purpose of fusion we have incorporated two sensors, one that monitors the header of the traffic packet and the other that monitors the packet content. The experiments were conducted with the simulated IDSs PHAD and ALAD (22). This choice of heterogeneous sensors in terms of their functionality is to exploit the advantages of fusion IDS (4). In addition, complementary IDSs provide versatility and similar IDSs ensure reliability. The PHAD being packet-header based and detecting one packet at a time, is totally unable to detect the slow scans. However, PHAD detects the stealthy scans much more effectively. The ALAD being content-based will complement the PHAD by detecting R2L (Remote to Local) and U2R (User to Root) attacks with appreciable efficiency.

5.2 Data set

The fusion IDS and all the IDSs that form part of the fusion IDS were separately evaluated with the same two data sets, namely 1) the real-world network traffic and 2) the DARPA 1999 data set. The real traffic within a protected University campus network was collected during the working hours of a day. This traffic of around two million packets was divided into two halves, one for training the anomaly IDSs, and the other for testing. The test data was injected with 45 HTTP attack packets using the HTTP attack traffic generator tool called libwhisker (23). The test data set was introduced with a base rate of 0.0000225, which is relatively realistic. The MIT-DARPA data set (IDEVAL 1999) (24) was used to train and test the performance of Intrusion Detection Systems. The network traffic including the entire payload of each packet was recorded in tcpdump format and provided for evaluation. The data for the weeks one and three were used for the training of the anomaly detectors and the weeks four and five were used as the test data. Each of the IDS was trained on distinct portions of the training data (ALAD on week one and PHAD on week three), which is expected to provide independence among the IDSs and also to develop diversity while being trained.
Even with the criticisms by McHugh (25) and Mahoney and Chan (26) against the DARPA dataset, the dataset was extremely useful in the IDS evaluation undertaken in this work. Since none of the IDSs perform exceptionally well on the DARPA dataset, the aim is to show that the performance improves with the proposed method. If a system is evaluated on the DARPA dataset, then it cannot claim anything more in terms of its performance on the real network traffic. Hence this dataset can be considered as the base line of any research (27). Also, even after 12 years of its generation, there are still a lot of relevant attacks in the data set for which signatures are not available in database of even the frequently updated signature based IDSs. The test data of the DARPA data set consisted of 190 instances of 57 attacks which included 37 probes, 63 DoS attacks, 53 R2L attacks, 37 U2R/Data attacks with details on attack types given in Table 1.

Attack Class	Attack Type
Probe	portsweep, ipsweep, queso, ntinfoscan, mscan, lsdomain, satan, illegal-sniffer
DoS	apache2, smurf, neptune, pod, mailbomb, back, teardrop, udpstorm, processtable, arppoison, tcpreset, crashiis, dosnuke, syslogd, land, selfping, warezclient
R2L	dict, guest, ftpwrite, xlock, xsnoop, httptunnel, framespoof, netbus, netcat, ppmacro, imap, named, ncftp, phf, sendmail, sshtrojan, snmpget
U2R/ Data	perl, xterm, eject, fdformat, ffbconfig, ps, loadmodule, casesen, nukepw, sechole, yaga, secret, ntfsdos, ppmacro, sqlattack

Table 1. Various attack types in DARPA'99 data set

The large observational data set were analyzed to find unsuspected relationships and was summarized in novel ways that were both understandable and useful for the detector evaluation. There are many types of attacks in the test set, many of them not present in the training set. Hence, the selected data also challenged the ability to detect the unknown intrusions. When a discrete IDS was applied to a test set, it yields a single confusion matrix. Thus, a discrete IDS produced only a single point in the ROC space, whereas scoring IDSs can be used with a threshold to produce different points in the ROC space.

5.3 Evaluation metrics

Let TP be the number of attacks that are correctly detected, FN be the number of attacks that are not detected, TN be the number of normal traffic packet/connections that are correctly classified, and FP be the number of normal traffic packet/connections that are incorrectly detected as attack. In the case of an IDS, there are both the security requirements and the usability requirements. The security requirement is determined by the TP_{rate} and the usability requirement is decided by the number of FPs because of the low base rate in the case of a network traffic.

The commonly used IDS evaluation metrics on a test data are the overall accuracy and F-score.

$$Overall\ Accuracy = \frac{TP + TN}{TP + FP + TN + FN}$$

Overall Accuracy is not a good metric for comparison in the case of network traffic data since the true negatives abound.

Precision is a measure of what fraction of test data detected as attack are actually from the attack classes.

$$Precision = \frac{TP}{TP + FP}$$

Recall is a measure of what fraction of attack class was correctly detected.

$$Recall = \frac{TP}{TP + FN}$$

There is a trade-off between the two metrics precision and recall. As the number of detections increase by lowering of the threshold, the recall will increase, while precision is expected to decrease. The recall-precision characterization of a particular IDS is normally used to analyze the relative and absolute performance of an IDS over a range of operating conditions. F-score, which is the harmonic mean of recall (R) and precision (P), scores the balance between precision and recall. The F-score is given by:

$$F\text{-}score = \frac{2 * P * R}{P + R}$$

The standard measures, namely precision, recall, and F-score are grounded on a probabilistic framework and hence allows one to take into account the intrinsic variability of performance estimation.

5.4 Experimental evaluation

The fusion element analyzes the IDS data coming from PHAD and ALAD distributed across the single subnet and observing the same domain. The fusion unit performed the aggregation of the IDS outputs for the purpose of identifying the attacks in the test data set. It used binary fusion by giving an output value of one or zero depending on the value of the aggregation of the various IDS decisions. The packets were identified by their timestamp on aggregation. A value of one at the output of the fusion unit indicated the record to be under attack and a zero indicated the absence of an attack.

The fusion IDS was initially evaluated with the DARPA 1999 data set. The individual IDSs chosen in this work are PHAD and ALAD, two research IDSs that are anomaly-based and having extremely low false alarm rate of the order of 0.00002. The other reason for the choice of PHAD and ALAD was that the are almost complementary in attack detection as evident fom table 2 and table 3. This helps in achieving best results from the fusion process. The

Attack type	Total attacks	Attacks detected	% detection
Probe	37	22	59%
DOS	63	24	38%
R2L	53	6	11%
U2R/Data	37	2	5%
Total	190	54	28%

Table 2. Types of attacks detected by PHAD at 0.00002 FP rate (100 FPs)

Attack type	Total attacks	Attacks detected	% detection
Probe	37	6	16%
DOS	63	19	30%
R2L	53	25	47%
U2R/Data	37	10	27%
Total	190	60	32%

Table 3. Types of attacks detected by ALAD at at 0.00002 FP rate (100 FPs)

analysis of PHAD and ALAD has resulted in a clear understanding of the individual IDSs expected to succeed or fail under a particular attack. On combining the two sensor alerts and removing the duplicates, an improved rate of detection is achieved as shown in table 4.

The performance in terms of F-score of PHAD, ALAD and the combination of PHAD and ALAD is shown in the tables 5, 6 and 7 respectively and figure 3 for various values of false positives by setting the threshold appropriately. In our experiment we are trying to maximize the true positive rate by fixing the false positive rate at α_0. α_0 determines the threshold T by

Attack type	Total attacks	Attacks detected	% detection
Probe	37	24	65%
DOS	63	39	62%
R2L	53	26	49%
U2R/Data	37	10	27%
Total	190	99	52%

Table 4. Types of attacks detected by the combination of ALAD and PHAD at 0.00004 FP rate (200 FPs)

FP	TP	Precision	Recall	Overall Accuracy	F-score
50	33	0.39	0.17	0.99	0.24
100	54	0.35	0.28	0.99	0.31
200	56	0.22	0.29	0.99	0.25
500	56	0.10	0.29	0.99	0.15

Table 5. F-score of PHAD for different choice of false positives

FP	TP	Precision	Recall	Overall Accuracy	F-score
50	42	0.45	0.21	0.99	0.29
100	60	0.37	0.31	0.99	0.34
200	66	0.25	0.34	0.99	0.29
500	72	0.12	0.38	0.99	0.18

Table 6. F-score of ALAD for different choice of false positives

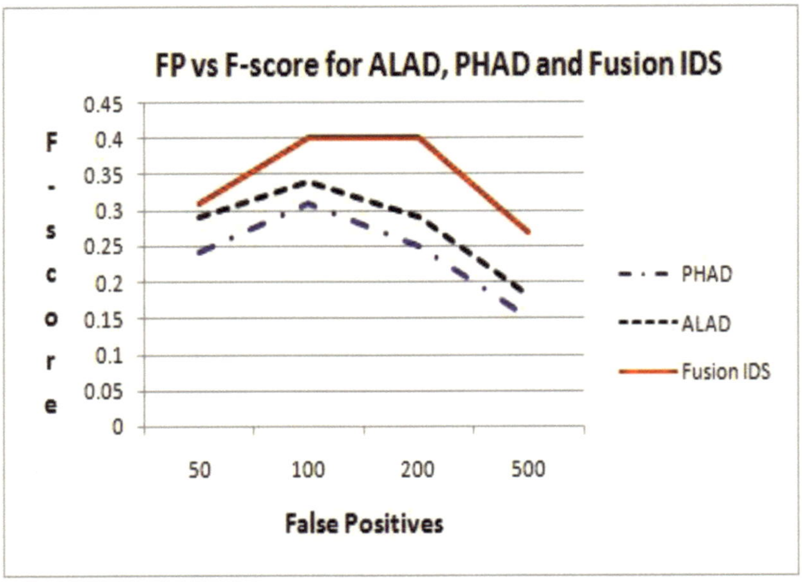

Fig. 3. F-score of PHAD, ALAD and fusion IDS for different choices of false positives

FP	TP	Precision	Recall	Overall Accuracy	F-score
50	44	0.46	0.23	0.99	0.31
100	73	0.42	0.38	0.99	0.40
200	99	0.33	0.52	0.99	0.40
500	108	0.18	0.57	0.99	0.27

Table 7. F-score and Detection Performance for different choice of false positives for fused IDS

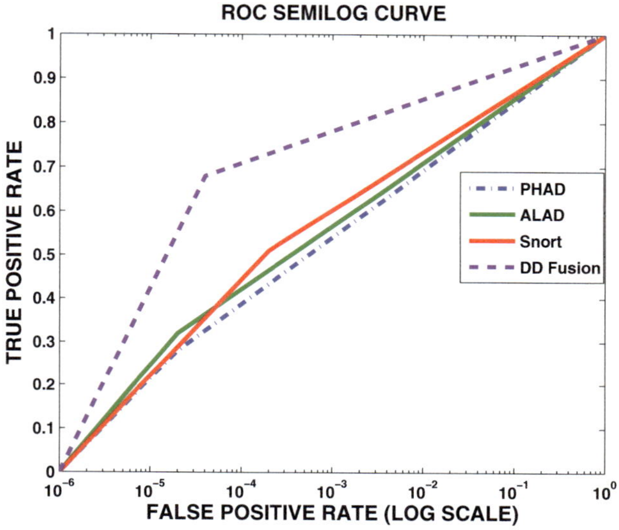

Fig. 4. ROC curve of PHAD, ALAD and fusion IDS

trial and error. We have noticed that within two or three trials in our case. This is done with the training data and hence it is done off line. The improved performance of the combination of the alarms from each system can be observed in table 7, corresponding to the false positives between 100 and 200, by fixing the threshold bounds appropriately. Thus the combination works best above a false positive of 100 and much below a false positive of 200. In each of the individual IDSs, the number of detections were observed at false positives of 50, 100, 200 and 500, when trained on inside week 3 and tested on weeks 4 and 5. The improved performance of fusion IDS compared to the two IDSs PHAD and ALAD is also illustrated with the ROC semilog curve shown in figure 4. The improved performance of the fusion IDS over some of the fusion alternatives using the real-world network traffic is shown in table 8 and figure 5.

6. Summary

Simple theoretical model is initially illustrated in this chapter for the purpose of showing the improved performance of fusion IDS. The detection rate and the false positive rate quantify the performance benefit obtained through the fixing of threshold bounds. Also, the more independent and distinct the attack space is for the individual IDSs, the better the fusion IDS performs.

The theoretical proof was supplemented with experimental evaluation, and the detection rates, false positive rates, and F-score were measured. In order to understand the importance

Detector/ Fusion Type	Total Attacks	TP	FP	Precision	Recall	F-score
PHAD	45	10	45	0.18	0.22	0.2
ALAD	45	18	45	0.29	0.4	0.34
OR	45	22	77	0.22	0.49	0.30
AND	45	9	29	0.24	0.2	0.22
SVM	45	21	44	0.32	0.47	0.38
ANN	45	21	61	0.26	0.47	0.28
Fusion IDS	45	22	32	0.41	0.49	0.45

Table 8. Comparison of the evaluated IDSs with various evaluation metrics using the real-world network traffic

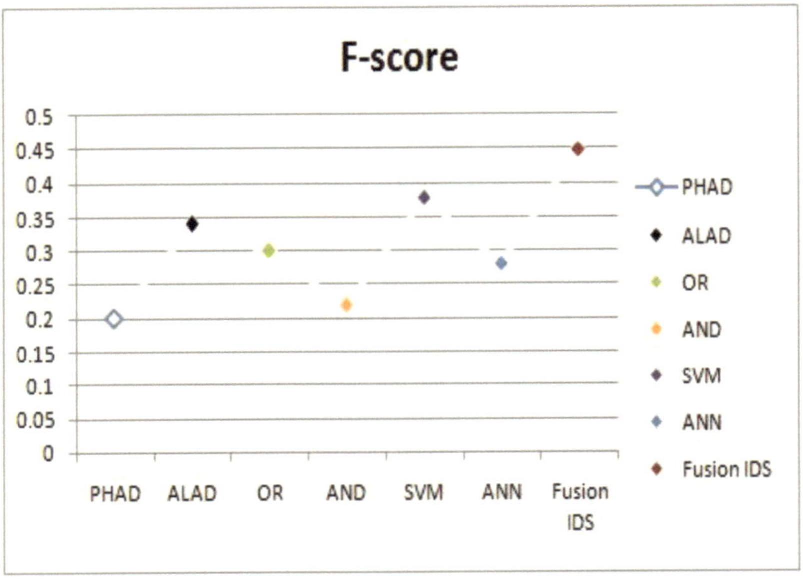

Fig. 5. F-score of the evaluated IDSs using the real-world network traffic

of thresholding, the anomaly-based IDSs, PHAD and ALAD have been individually analyzed. Preliminary experimental results prove the correctness of the theoretical proof. The chapter demonstrates that our technique is more flexible and also outperforms other existing fusion techniques such as OR, AND, SVM, and ANN using the real-world network traffic embedded with attacks. The experimental comparison using the real-world traffic has thus confirmed the usefulness and significance of the method. The unconditional combination of alarms avoiding duplicates as shown in table 4 results in a detection rate of 52% at 200 false positives, and F-score of 0.4. The combination of highest scoring alarms as shown in table 7 using the DARPA 1999 data set results in a detection rate of 38% and threshold fixed at 100 false positives, and F-score of 0.4.

7. References

[1] C. Siaterlis and B. Maglaris, Towards Multisensor Data Fusion for DoS detection, ACM Symposium on Applied Computing, 2004

[2] H. Wu, M. Seigel, R. Stiefelhagen, J. Yang, Sensor Fusion using Dempster-Shafer Theory, IEEE Instrumentation and Measurement Technology Conference, 2002
[3] C. Thomas and N. Balakrishnan, Advanced Sensor Fusion Technique for Enhanced Intrusion Detection, IEEE International Conference on Intelligence and Security Informatics, Jun. 2008
[4] T. Bass, Multisensor Data Fusion for Next Generation Distributed Intrusion Detection Systems, IRIS National Symposium, 1999
[5] G. Giacinto, F. Roli, and L. Didaci, Fusion of multiple classifiers for intrusion detection in computer networks, Pattern recognition letters, 2003
[6] L. Didaci, G. Giacinto, and F. Roli, Intrusion detection in computer networks by multiple classifiers systems, International Conference on Pattern recognition, 2002
[7] Y. Wang, H. Yang, X. Wang, and R. Zhang, Distributed intrusion detection system based on data fusion method, Intelligent control and automation, WCICA 2004
[8] W. Hu, J. Li, and Q. Gao, Intrusion Detection Engine on Dempster-Shafer's Theory of Evidence, Proceedings of International Conference on Communications , Circuits and Systems, vol.3, pp. 1627-1631, Jun 2006
[9] A. Siraj, R.B. Vaughn, and S.M. Bridges, Intrusion Sensor Data Fusion in an Intelligent Intrusion Detection System Architecture, Proceedings of the 37th Hawaii international Conference on System Sciences, 2004
[10] S.C.A. Thomopolous, R. Viswanathan, and D.K. Bougoulias, Optimal distributed decision fusion, IEEE Transactions on aerospace and electronic systems, vol. 25, No. 5, Sep. 1989
[11] C. Thomas and N. Balakrishnan, Selection of Intrusion Detection Threshold bounds for effective sensor fusion, Proceedings of SPIE International Symposium on Defense ans Security, vol.6570, Apr 2007
[12] R. Perdisci, G. Giacinto, and F. Roli, Alarm clustering for intrusion detection systems in computer networks, Engg. applications of Artificial intelligence, Elsevier publications, March 2006
[13] A. Valdes and K. Skinner, Probabilistic alert correlation, Springer Verlag Lecture notes in Computer Science, 2001
[14] O.M. Dain and R.K. Cunningham, Building Scenarios from a Heterogeneous Alert Stream, IEEE Workshop on Information Assurance and Security, 2001
[15] F. Cuppens and A. Miege, Alert correlation in a cooperative intrusion detection framework, Proceedings of the 2002 IEEE symposium on security and privacy, 2002
[16] B. Morin and H. Debar, Correlation of Intrusion Symptoms : an Application of Chronicles, RAID 2003
[17] H. Debar and A. Wespi, Aggregation and Correlation of Intrusion-Detection Alerts, RAID 2001
[18] C. Thomas and N. Balakrishnan,Improvement in Attack Detection using Advances in Sensor Fusion, IEEE Transactions on Information Forensics and Security, vol.4, No.3, Sep.2009
[19] M.V. Mahoney and P.K. Chan, Detecting Novel attacks by identifying anomalous Network Packet Headers, Florida Institute of Technology Technical Report CS-2001-2
[20] M.V. Mahoney and P.K. Chan, Learning non stationary models of normal network traffic for detecting novel attacks, SIGKDD, 2002
[21] R.R. Tenney and N.R. Sandell, Detection with distributed sensors, IEEE Transactions on Aerospace and Electronic Systems, vol. 17, No.4, Jul 1981

[22] M.V. Mahoney, A Machine Learning approach to detecting attacks by identifying anomalies in network traffic, PhD Dissertation, Florida Institute of Technology, 2003
[23] rfp@wiretrip.net/libwhisker
[24] DARPA Intrusion Detection Evaluation Data Set, http://www.ll.mit.edu/IST/ideval/data/data_index.html
[25] J. McHugh, Testing Intrusion Detection Systems: A Critique of the 1998 and 1999 DARPA IDS evaluations as performed by Lincoln Laboratory, ACM Transactions on Information and System Security, vol.3, No.4, Nov. 2000
[26] M.V. Mahoney and P.K. Chan, An analysis of the 1999 DARPA /Lincoln Laboratory evaluation data for network anomaly detection, Technical Report CS-2003-02
[27] C. Thomas and N. Balakrishnan, Usefulness of DARPA data set in Intrusion Detection System evaluation, Proceedings of SPIE International Defense and Security Symposium, 2008

5

Data Association Techniques for Non-Gaussian Measurements

Stephen C. Stubberud[1] and Kathleen A. Kramer[2]
[1]*Oakridge Technology*
[2]*University of San Diego*
USA

1. Introduction

In target tracking, the primary algorithm for estimating the target state has been the Kalman filter(Blackman, 1986) and (Blackman and Popoli, 1999). This workhorse algorithm and its nonlinear implementation, the extended Kalman filter (EKF), provide both an estimate of the target kinematic state, e.g. position and velocity, and an estimate of the error covariance associated with the target state (Brown, 1983). The linear Kalman filter is given as

$$\mathbf{K}_k = \mathbf{P}_{k|k-1}\mathbf{H}^T(\mathbf{H}\mathbf{P}_{k|k-1}\mathbf{H}^T + \mathbf{R}_k)^{-1} \tag{1a}$$

$$\mathbf{x}_{k|k} = \mathbf{x}_{k|k-1} + \mathbf{K}_k(\mathbf{z}_k - \mathbf{H}\mathbf{x}_{k|k-1}) \tag{1b}$$

$$\mathbf{P}_{k|k} = (\mathbf{I} - \mathbf{K}\mathbf{H})\mathbf{P}_{k|k-1} \tag{1c}$$

$$\mathbf{x}_{k+1|k} = \mathbf{F}\mathbf{x}_{k|k} \tag{1d}$$

$$\mathbf{P}_{k+1|k} = \mathbf{F}\mathbf{P}_{k|k}\mathbf{F}^T + \mathbf{Q}_k \tag{1e}$$

where **x** is the track estimate, **H** is the output-coupling matrix that maps the track space to the measurement space, **F** is the state-coupling matrix that models the target motion, **z** is the measurement from the sensor, **P** is the error covariance, **K** is the Kalman gain that weights the residual or innovation, **R** is the measurement error covariance, and **Q** is the process noise. The subscript k is the scan, $k\,|\,k$ indicates the update from the new measurement, and $k{+}1\,|\,k$ denotes prediction. The basic premise of the Kalman filter is that the measurements are Gaussian random variables and the target motion's error or process noise is also Gaussian. For this reason, the Mahalanobis distance (Blackman, 1986) and (Mahalanobis, 1936)

$$(\mathbf{h}(\mathbf{x}_1) - \mathbf{z}_2)^T(\mathbf{H}\mathbf{P}_1\mathbf{H}^T + \mathbf{R}_2)^{-1}(\mathbf{h}(\mathbf{x}_1) - \mathbf{z}_2) \tag{2}$$

has been used as a measure to determine how "close" a measurement is to a given track. The inner product measure of Eq. (2) weights the residual in measurement space by the error covariance:

$$\mathbf{H}\mathbf{P}_1\mathbf{H}^T + \mathbf{R}_2 \tag{3}$$

Thus, a large covariance error reduced the effect of large differences. This chi-squared random variable has provided the primary scoring algorithm for the data association step in the tracking problem as seen in the Bowman model (Steinberg, Bowman, and White, 1999) in Fig. 1.

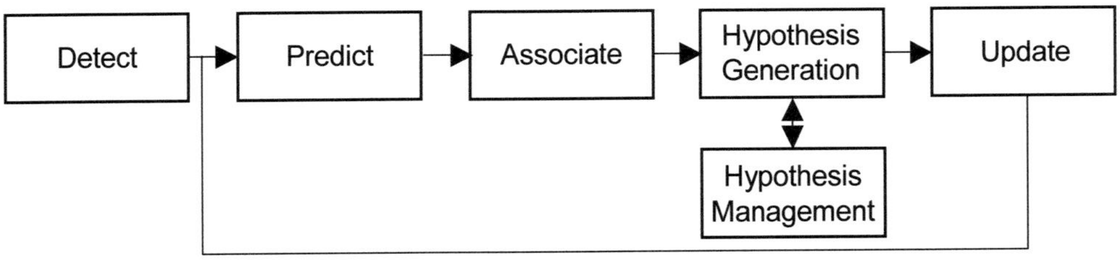

Fig. 1. Target Tracking Functional Flow.

While the Kalman filter has its strengths, the Gaussian assumption is not always accurate. As seen in Figure 2, a simplified model of a passive acoustic measurement is not well approximated by a Gaussian model. Near-field measurements of such a sensor have greater errors as shown in Figure 3. With terrain and mapping information, target limitations and sensor blockage (Figure 4) cannot be mapped without changing the underlying Gaussian assumption of the Kalman filter as done in (Yang and Blasch, 2009).

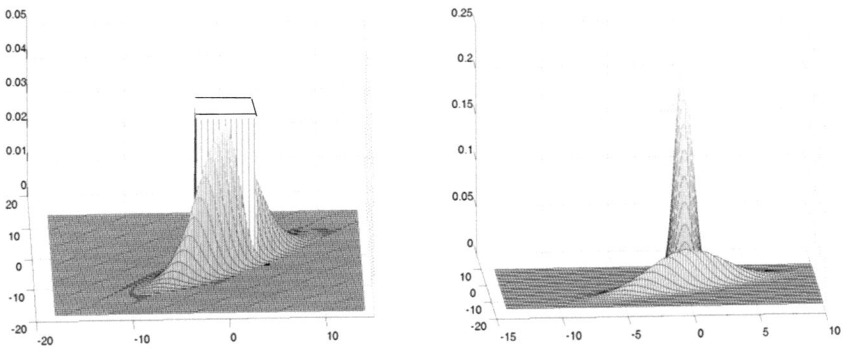

(a) (b)

Fig. 2. a) Measurement Distributions - One with Uniform Distribution and One with Gaussian b) Measurement Distributions - With Gaussian Distribution Used to Approximate a Uniform Distribution

A number of techniques have been derived to handle these issues. Some of the basic techniques include the use of Gaussian sums (Alspach, 1970) and (Alspach and Sorenson, 1972) to approximate the non-Gaussian sensor measurements. The so-called unscented Kalman filer (UKF) (Haykin, 2002) and the particle filter (Arulampalam et al, 2002) use samples of the probability density functions to combine the random variables of the track and measurement. In (Stubberud, 2003) and (Stubberud, 2000), a sequential estimator that can incorporate constraints on the target state without a Gaussian assumption was developed. The single Mahalanobis distance of Eq. (2) is not valid for the Gaussian sum and completely invalid for the other techniques.

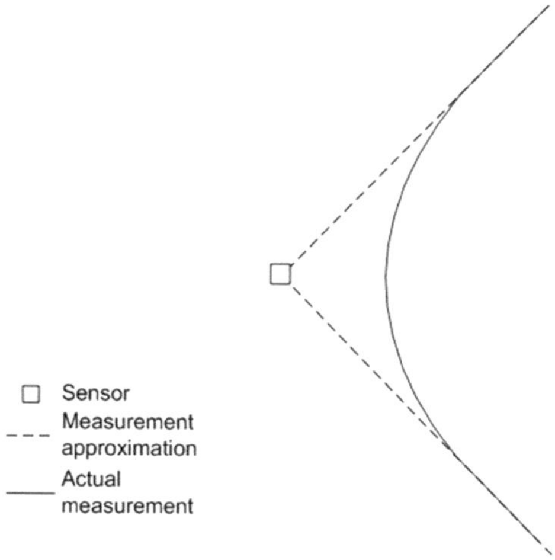

Fig. 3. Near field measurement error

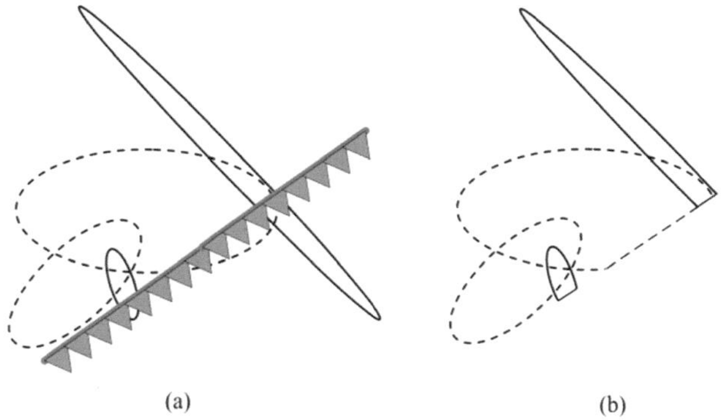

Fig. 4. Uncertainty error ellipses are affected by sensor blockage. a) Gaussian ellipses and blockage indicated; b) resulting alteration of ellipse.

Also, target tracking has grown beyond simple kinematic tracking. The use of classification information has also been incorporated into target track states. Usually, a Bayesian taxonomy (Pearl, 1988) or a Dempster-Shafer algorithm (Dempster, 1967) and (Shafer, 1976) has been employed to provide the scoring for the classification association of the measurement to the target state. The kinematic score is then combined with the classification score using a technique defined by the implementer.
Besides the improvement in the tracking algorithms, data fusion has grown well beyond the target tracking application. As defined in (Steinberg, Bowman, and White, 1999), fusion now covers a number of interpretations of information from a wide variety of sources. Applications are more varied than the military surveillance and targeting that founded the research. However, as discussed in (Stubberud, Shea, and Klamer, 2003) and (Llinas et al,

2004), the association step is still important. Techniques that do not rely on a Gaussian random variable are necessary.
In this chapter, an association technique that maps data into image representation is presented. The images are compared using a phase-only filter (POF) technique to provide a correlation score. The data can be from any source. Each source may contain an uncertainty measure or not. A single association score results.
The development of this effort begins with a previously developed fuzzy logic kinematic association scoring technique that did not rely on the Gaussian assumption. This approach provides the foundation for incorporation of uncertainty into the image representation. The fuzzy approach discussion is followed by the original image correlation technique without uncertainty. The uncertainty incorporation is then developed and example of tracking with kinematics and classification is provided

2. Data association using fuzzy logic

With the desire for more accurate target tracking (Smith and Singh, 2006) and the incorporation of terrain information into tracking systems (Shea et al, 2000) the use of Gaussian models for target tracks and their associated uncertainties became less desirable. While new methods for incorporating non-Gaussian measurements into track states were being developed, different metrics and association routines were being developed for measurements and tracks that were modeled as generic probability density functions. One such method used in some tracking system is the Dempster-Shafer method. This technique does not require any specific probability density function. It is a powerful tool but complex to implement. Other approaches that have been researched use fuzzy logic.
One fuzzy logic approach for association was developed to emulate the chi-squared metric (Stubberud and Lugo, 1999). The fuzzy logic approach used a linguistic model of two interpretations of the chi-squared metric for Gaussian tracks and measurements. Using this basic concept of fuzzy logic (Kosko, 1992), two interpretations of the chi-squared metric were used. The first, a mathematical description, was that the distance between the means was relative based on the size of the covariances. A layered fuzzy logic approach (see Figure 5) was developed where the size of the error covariances of the target and the measurement were used to generate the parameters of the membership functions of the antecedent with regards to the size of the residual.

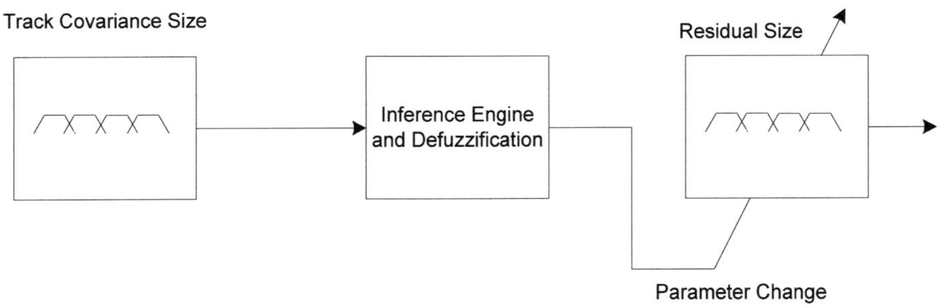

Fig. 5. Layered fuzzy logic is used to reduce computational explosions by making the parameters of membership functions dependent on inputs as opposed to adding to the inference engine.

The second interpretation, a graphical interpretation, determined amount of overlap that the 2-sigma ellipses for the measurements created relative to the maximum amount of overlap possible. Then the three inputs, residual size and the two percentages of overlap, are operated on by the inference engine, as shown in Figure 6, and that results in the association score. This approach was shown to have comparable behavior to the chi-squared metric in two standard challenge problems, crossing targets and closely spaced parallel targets.

The development of the fuzzy association scheme was then extended to handle the two cases where measurements and/or tracks that had uniform probability density functions, (Stubberud and Kramer, 2006) and (Stubberud and Kramer, 2007). The first case considered two uniformly distributed representation. The association is performed using the graphical interpretation of overlapping areas. As seen in Figure 7, when the uniform distributions are being associated, the scoring can be looked at as the graphical interpretation. A percentage of overlap between the two density functions can be interpreted to give a score between 0 and 1. The fuzzy system operates on the overlapping region.

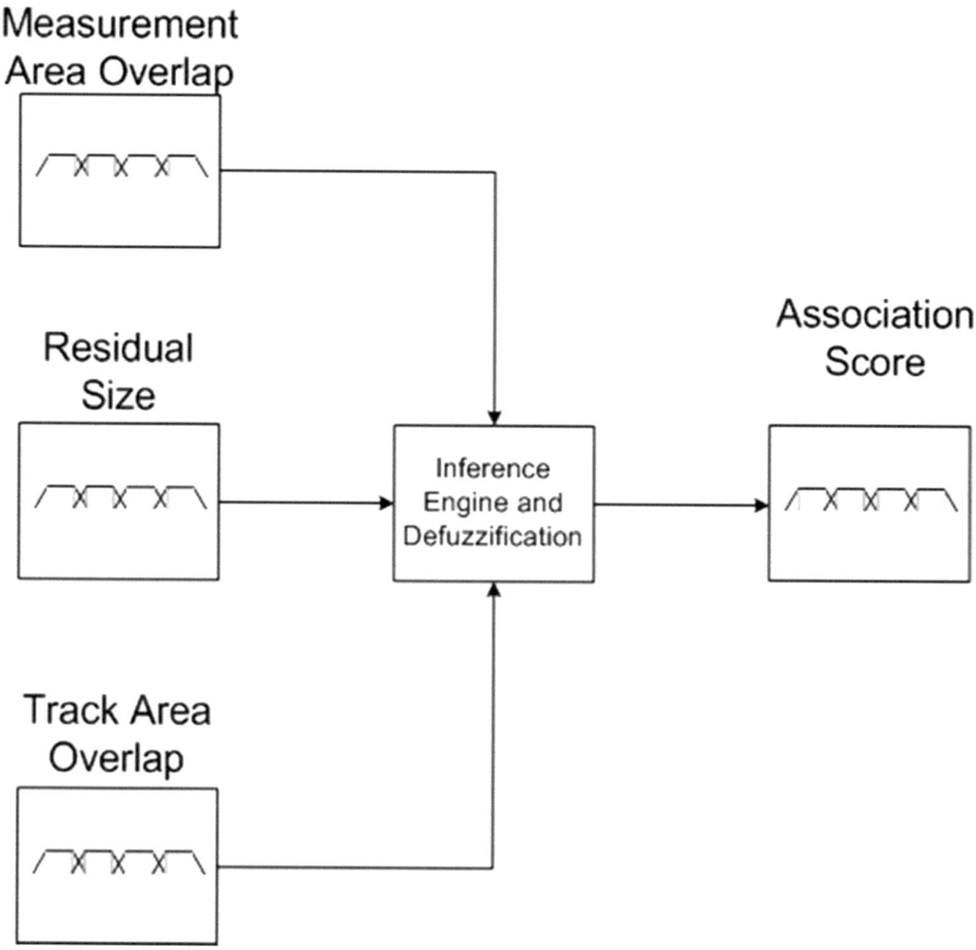

Fig. 6. Association score is generated using three inputs: overlap of the measurement uncertainty, overlap of the track uncertainty, and the residual size.

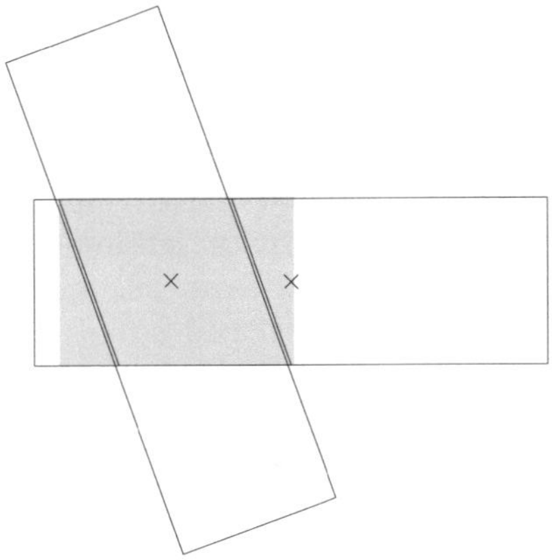

Fig. 7. Two uniform distributions show that overlap is the primary interpretation of their association.

The area of the overlapping region is compared to the possible percentage of overlap and a score is determined. The antecedent functions for each percentage is defined as shown in Figure 8.

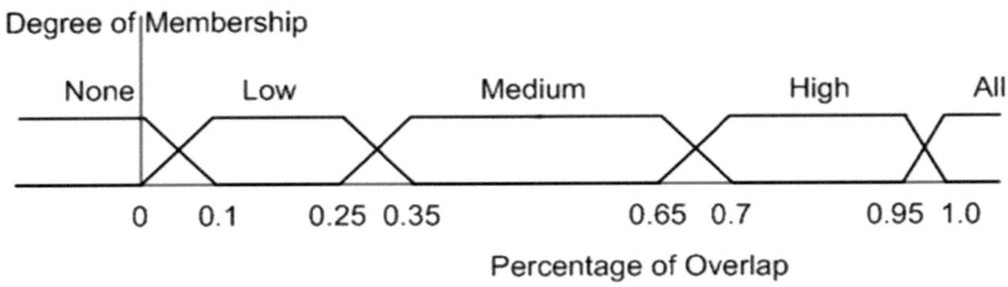

Fig. 8. Example antecedent functions

The consequence functions shown in Figure 9 are used to provide the association score.

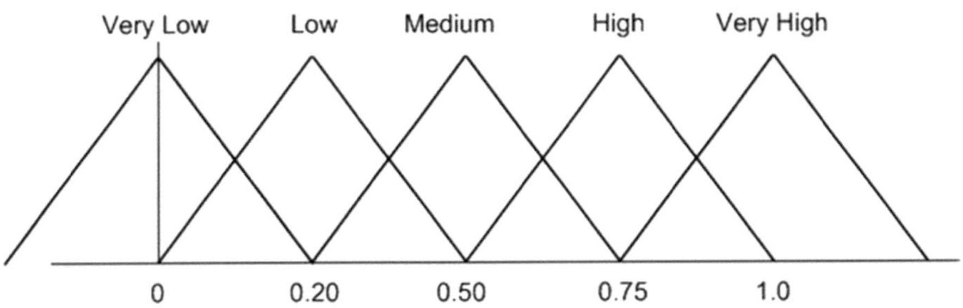

Fig. 9. Example consequence functions

In the second case, a Gaussian track and uniform measurement (or vice versa) was considered. The layered fuzzy logic approach with regards to the residual distance is again used to create the membership functions used to input information to the inference engine. The distance measure is the average distance from the Gaussian mean to the vertices of the uniform distribution and the mean of the uniform distribution. Figure 10 shows the distances used to form the measure for one example.

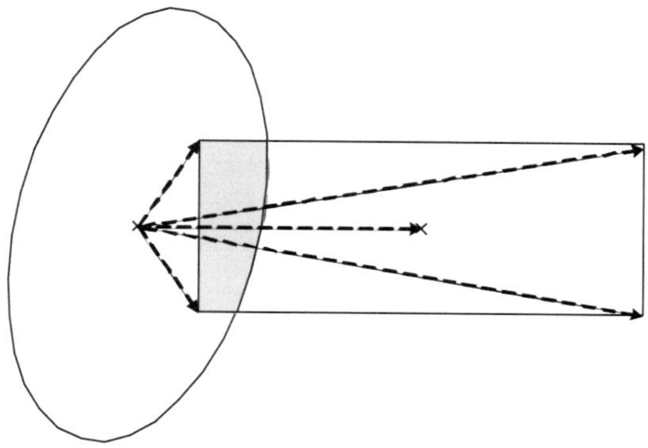

Fig. 10. Association of a Gaussian to a uniform distribution is dependent on both overlap and a modified residual.

The kinematic solution for fuzzy association was applied to other classes of probability density functions for research purposes. These efforts led to the concept of generating rule bases for different applications of association such as target classification and threat assessment. In these applications, the values can be discrete, continuous, or both. The associated uncertainties which are represented by the covariance matrix in the Gaussian representations often are not even measured values. For such applications, the use of fuzzy logic is quite effective. For the Level 1 target tracking application, classification is an important component of sensor reports and track description.

For classification information to be incorporated into the target track, the measurements can exist in two forms: target class and attribute information. For the target classification type measurements, two typical formats include a class and an associated quality score or a set of possible classes with associated probabilities (Pearl, 1988) or levels of evidence accrual (Stubberud, Kramer, and Geremia, 2007). For the former format, reports are provided similar to those exemplars listed in Table 1. The report is a pre-defined classification of the target. This is considered the measurement. The quality score can be a covariance as with Report 1, a probability as with Report 2, or generalized quality level as given with Report 3. The second reporting scheme provides a number of potential classes that the target may be, as seen in Table 2. Each possible class has an associated probability or degree of evidence. The reports look as a discrete probability density function. The probabilities should incorporate the quality in their generation. The evidence accrual results each have an associated quality score with them as shown in the table. In application, different classification algorithms have different methods of reporting results. Therefore, a centralize fusion system must be able to associate and combine the different reporting schemes. For target classification association, when the measurement reports are provided as a target class the fuzzy association scheme is implemented as seen in Figure 11.

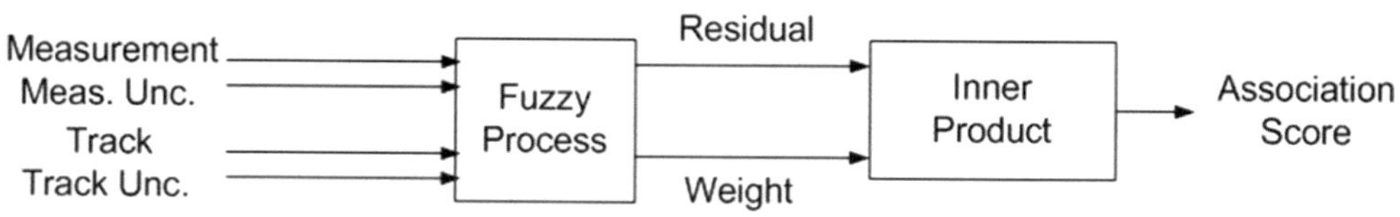

Fig. 11. Classification Association Process Flow

Target	Class	Uncertainty/Quality
1	2 (APC)	2.5
2	Tank	0.62
3	Truck	Good

Table 1. Exemplars of Classification Reporting Schemes and Their Uncertainties

Target Track #	Class 1 (Tank)	Class 2 (APC1)	Class 3 (APC2)	Class 4 (APC3)	Class 5 (Truck)
1	.1	.4	.3	.1	0
1 (Unc)	NA	NA	NA	NA	NA
2	.8	.5	.5	0	0
2 (Unc)	.01	.2	.01	0	0

Table 2. Exemplars of Probability and Evidence Accrual Based Reporting Schemes

The actual class is mapped through a fuzzy logic system to map the various classes into related values. For example, tank can be mapped into several values. Three such values are *armored vehicle, large vehicle,* and *tread vehicle.* An armored personnel carrier (APC) could map into *armored vehicle, medium vehicle* or *large vehicle,* and *tread vehicle* or *wheeled vehicle.* The fuzzy mapping would therefore a set of consequence membership functions for each value. These values then create a vector similar to the position and velocity vector for kinematics:

$$x_{class} = \begin{bmatrix} Tank \\ APC1 \\ APC2 \\ APC3 \\ APC4 \\ Truck \end{bmatrix} \quad (4)$$

The vector can be from one element to over a hundred elements that measure each attribute that various sensors can provide. The vector of the measurement and its counterpart for the target track then can create a residual vector

$$z_{class} = x_{class}^{Meas} - x_{class}^{Track} \quad (5)$$

Based on the different classification schemes six different reporting schemes exist as defined in Table 3. In each case, the following fuzzy logic association scheme is used as the basis with minor modifications in its implementation. Figure 12 shows the functional flow of this basis. As with the kinematic fuzzy association technique, the foundation is to use the Mahalanobis distance to build the algorithm. A residual of each classification vector exists from above. The next step is to generate a comparable uncertainty score. Unlike the Gaussian generation of the covariance, the score here will be a weighting, $\mathbf{w}^{-1}$, for the inner product

$$assoc\_score = z_{class}^{T} \mathbf{w}^{-1} z_{class} \quad (6)$$

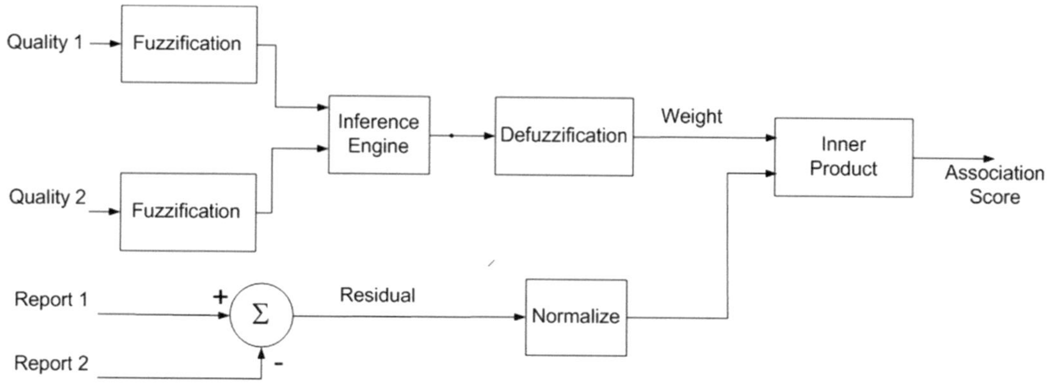

Fig. 12. Classification Uncertainty Association Incorporation

If the measurement or track is defined by a level of quality, the consequence membership function of Figure 13 is used. As with all previous defuzzification steps, the authors have used a center of gravity method. This is not a requirement but simply a preference. For evidence accrual technique, the associated covariance for the given class mapped to the consequence function using the antecedent membership function and the associated inference engine in Figure 14. The probability reports have a covariance set as a *good* covariance. The weighting score is generated using the fuzzy process defined in Figure 12.

Association	Report 1	Report 2
Type 1	Class/Quality	Class/Quality
Type 2	Class/Quality	Probability
Type 3	Class/Quality	Evidence
Type 4	Probability	Probability
Type 5	Probability	Evidence
Type 6	Evidence	Evidence

Table 3. Possible Association Combinations

For association Type 1 defined in Table 3, this is the process to compute the weight for Eq. (6) which is then the association score. For Type 2, the association score is generated for each potential class and multiplied by the probability of each class

$$\sum_{i=1}^{num\_class} p_i \cdot assoc\_score_i \tag{7}$$

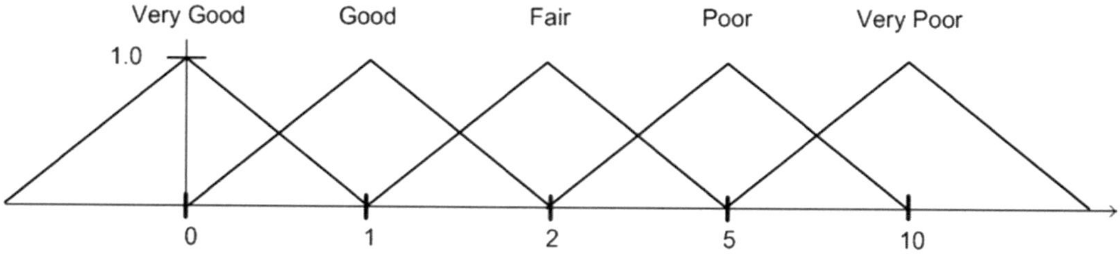

Fig. 13. Consequence Membership Function for Quality Mapping

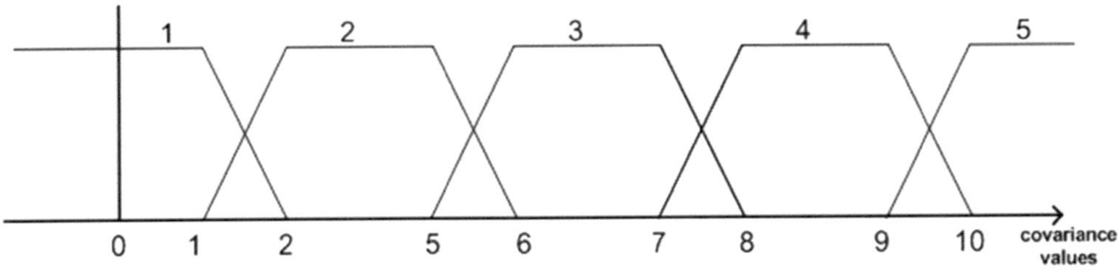

1	2	3	4	5
VP	P	F	G	VG

Fig. 14. Antecedent Function and Inference Engine for Numeric Quality Mapping

Type 3 is similar to Type 2 except that the probabilities used in Eq. (7) are replaced by a normalized evidence score for each class

$$p_i = \frac{e_i}{\sum_{i=1}^{num\_class} e_i} \tag{8}$$

Type 4 is generated by performing the association scores for each permutation and weighting them appropriately with the two sets of probabilities

$$\sum_{i=1}^{num\_class} \sum_{i=1}^{num\_class} p_j \cdot p_i \cdot assoc\_score_i \tag{9}$$

Type 5 replaces one set of probability scores from the Type 4 technique with one set of normalized evidence scores from Eq. (8). Type 6 association is the same as Type 4 association with the exception that the probabilities are replaced normalized evidence scores.

3. Optically inspired correlation

Optical computation is the process of using the physics of light, lenses, and filters to provide specific computations at the fastest speeds possible, the speed of light. By modeling light's behavior with mathematical functions, these functions can be performed on dedicated hardware platforms. One such function is that of optical correlation (Kumar, Mahalanobis, and Juday, 2005). The process is shown in Figure 15.

Coherent light is passed through a thin film Red-Green-Blue (RGB) screen called a spatial light modulator (SLM). On one of the set of pixels (assume Red) or channel, an image is placed. The coherent light projects this image through lens onto a second SLM. The new image is the two-dimensional (2D) Fourier transform or fast Fourier transform (FFT) of the image on the first SLM. On the second channel, the 2D-FFT of a comparison image is placed and the pixels are individually multiplied. The resulting multiplied image is placed in the third channel and projected by the coherent light through a second lens onto a third lens. This provides the inverse 2D-FFT of the combined image which is actually the correlation

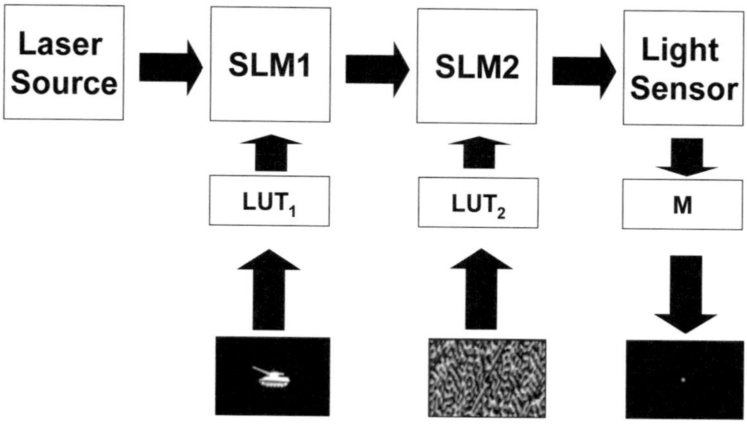

Fig. 15. Optical correlation functional flow using an optical correlation device.

function. (This will be mathematically described below.) For clarity, the input image at the first SLM will be the measured or test image. The image whose 2D-FFT is placed at the second SLM is the reference image. As seen in Figure 16, if the reference image is centered when transformed so that the correlation peak will be centered about the location of the reference image in the test image. In the case of Figure 16, two copies of the reference image exist in the test image and, thus, two correlation peaks exist. Both are centered about their locations in the test image. This implies that optical correlation can be used for both fast processing and parallel processing (multiple correlation comparisons simultaneously).
One problem with the optical correlator is that the SLMs limit the range of values that the image amplitude can attain. The pixels are integer values between 0 and 255. The image at the last SLM is only the real part of the final transform. Also, in some optical correlators, the amplitude of the test image's 2D-FFT is normalized to 1, (Kumar, Mahalanobis, and Juday, 2005). This is a physical limitation in optical correlators, especially early versions. The normalization of amplitude implied that only the phase information was used. Despite this apparent limitation, this so-called phase-only filter has actually provided improved correlation results.

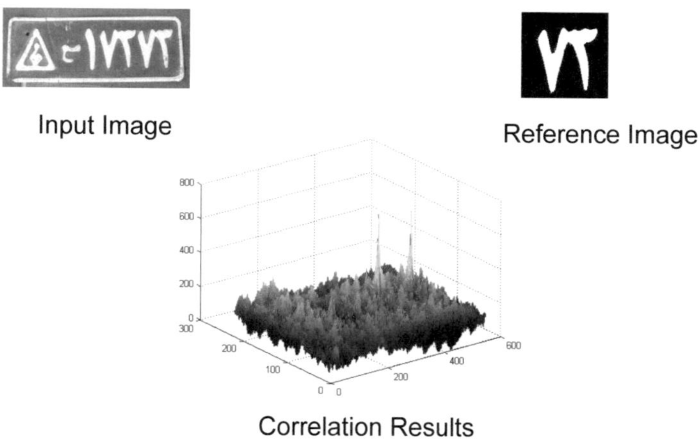

Fig. 16. Two correlation peaks are created aligned with the location of the reference image in the input image.

The mathematics of the optical physics is straightforward. The standard correlation function between two images **F** and **G**, is defined as

$$\begin{aligned} &\sum\sum f^*(x,y)f(x+X,y+Y) \\ &=\sum\sum F^*(k_x,k_y)F(k_x,k_y)e^{-j(k_xX+k_yY)} \\ &= \Im^{-1}\left\{\left|F(k_x,k_y)\right|^2\right\} \end{aligned} \tag{3}$$

As in basic signal processing (Oppenheim, Willsky, and Hamid, 1996), the spatial or time domain can be computed using frequency domain techniques. The frequency domain version of the correlation function is defined as

$$\begin{aligned} &\sum\sum g^*(x,y)f(x+X,y+Y) \\ &=\sum\sum G^*(k_x,k_y)F(k_x,k_y)e^{-j(k_xX+k_yY)} \\ &=\mathrm{Re}\left\{\Im^{-1}\left\{G^*(k_x,k_y)F(k_x,k_y)\right\}\right\} \end{aligned} \tag{4}$$

where F is the Fourier transform. With amplitude information normalized, the POF form of the correlation function based on the optics is

$$\begin{aligned} &\sum\sum G^*(k_x,k_y)\frac{F(k_x,k_y)}{\left|F(k_x,k_y)\right|}e^{-j(k_xX+k_yY)} \\ &=\mathrm{Re}\left\{\Im^{-1}\left\{G^*(k_x,k_y)\frac{F(k_x,k_y)}{\left|F(k_x,k_y)\right|}\right\}\right\} \end{aligned} \tag{5}$$

With the advent of faster computers and the desire to avoid using specialized hardware, the optical correlator is rarely used but the mathematics of the optical physics provides a powerful tool for the correlation problem which can be applied to the data fusion association problem.

3.1 Amplitude tile

For the Level 1 fusion problem, the image used for correlation is referred as a tile. The tile is a two-dimensional image that maps data about the target track and the measurement reports into an amplitude plot. Figure 17 presents an example of a amplitude tile. In this case, the tile is 50 x 200 pixel image. The attributes of the target track vector or the measurement vector are the vertical components. The horizontal components are the time reports. Each 10 x 10 block represents a given attribute at a given time. The rows or attributes are defined as:

- target latitude
- target longitude
- target altitude
- target speed

- target course heading
- target classification

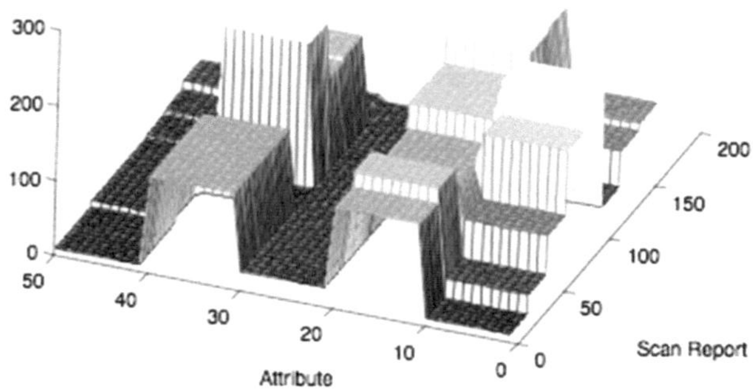

Fig. 17. The amplitude tile creates blocks valued between 0 and 255.

The columns are these attributes at different times. For amplitude tile, the intensity of each 10x10 block is a value between 0 and 255. The range of values for position is limited by the region of interest. This implies that, if the region of interest is for air targets between latitude 32°N and 40°N and 25°E and 33°E, each degree change for latitude the intensity would change by 31.875 in amplitude as would each change in intensity for longitude. Similar ranges could be created for the other attributes.

3.2 Phase tile

The image tiles can be created in a different format. This second 10x10 block format is that each pixel of the (m,n)th block is generated by

$$\sin\left(\omega_m i_{mn} + \phi_{mn}\right) \cdot \sin\left(\omega_m j_{mn} + \phi_{mn}\right)$$

where m is the number of attributes and n is the number of time steps. The values of i and j represent the individual pixel within the 10x10 block. An example of a tile from this phase generation is shown in Figure 18.

The phase variations range between 0 and 180 degrees depending on the attribute value. The selection of the phase would be made similar to the amplitude change above. Each attribute has its own frequency as well. This is done to separate each attribute more clearly in the frequency domain.

4. Image-based association with uncertainty and classification information

The image correlation technique, described in Section 3, has been demonstrated to work comparably to the techniques that use the chi-squared metric based on the Mahalanobis distance (Stubberud and Kramer, 2010a) and (Stubberud and Kramer, 2009). The incorporation of uncertainty into the algorithm began with the use of a sample set of points in the uncertainty ellipse (Stubberud and Kramer, 2010b). An iterative process of association scores were generated and weighted similar to the technique that was developed in this chapter for the fuzzy logic association applied to the classification problems where evidence or probabilities described the measurement. The technique is easily extended to uniform distributions as well as other distributions. In this section, an uncertainty model is introduced that remaps the phase distribution for the uncertainty model using fuzzy logic techniques derived from the methods of Section 2. The section concludes with the incorporation of classification information where the Gaussian assumption is not valid.

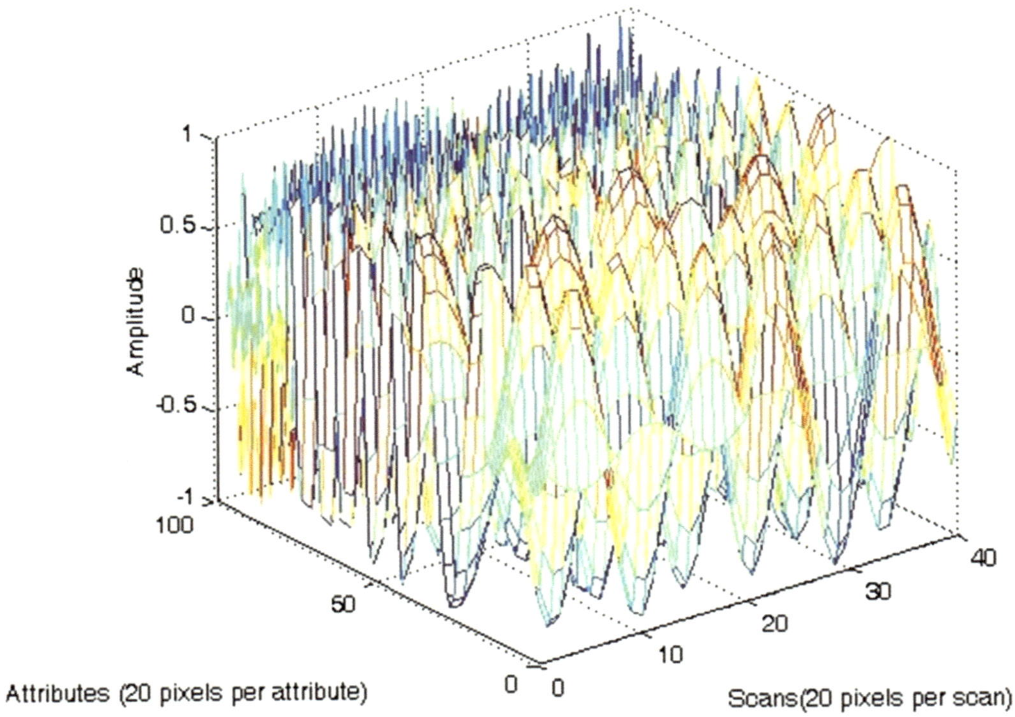

Fig. 18 The phase tile maps sinusoids with phases between 0 and 180 degrees.

4.1 Fuzzy logic applied to phase variation for tile generation

In (Stubberud and Kramer, 2009) a fuzzy logic approach to the data association problem was developed. There, the technique used multi-layered fuzzy logic to create a score by implementing a linguistic interpretation of the chi-squared metric. Closeness was based on the weighted distance between the two track elements. It also considered the overlap of the error ellipses. In this approach, a similar view of the error ellipses is used. Instead of using complex mathematical distances which would be adverse to the simplicity of the image correlation technique, a basic fuzzy logic approach is developed to adjust the tile creation based on the uncertainty ellipse size and orientation.

The first variable to be mapped is that of the orientation of the error ellipses. The fuzzy memberships are defined as five potential angles of the error ellipse relative to the longitude (x-axis). As seen in Figure 19, the memberships are 0°, 45°, 90°, 135°, 180°. The alignments provide two inputs to the fuzzy decision process described in Figure 20. The first is a weighting factor that is generated using the crisp value of the antecedent functions. These weights are used in the defuzzification process of the second defuzzification step in Figure 20. The second input is the different rules that are activated from the inference engines in

Tables 4 and 5 which relate to longitude component of the target track (or the x-axis) or the latitude component of the track (x-axis). These rules from the first stage determine vary the inputs to the inference engine the second stage (Table 6). The variations in the inputs are defined in Table 7. The input A indicates the average size of the minor or major axes of the uncertainty ellipse of Track 1 and B indicates the average of minor or major axes of Track 2. The table also indicates that multiple instantiations of the fuzzy inference engine are used in several cases.

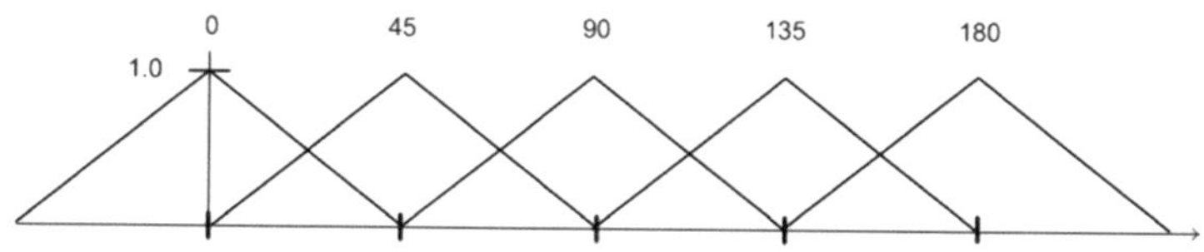

Fig. 19. The antecedent triangular membership functions used to determine the alignment of the uncertainty ellipses.

Degrees Longitude	0	45	90	135	180
Effect	1	2	3	2	1

Table 4. Inference Engine Major Axis Alignment of Covariance Effect on Longitude (x-axis)

Degrees Latitutde	0	45	90	135	180
Effect	3	2	1	2	3

Table 5. Inference Engine Major Axis Alignment of Covariance Effect on Latitude (y-axis)

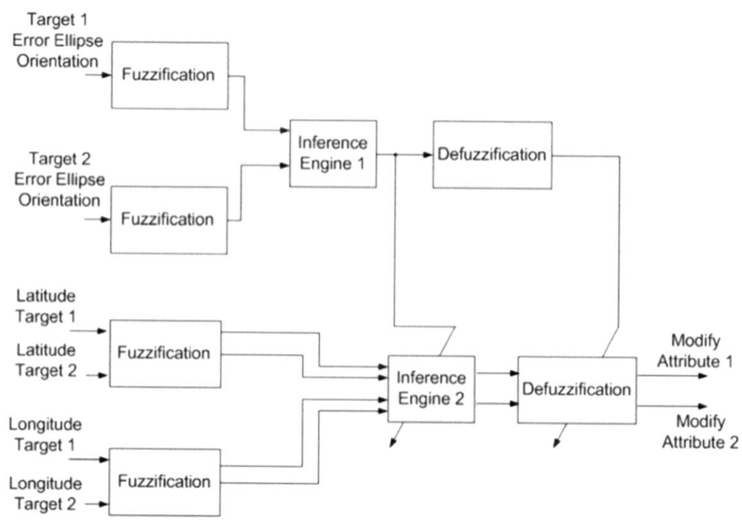

Fig. 20. The fuzzy logic process to vary the tile generation used in image correlation track association process.

		B			
		Small	Medium1	Medium2	Large
A	Small	1	2	3	4
	Medium1	2	3	4	5
	Medium2	3	4	5	6
	Large	4	5	6	7

Table 6. Inference Engine of Averaged Aligned Error Covariance Axis Sizes

The second process is based on the size of the axis of interest of the uncertainty ellipse. The actual uncertainty in the directions of latitude and longitude are mapped to membership functions in Figure 21. The elements of the membership functions are small, medium_one, medium_two, and large. Since the uncertainty measure is global, i.e., averaged over the entire time window, the results will affect all of the time subtiles of the entire attribute similarly. The proposed inference engine, defined in Table 4, vary depending on the alignment rule from the first step. If multiple instantiations are desired, then the output of each instantiation is weighted by the noted factor in Table 5. An NA indicates that those rules are not used and result in a zero value for that component.

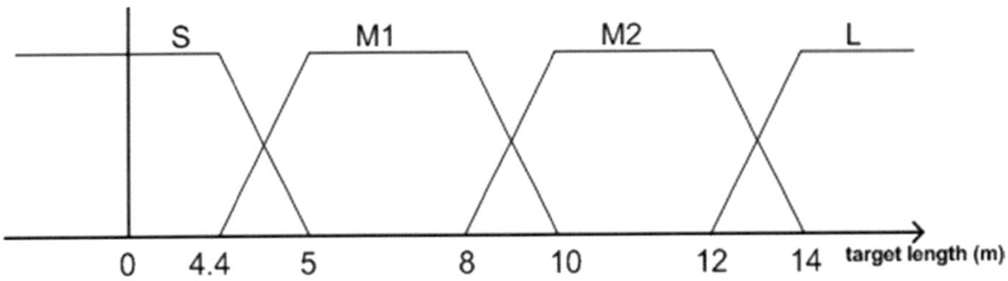

Fig. 21. The antecedent triangular membership functions used to determine the size of the uncertainty ellipses axis.

Alignment			Component							
T1	T2	Wt.	*A1*	*B1*	*A2*	*B2*	*A3*	*B3*	*A4*	*B4*
1	1	1	Maj	Maj	NA	NA	NA	NA	NA	NA
1	2	0.5	Maj	Min	Maj	Min	NA	NA	NA	NA
1	3	1	Maj	Min	NA	NA	NA	NA	NA	NA
2	1	0.5	Maj	Maj	Maj	Maj	NA	NA	NA	NA
2	2	0.25	Maj	Maj	Maj	Min	Min	Maj	Min	Min
2	3	0.5	Maj	Min	Min	Min	NA	NA	NA	NA
3	1	1	Min	Maj	NA	NA	NA	NA	NA	NA
3	2	0.5	Min	Maj	Min	Min	NA	NA	NA	NA
3	3	1	Min	Min	NA	NA	NA	NA	NA	NA

Table 7. Alignment Effects on the Covariance Inference Engine and Process

To simplify the development in this paper, the association score is generated on only the position data. The extension to include velocity is simply to repeat the process on the velocity measures and uncertainties.
The consequence function is the same as that seen in Figure 22. The defuzzified value is used to reduce the phase range.

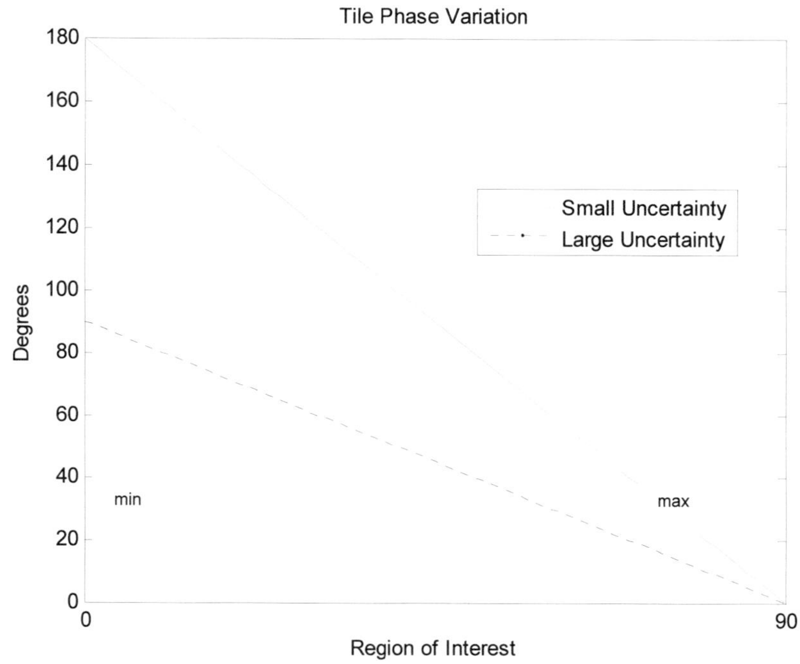

Fig. 22. An Example of the Change in Range of the Phase Variation Based on Track Uncertainty.

4.1.1 Incorporating uncertainty as a quantization of the phase separation

To vary the phase difference in generating the subtiles using uncertainty, at each time instance, an estimate of the effects of both track's uncertainty on the association score is made. For each attribute, a modification score using the fuzzy logic process described in Section 4 is made. This modification score is computed for each attribute at each report time or scan for both tracks.
The modification score changes the range of phase values used to adjust the phase of each attributes value. As seen in Figure 22, the linear range of values is reduced as the uncertainty becomes larger. The nominal range is from 0° to 180°. As the range shrinks the variation in the phase for changes becomes less. As would be the case with the chi-squared metric, larger differences can be tolerated.

4.1.2 Example and results

To analyze the performance of the new image-correlation association that incorporates uncertainty, a scenario with two targets and four sensor systems was generated to compare

the incorporation of uncertainty in a global manner to image-correlation data association routine. The two targets' trajectories are depicted in Figure 23.
The target on top is considered Target 2. The three sensors platforms, which are stationary throughout the scenario, are also shown. The sensors are identical range-bearing sensors, i.e., a radar or active sonar, each with Kalman filter kinetic target tracking systems on them. The association scores are generated to compare the position-only data form each track from the first three sensor systems.
Figure 24 depicts the difference between the track of Target 2 from both Sensor 1 and Sensor 3 and the track for Target 1 from Sensor 1. In Figure 25, the association scores between the two tracks of Target 2 are shown for the original image-correlation technique without uncertainty and the new uncertainty-incorporation method.

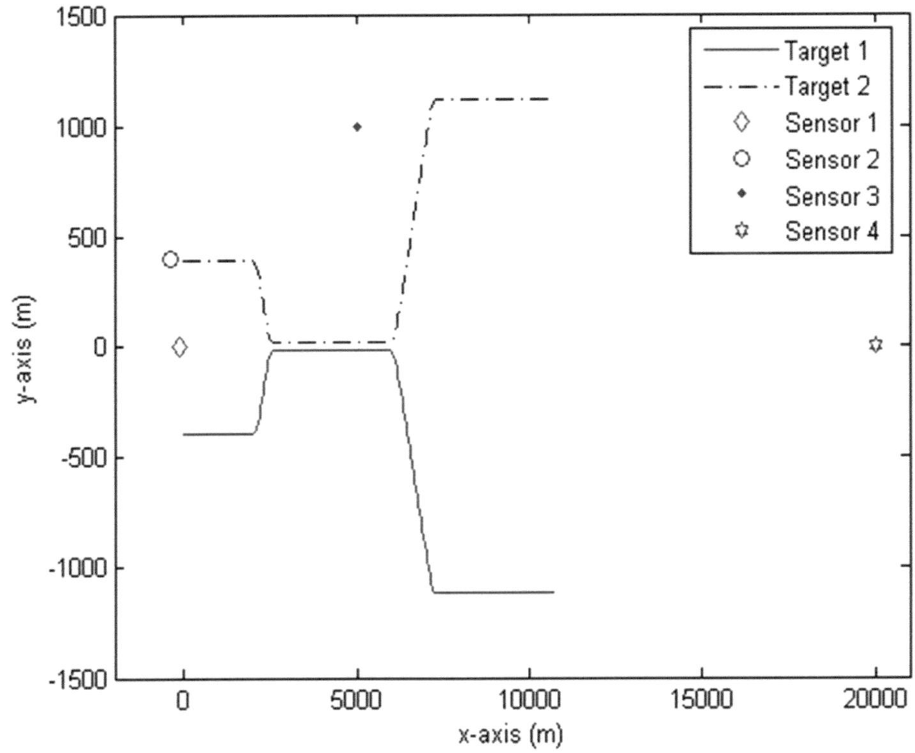

Fig. 23. Two target example scenario.

The incorporation of uncertainty reduces the variation in the scores. This is a benefit of the use of uncertainty similar to the chi-squared metric. Also, the uncertainty method reduces the loss of association score as the target gets further from the sensors. A similar comparison of the association scores for both techniques between Sensor 1's Target 1 track and Sensor 3's Target 2 track is shown in Figure 26. In this comparison, the uncertainty association does better in that it produces lower scores for the differences between the positions of the target tracks.

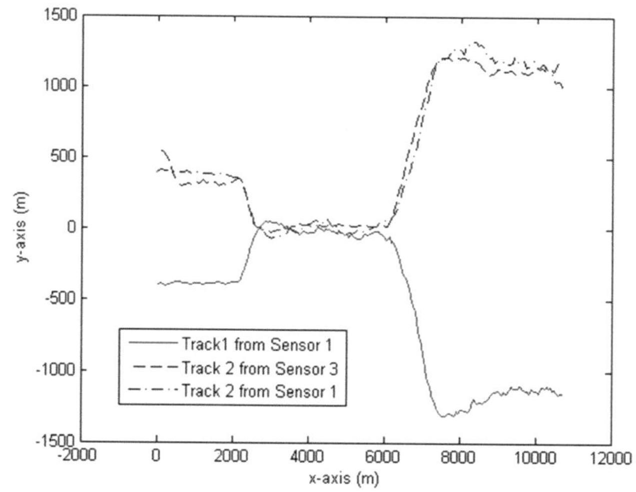

Fig. 24. Track results for Target 1 and Target 2 for Sensor 1 compared to track results for Target 2 for Sensor 3.

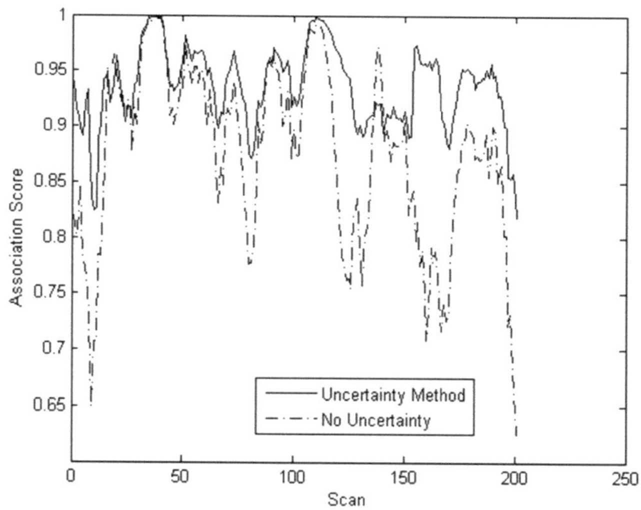

Fig. 25. Association score comparison of results with and without incorporation of uncertainty.

Figure 27 provides a more detailed analysis of uncertainty association scores of the tracks for Target 2 from Sensor 1 and Sensor 3. The reports of Scan 10 and Scan 195 are marked. By comparing the results to the actual tracks seen in Figure 26, it is apparent that the scores follow the differences in the tracks. At Scan 10, a full tile is generated for the first. (Prior to this, ten scans of data for each attribute were not available.) At the tenth scan, the target tracks were separated in position in the y-coordinate of the first leg of the trajectory. Since the sensors were fairly close to the targets, the uncertain ty was "small" and, therefore, the association score became poor. For the most part after that, the association score settled near or above 0.9 until the end. By this time, around Scan 195, the uncertainty from both tracks started to become more aligned. This means that the small range error would bring the x-coordinate uncertainty in the track down. The x-coordinate lag Sensor 1's track seen in the last two legs of the Target 2 trajectory now have a greater impact on the score. This drives the association score downwards.

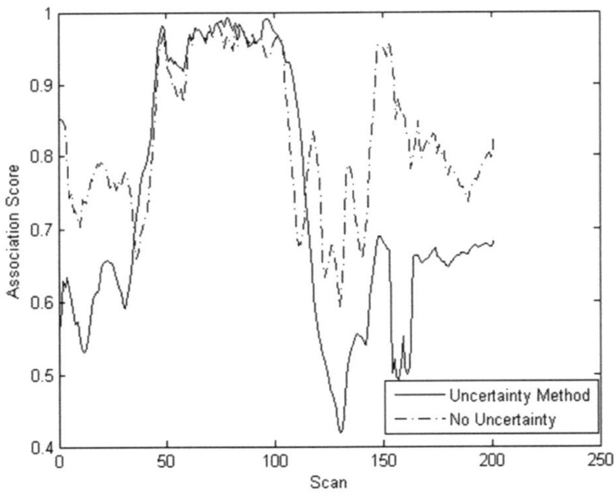

Fig. 26. Association score comparison of results with and without incorporation of uncertainty.

4.2 Incorporating classification into image-based association

The incorporation of classification is a two-step process. The first step is the definition of the new attributes for the tile. The second is the incorporation of uncertainty.

In Section 2, the classification of a target was mapped into a set of subclasses and those were mapped into numeric values using fuzzy logic. In this development, a similar approach is used. The use of the subclasses creates more attributes for the tile generation. As with the exemplar above, three attributes are used for each class *armor_type*, *wheel/tread* , and *vehicle_size*. Each attribute would have a 20x20 pixel subtile for each scan or report of data. Each class would be set to its own phase predetermined by the nearness of the relationship. Thus, as seen in Table 8, a number phase differences vary based on their relationships to each other.

To incorporate uncertainty, two techniques are used. The first is the probability of an attribute. In the example, and a generic APC can be either a wheeled or tread vehicle. If the specific type of APC is not know then a probability of either type would exist. For example, assume there is a 0,6 probability that an APC would have a tread. The subtile would then have 12 pixels of its horizontal set to tread while 8 pixels would be set as wheeled (see Figure 28). Every change as the classification information becomes better would change the composition of the subtile for the given scan as further shown in Figure 28 The quality uncertainty of each report would then be generated using the redistribution of the phases by shrinking or expanding the range of values in Table 8 using the technique described in Section 4.1. For example, a large uncertainty (or poor quality) measurement/track pair would have phase ranges similar to those in Table 9.

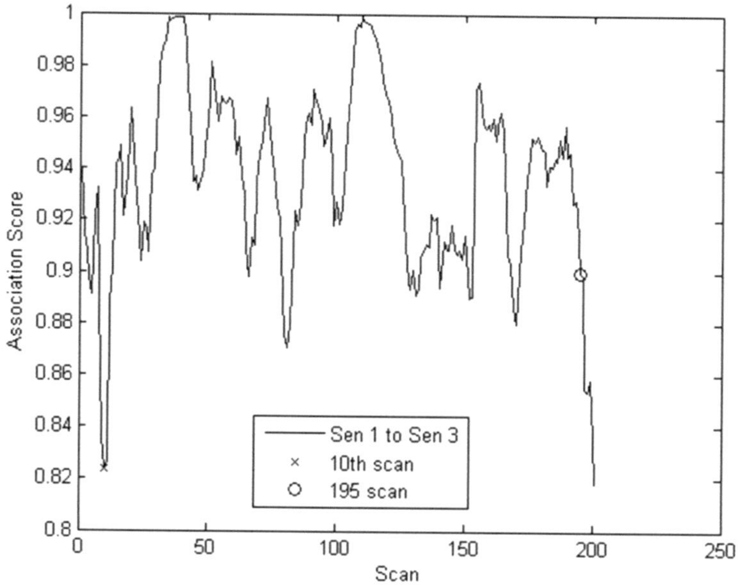

Fig. 27. Association scores using the uncertainty method for the tracks for Target 2 from Sensor 1 and Sensor3

Class	Attribute Phase Value (degrees)			
	Val. 1	Val. 2	Val. 3	Val.4
Armor_Type	0	45	170	NA
Wheel/ Tread	0	180	NA	NA
Vehicle_Size	0	10	45	90

Table 8. Classification Attribute Mapping To Phase Differences

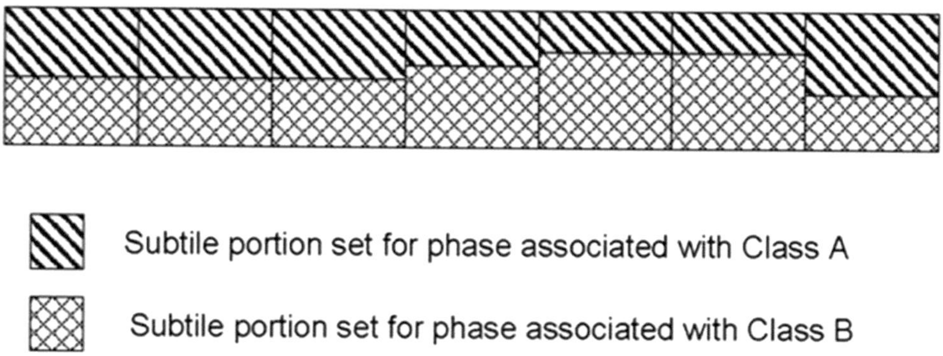

Fig. 28. Subtile separation based on probability score of different attributes.

Class	Attribute Phase Value (degrees)			
	Val. 1	Val. 2	Val. 3	Val.4
Armor_Type	0	12	90	NA
Wheel/ Tread	0	45	NA	NA
Vehicle_Size	0	5	20	45

Table 9. Classification Attribute Mapping To Phase Differences With Large Uncertainties

4.2.1 Example and results

The aforementioned example from Section 4.1.2 is again used. In this problem, the target classification is given as in Table 10 for the measurement and the track. The table details probability for each of the element of the attribute. The results of the classification algorithm for Target 2 from Sensor 1 and Sensor 3 are shown in Table 11. The classification differences have significant impact on the association scores as expected.

Class (Target #)	Report (Uncertainty)										
Armor_Type (1)	1 (.3)	1 (.2)	1 (.2)	1 (.2)	1 (.2)	1 (.2)	1 (.2)	1 (.2)	1 (.2)	1 (.2)	1 (.2)
Wheel/ Tread (1)	1 (7)	2 (7)	2 (6.5)	2 (6.8)	1 (6.9)	1 (6.4)	1 (6.2)	1 (6.1)	2 (6.4)	2 (6.5)	1 (6.6)
Vehicle_Size (1)	1 (2.3)	1 (2.2)	1 (2.2)	1 (2.1)	1 (2)	1 (1.9)	1 (1.9)	1 (2)	1 (1.8)	1 (1.8)	1 (1.7)
Armor_Type (2)	2 (.3)	2 (.2)	2 (.2)	2 (.2)	2 (.2)	2 (.2)	2 (.2)	2 (.2)	2 (.2)	2 (.2)	2 (.2)
Wheel/ Tread (2)	2 (7)	2 (6.4)	2 (6.4)	2 (6.4)	2 (6.4)	1 (6.4)	2 (6.4)	1 (6.4)	2 (6.4)	2 (6.4)	2 (6.4)
Vehicle_Size (2)	2 (2.3)	2 (2.3)	2 (2.4)	3 (2.7)	3 (2.8)	3 (2.9)	2 (2.7)	2 (2.6)	2 (2.5)	1 (2.8)	2 (2.9)

Table 10. Classification Reports For Target 2/Sensor 1 And Target 2/Sensor 3

Association Score			
Scan 8	Scan 9	Scan 10	Scan 11
0.59	0.56	0.56	0.52

Table 11. Association Score Between Target 2/Sensor 1 and Target 2/Sensor2

5. Conclusion

The use of Gaussian distribution in target tracking is becoming less prevalent that in the past. New association techniques are needed for the fusion process. The approaches need to look at previously unused technologies or techniques used for other applications in the past. In this chapter, new techniques for data association were addressed and applied to current fusion problems that combine kinematics and classification as well as address nonGaussian uncertainty measures.

6. References

Alspach, D.L., A Bayesian Approximation Technique for Estimation and Control of Time Discrete Stochastic Systems, Ph.D. Dissertation, University of California San Diego, 1970.

Alspach, D. & Sorenson, H., (1972). "Nonlinear Bayesian Estimation Using Gaussian Sum Approximations," *IEEE Transactions on Automatic Control*, Vol. 17 , No. 4 , pp. 439 - 448, Aug., 1972.

Arulampalam, M.S., Maskell, S., Gordon, N., and Clapp, T. (2002). A Tutorial on Particle Filters for Online Nonlinear/Non-Gaussian Bayesian Tracking, IEEE Transactions on Signal Processing, Vol. 50, No. 2, (February 2002), pp. 174 - 188, ISSN: 1053-5872

Blackman, S. (1986). Multiple-Target Tracking with Radar Applications, Artech House, ISBN 0890061793, Norwood, MA, USA

Brown, R.G. (1983). Introduction to Random Signal Analysis and Kalman Filtering, John Wiley and Sons, ISBN: 978-0471087328, New York, NY, USA

Dempster, A. (1967). Upper and Lower Probabilities Induced by a Multivalued Mapping, Annals of Mathematical Statistics, Vol. 38, No. 2, (April 1967), pp. 325 - 339, ISSN 0003-4851

Haykin, S. (2002). Kalman Filtering and Neural Networks, Prentice-Hall, ISBN 0471369985, Englewood Cliffs, NJ, USABlackman, S. (1986). Multiple-Target Tracking with Radar Applications, Artech House, ISBN 0890061793, Norwood, MA, USA

Kosko, B. (1992). Neural Networks and Fuzzy Systems: A Dynamical Systems Approach to Machine Intelligence, Prentice Hall, ISBN 0136114350, Englewood Cliffs, New Jersey, USA

Kumar, B.V.K.V., Mahalanobis, A., and Juday, R.D. (2005). Correlation Pattern Recognition, Cambridge University Press, ISBN:0-521-57103-0, Cambridge, UK

Llinas, J.; Bowman C.; Rogova, G.L.; Steinberg, A.; Waltz, E. & White, F. (2004). Revisions and Extensions to the JDL Data Fusion Model II, Proceedings of Fusion 2004, pp.1218-1230, ISBN 91-7056-117-6, Stockholm, Sweden, July 2004.

Mahalanobis, P C (1936). On the Generalised Distance in Statistics, Proceedings of the National Institute of Sciences of India, Vol. 2, No. 1, pp. 49–55.

Oppenheim, A.V., Willsky, A.S., and Hamid, S. (1996). Signals and Systems, 2nd Edition, Prentice Hall, ISBN: 9780138147570, Upper Saddle River, NJ, USA

Pearl, J. (1988). Probabilistic Reasoning in Intelligent Systems: Networks of Plausible Inference, Morgan Kaufmann Publishers, ISBN 1558604790, San Mateo, CA, USA

Shafer, G. (1976). A Mathematical Theory of Evidence, Princeton University Press, ISBN 978-0691100425, Princeton, N.J., USA

Shea P.J., Zadra, T., Klamer, D., Frangione, E., Brouillard, R. and Kastella, K. (2000). Precision Tracking of Ground Targets, Proceedings of IEEE Aerospace Conference, Vol. 3 Big Sky, Montana, ISBN: 0-7803-5846-5, pp: 473 - 482, March, 2000

Smith, D. and Singh, S. (2006). Approaches to Multisensor Data Fusion in Target Tracking: A Survey, IEEE Transactions on Knowledge and Data Engineering, Vol. 18, No. 12, (December 2006), pp. 1696 - 1710, ISSN: 1041-4347

Steinberg A.; Bowman, C. & White, F. (1999). Revisions to the JDL Data Fusion Model, Proc. of the SPIE Sensor Fusion: Architectures, Algorithms, and Applications III, pp. 430-441, ISBN 0-8194-3193-1, Orlando, FL, USA, April 1999.

Stubberud, A.R. (2000). "Constrained Estimate Of The State Of A Time-Variable System," Proceedings of the 14th International Conference on Systems Engineering, Coventry, UK, pp. 513 - 518, September, 2000.

Stubberud, A.R. (2003)."Sequential target tracking using constrained nonlinear estimators," Proceedings of the 16th International Conference on Systems Engineering, Coventry, UK, ISBN 0-905949-91-9, pp. 652 - 656, September, 2003.

Stubberud, S. and Kramer, K. (2006). Data Association for Multiple Sensor Types Using Fuzzy Logic, IEEE Transactions on Instrumentation and Measurement, Vol. 55, No. 6, (December 2006), pp: 2292 - 2303, ISSN: 0018-9456

Stubberud, S. C., Kramer, K. (2007). Fuzzy Logic Based Data Association with Target/Sensor Soft Constraints, Proceedings of the 2007 IEEE International Symposium on Intelligent Control, Singapore, ISBN: 978-1-4244-0441-4, pp. 620 - 625, October 2007

Stubberud, S. and Kramer, K. (2009). Phase vs. Amplitude: Analysis of Optically-Based Correlation For Track Association, Proceedings of the IEEE International Conference of Systems Engineering. Coventry, UK, pp. 429-433, September 2009.

Stubberud, S. and Kramer, K. (2010). Track-to-Track Association Using a Phase-Only Filter, Signal Processing, Sensor Fusion, and Target Recognition XIX - Proceedings of SPIE, Vol. 7697, Orlando, FL, ISBN: 9780819481610. pp. 733605.1-12, April 2010

Stubberud, S. and Kramer, K. (2010). Incorporation of Measurement Uncertainty into Optically-Inspired Measurement-To-Track Data Association, Proceedings of the IEEE International Instrumentation and Measurement Technology Conference, San Antonio, TX, ISBN: 978-1-4244-2832-8. pp. 18-23, May 2010

Stubberud, S. C., Kramer, K., Geremia, J. A. (2007). Feature Object Extraction: Evidence Accrual For The Level 1 Fusion Classification Problem. IEEE Transactions on Instrumentation and Measurement, Vol. 56, No. 6, (December 2007), pp. 2705-2716, ISSN: 0018-9456.

Stubberud, S. and Lugo, K. (1999). A Fuzzy Logic Approach For Data Association, Sensor Fusion: Architectures, Algorithms, and Applications III, Proceedings of SPIE, vol. 3719, Orlando, FL, ISBN: 9780819431936, pp.242 - 249, April, 1999

Stubberud, S.; Shea, P.J. & Klamer, D. (2003). Data Fusion: A Conceptual Approach to Level 2 Fusion (Situational Awareness), Proceedings of SPIE, Aerosense03, pp. 455-462, ISBN 0-8194-4956-3, Orlando, FL., April 2003.

Yang, C. and Blasch, E. (2009). Kalman Filtering With Nonlinear State Constraints, IEEE Transactions on Aerospace and Electronic Systems, Vol. 45, No. 1, (January 2009), pp. 70-84, ISSN 0018-9251

6

Sensor Fusion Techniques in Navigation Application for Mobile Robot

Surachai Panich[1] and Nitin Afzulpurkar[2]
[1]*Srinakharinwirot University*
[2]*Asian Institute of Technology*
Thailand

1. Introduction

Generally, the mobile robots are researched and used in many applications. The fundamental method of navigation is localization, which includes aspects such as sensing, mapping, localization, planning and control. In order to we build the navigation system for mobile robot operating in unmodified environments, we must know the mobile robot's position in reference frame. Typically reference frame is a Cartesian or we can specify the landmark as reference position. The dead reckoning based on odometry is widely to identify and classify landmarks. Dead reckoning systems estimate the position of the mobile robot with respect to an initial point by measuring linear and angular velocities or accelerations. There are also hybrid models, which have sensor systems of both dead reckoning and landmark recognition. Dead reckoning is the estimate of current position based on a previously determined position or estimated speeds from a known past position over time. This can also determine the future position by projecting an ordered course and speed of advance from a known present position. A disadvantage of dead reckoning is that new positions are calculated solely from previous positions. The errors of the process are cumulative, so the error in the position fix grows with time (Hans-Joachim von der Hardt & Didier Wolf Rene Husson, 1996; Hye Ri Park et al., 2009). Odometry is widely used in mobile robots to estimate their position relative to reference frame. This method estimates the position over time with errors accumulation due to the integration of velocity measurements (Cristina Tar in Sauer and Hermann Brugger, 2001; Houshangi, N. and Azizi, F., 2006; Agostino Martinelli, 2002). The navigation system for mobile robot based on odometry has typically problem with inaccuracy. Because of the odometry problems navigation systems for mobile robot application cannot work solely on dead reckoning. This is the reason that the navigation system needs the addition of sensory information. Additional sensors can be defined as information provided to assist the accomplishment of a task by a system. Practically, sensor fusion refers to the integration process where the different sources of sensory information are combined into one representational format. Typically in navigation systems sensory information can be fused from complementary sensors, redundant sensors or even from a single sensor over a period of time. The advantage of sensory fusion provides uncertainty and noise reduction, failure toleration and extended flexibility of sensor ability.

2. Sensor system in navigation application

The navigation system plays very an important role and challenging competence for mobile robot. In navigation application, a mobile robot must interpret its sensors data to extract environment information, with which the mobile robot can determine its position. After mobile robot can localize its position, it must decide how to act to achieve its purpose. Lastly, the mobile robot must control its drive system to achieve the desired path. The localization of mobile robot's position has significant research,so in this chapter we introduce mainly the navigation system for the successful localization with relative and absolute instruments only. The increment distance of the robot's movement sensed with wheel encoders is integrated to calculate position. Because measurement errors from encoders are integrated, the position error is accumulated over time. Using additional sensors can help to reduce the cumulative errors. There are many error sources of differential encoder system from environmental factors, such as misalignment of the wheels, uncertainty of wheel diameter, variation in the contact point between wheel and floor and slipping.

2.1 The differential encoder system in navigation application

The basic concept of differential encoder system is the transformation from wheel revolution to linear translation on the floor. This transformation is affected with errors by wheel slippage, unequal wheel diameter, inaccurate wheel base distance, unknown factor and others. The real increment translation of wheel encoder is prone to systematic and some non-systematic error. Optical incremental encoders are widely used in industries for shaft angular displacement measurement. Normally the encoder is mounted on the motor shaft.

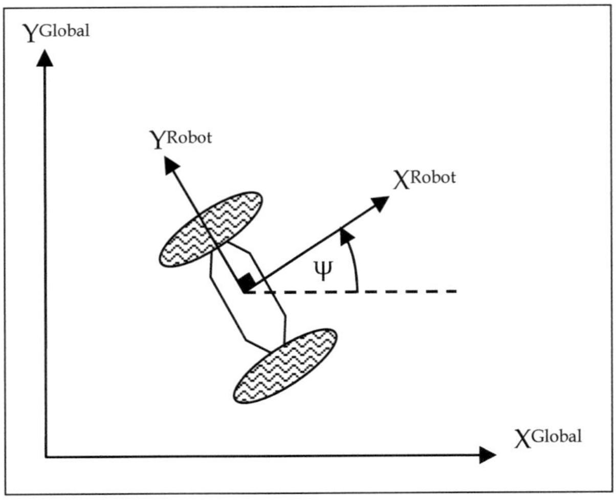

Fig. 1. Mobile robot navigation in global coordinate

We can estimate the position and heading angles of the mobile robot using the feedback information from the two encoders on the left and right wheels of mobile robot shown in Fig.1. The distance and heading increment can be obtained as follows.

$$\Delta d_k^{encoder} = \frac{\Delta d_{R,k}^{encoder} + \Delta d_{L,k}^{encoder}}{2} \tag{1}$$

$$\Delta \psi_k^{encoder} = \frac{\Delta d_{R,k}^{encoder} - \Delta d_{L,k}^{encoder}}{B_k} \tag{2}$$

Then the position can be estimated as

$$X_{k+1} = X_k + \Delta d_{x,k}^{encoder} \tag{3}$$

$$Y_{k+1} = Y_k + \Delta d_{y,k}^{encoder} \tag{4}$$

$$\psi_{k+1} = \psi_k + \Delta \psi_k^{encoder} \tag{5}$$

where

$$\Delta d_{x,k}^{encoder} = \Delta d_k^{encoder} * \cos \psi_k$$

$$\Delta d_{y,k}^{encoder} = \Delta d_k^{encoder} * \sin \psi_k$$

From equations (1) - (5) hold true when wheel revolutions can be translated accurately into linear displacement relative to the floor. We can calculate the integration of incremental motion information over time using the information from shaft encoders. But since the encoder measurement is based on the wheel shaft, it leads inevitably to unbounded accumulation of errors if wheel slippage occurs. Specifically, orientation errors will cause large lateral position errors, which will increase proportionally with the distance traveled by the robot. The differential encoder as the basic system can read the sensor information from encoders mounted at the left and right wheel shaft. The encoder information will be converted to $\Delta \psi_k^{encoder}$, $\Delta d_{x,k}^{encoder}$ and $\Delta d_{y,k}^{encoder}$ by equation (1) and (2).

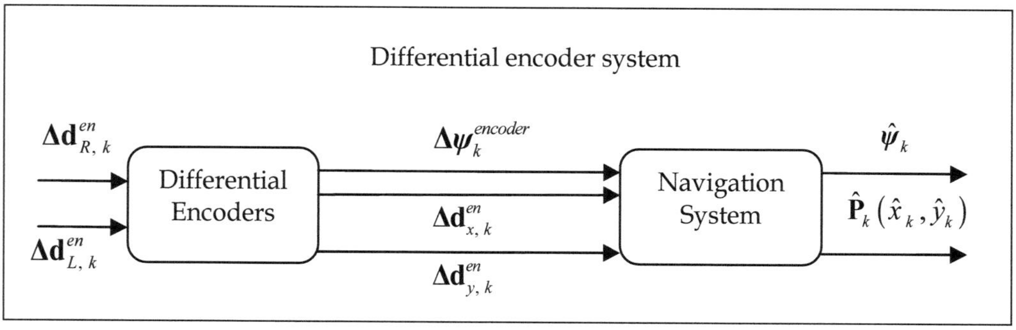

Fig. 2. Block diagram of the navigation system with the basic differential encoder system

For the differential encoder system the position and orientation can be calculated by equation (3), (4) and (5). In Fig.2 shows the simple block diagram of navigation system with two encoders mounted at the wheel shaft.

2.2 Complementary sensors for differential encoder system

Although the differential encoder system for mobile robot is practice in navigation application, the mobile robot's position is estimated over time with unbound errors accumulation due to the integration of sensor information. To construct a reliable navigation system, sensor fusion is considered to improve the position estimation generated by differential encoder system. Many researchers worked on sensor fusion with variety of sensors.

2.2.1 Gyroscope

Gyros measure rotational rate, which can be integrated to yield orientation. Typically, Gyroscope is mounted at the mobile robot. Many researchers work on fusion between encoder and gyroscope (Komoriya, K. and Oyama, E., 1994; Houshangi, N. and Azizi, F., 2006)

2.2.2 Compass

Compass is a navigational instrument for determining direction relative to the Earth's magnetic poles. It can be used to calculate heading and provided a much improved navigational capability. A compass is any magnetically sensitive device capable of indicating the direction of the magnetic north of a planet's magnetosphere. In navigation compass is introduced to capture the navigating direction information including direction angle and angular velocity (Yikang Yang, 2002).

2.2.3 Ultrasonic sensor

Ultrasonic sensor generates an ultrasonic pulse in a particular direction. If there is an object in the range of this pulse, part or all of the pulse will be reflected back to the transmitter as an echo and can be detected through the receiver. By measurement of the difference in time between the transmitted pulse and the received echo, it is possible to determine object's range. In navigation system, ultrasonic sensors play an important role for automatic guidance (Miao Yu and Li Shu-qin, 2010; Tao Guo and Silamu, W., 2009).

2.2.4 Camera

Camera is used to determine the relationship between the three-dimensional geometric location of one point and its corresponding point in the image. In many applications for navigation system, camera is utilized to detect landmarks and match them to pre-provided map information for localization. They implement matching process using environmental objects such as walls or obstacles. (Zhang, J., et. al, 2010; Chen, Yen-Sheng, et. al, 2010)

2.2.5 Accelerometer

Accelerometer measures static acceleration forces such as gravity, allowing it to be used as a tilt sensor (Surachai, P., 2009).

2.2.6 Light intensity sensor

Light intensity sensor system measures light in analog and convert to digital values. In a Cartesian coordinate system the robot's current position can be determined using the differences between opposite light intensity sensors in X and Y direction due to reference frame (Surachai, P., 2009).

3. Sensor fusion algorithm

Consider an algorithm to comprise of two sensors, which are sensor A and B. The data processing algorithm, which can combine the information from sensor A and B generates the proper estimate. The measurements of sensor A and B are given as

$$Z_{A,k} = x + \delta_A \tag{6}$$

$$Z_{B,k} = x + \delta_B \tag{7}$$

where x is true value and δ is measurement error.
The estimate of x is determined using the combination of information from sensor A and B and defined as

$$\hat{X}_k = w_{A,k} * Z_{A,k} + w_{B,k} * Z_{B,k} \tag{8}$$

where $w_{A,k}$ and $w_{B,k}$ are optimal quantity computed by variances.
To fuse information of two sensors corrupted by error, the optimal estimation (Gelb, 1974) can be expressed as follows

$$\hat{X}_k = \left(\frac{\sigma^2_{B,k}}{\sigma^2_{A,k} + \sigma^2_{B,k}} \right) * Z_{A,k} + \left(\frac{\sigma^2_{A,k}}{\sigma^2_{A,k} + \sigma^2_{B,k}} \right) * Z_{B,k} \tag{9}$$

where $Z_{A,k}$ is the measurement from sensor A with error variance ($\sigma^2_{A,k}$) and $Z_{B,k}$ is the measurement from sensor B with error variance ($\sigma^2_{B,k}$).
Generally, the differential encoder system generates incremental distance and heading and with integration of them over time the position and orientation are estimated. In this work the differential encoder is used as basic system in navigation purpose for mobile robot. To improve the accuracy of basic drive system the complementary systems with additional sensors are combined by sensor fusion algorithm. The complementary system can give the information both increment position and orientation and also both absolute position and orientation. This complementary information will be combined with information generated by differential encoder system with fusion algorithm.

3.1 Sensor fusion between differential encoder as basic system of mobile robot and gyroscope as relative complementary system with incremental rotation

The creation of relative complementary for differential encoder sytem is the compensation of incremental translation and rotation. With the compensation of additional relative sensor, it can reduce error directly generated by encoders. Gyroscope is widely used to compensate error in many applications. Many researchers on robotic field work on about gyroscope sensor (Komoriya & Oyama, 1994; Karthick Srinivasan and Jason Gu, 2007). The dominant gyroscope random errors are the random bias and the scale factor error. Generally, the gyroscope output signal is not zero even though there is no input and the effect of the earth rotation is neglected. The bias error of gyroscope is the signal output from the gyroscope when it is not any rotation. The bias error tends to vary with temperature and over time. The scale factor relates the output of the gyroscope to rotation angle of corresponding gyroscope about its input axis. By error of the scale factor in gyroscope it means the deviation between the actual scale factor and the nominal scale factor. The output beat frequency changes with the changing of the scale factor when the input rate is the same, which affects precision of gyro directly. This chapter introduces relative complementary system of incremental rotation $\Delta\psi_k^G$ with gyroscope sensor. The output of gyroscope is used to compensate the rotation increment. The commercial gyroscope CRS-03 is selected working in measurement area ± 200°/s and $\Delta\psi_k^G$ = 3.2°±0.01° as shown in Fig.3. It is used to combine with differential encoder system for error reduction of incremental rotation as shown in Fig.4.

Fig. 3. Gyroscope CRS03

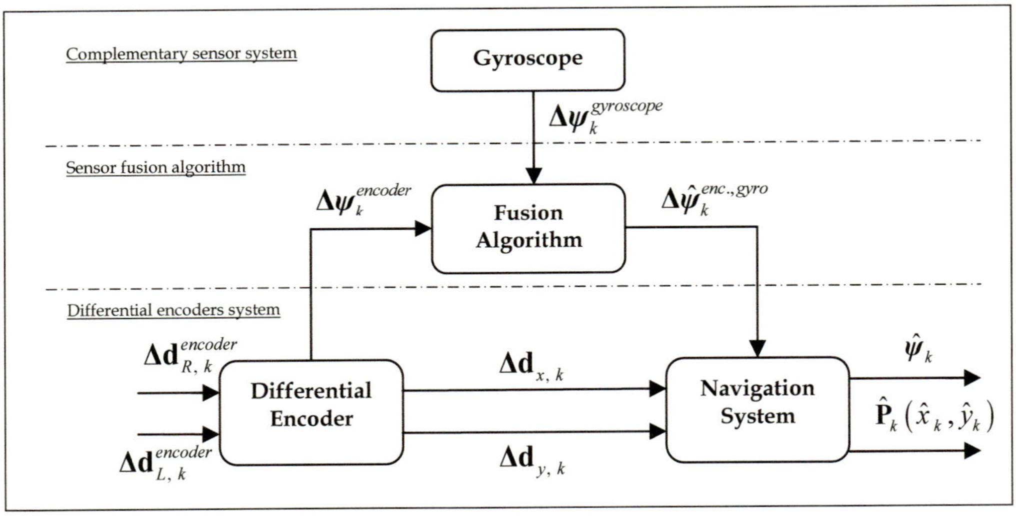

Fig. 4. Block diagram of the navigation system with the basic differential encoder system compensated with gyroscope.

The sensor fusion algorithm in this chapter we use the simple algorithm by equation (9), so the formular of fusion algorithm is given as

$$\Delta\hat{\psi}_k^{enc.,gyro.} = \left(\frac{\sigma^2_{\Delta\psi_k^{gyroscope}}}{\sigma^2_{\Delta\psi_k^{encoder}} + \sigma^2_{\Delta\psi_k^{gyroscope}}} \right) * Z_{\Delta\psi_k^{encoder}} + \left(\frac{\sigma^2_{\Delta\psi_k^{encoder}}}{\sigma^2_{\Delta\psi_k^{encoder}} + \sigma^2_{\Delta\psi_k^{gyroscope}}} \right) * Z_{\Delta\psi_k^{gyroscope}} \quad (10)$$

where

$\Delta\hat{\psi}_k^{enc.,gyro.}$ is the estimated incremental rotation from sensor fusion algorithm,

$Z_{\Delta\psi_k^{encoder}}$ is the measurement of incremental rotation read from encoder,

$Z_{\Delta\psi_k^{gyroscope}}$ is the measurement of incremental rotation read from gyroscope,

$\sigma^2_{\Delta\psi_k^{encoder}}$ is variance of the measurement of incremental rotation read from encoder and

$\sigma^2_{\Delta\psi_k^{gyroscope}}$ is variance of the measurement of incremental rotation read from gyroscope.

3.2 Sensor fusion between differential encoder as basic system of mobile robot and tilt sensor as relative complementary system with incremental translation

The tilt sensor, an accelerometer or inclinometer is one of the most common inertial sensors, which is dynamic sensor and measures static acceleration forces such as gravity. There are advantages by using of an accelerometer that is opposed by using inclinometer such as inclinometers can only detect when the tilt has exceeded some thresholding angle. This is equivalent to inertial acceleration minus the local gravitational acceleration, where inertial acceleration is understood in the Newtonian sense of acceleration with respect to a fixed reference frame, which the Earth is often considered to approximate. An accelerometer at rest on the Earth's surface will actually indicate 1 g upwards along the vertical axis. To obtain the inertial acceleration, this gravity offset must be subtracted.

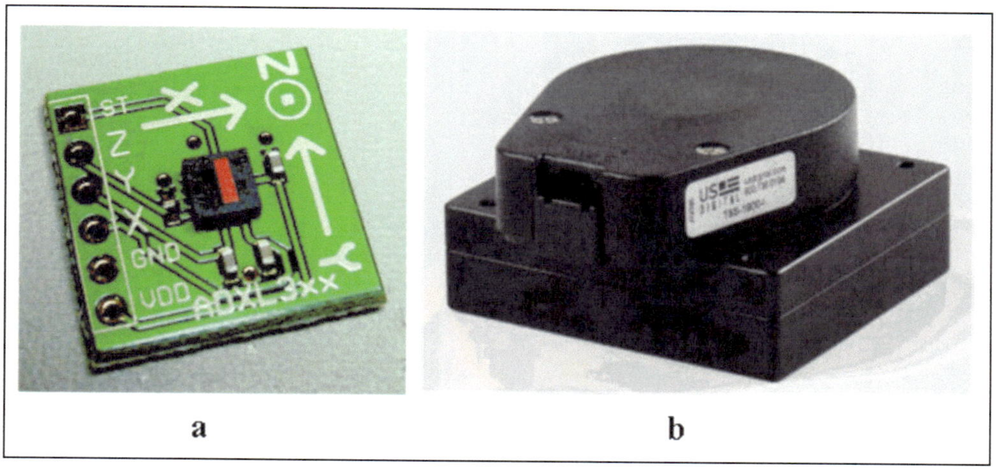

Fig. 5. Tilt sensor a) Accelerometer ADXL-SERIES, b)Incremental Inclinometer from US Digital Company

Accelerometers are available that can measure acceleration in one, two, or three orthogonal that implement capacitive sensing output a voltage dependent on the distance between two planar surfaces. The accelerometer and inclinometer as shown in Fig.5(a) and Fig.5(b) are capable of measuring both positive and negative tilt. The tilt sensor can generate the information, which can be converted to distance increment in X and Y coordinate $\Delta d_{x,k}^{Tilt}$ and $\Delta d_{y,k}^{Tilt}$ (Surachai, P., et. al, 2009; Surachai, 2010a). The information from encoders and tilt sensor can be combined by equation (11) and (12).

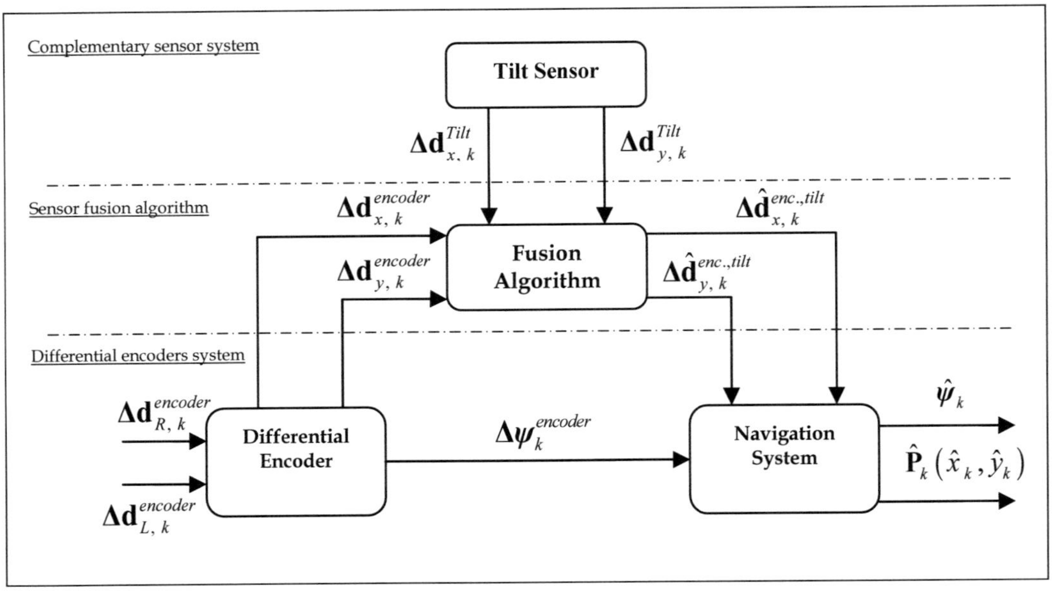

Fig. 6. Block diagram of the navigation system with the basic differential encoder compensated with tilt sensor.

In X coordinate:

$$\Delta\hat{d}^{enc.,tilt}_{x,k} = \left(\frac{\sigma^2_{\Delta d^{tilt}_{x,k}}}{\sigma^2_{\Delta d^{encoder}_{x,k}} + \sigma^2_{\Delta d^{tilt}_{x,k}}} \right) * Z_{\Delta d^{encoder}_{x,k}} + \left(\frac{\sigma^2_{\Delta d^{encoder}_{x,k}}}{\sigma^2_{\Delta d^{encoder}_{x,k}} + \sigma^2_{\Delta d^{tilt}_{x,k}}} \right) * Z_{\Delta d^{tilt}_{x,k}} \tag{11}$$

In Y coordinate:

$$\Delta\hat{d}^{enc.,tilt}_{y,k} = \left(\frac{\sigma^2_{\Delta d^{tilt}_{y,k}}}{\sigma^2_{\Delta d^{encoder}_{y,k}} + \sigma^2_{\Delta d^{tilt}_{y,k}}} \right) * Z_{\Delta d^{encoder}_{y,k}} + \left(\frac{\sigma^2_{\Delta d^{encoder}_{y,k}}}{\sigma^2_{\Delta d^{encoder}_{y,k}} + \sigma^2_{\Delta d^{tilt}_{y,k}}} \right) * Z_{\Delta d^{tilt}_{y,k}} \tag{12}$$

where

$\Delta\hat{d}^{enc.,tilt}_{x,k}$, $\Delta\hat{d}^{enc.,tilt}_{y,k}$ are estimated incremental distance from fusion algorithm in X and Y coordinate,

$Z_{\Delta d^{encoder}_{x,k}}$, $Z_{\Delta d^{encoder}_{y,k}}$ are the measurement of incremental distance from encoder in X and Y coordinate,

$Z_{\Delta d^{tilt}_{x,k}}$, $Z_{\Delta d^{tilt}_{y,k}}$ are incremental distance from tilt sensor in X and Y coordinate,

$\sigma^2_{\Delta d^{encoder}_{x,k}}$, $\sigma^2_{\Delta d^{encoder}_{y,k}}$ are variance of incremental distance from encoder in X and Y coordinate and

$\sigma^2_{\Delta d^{tilt}_{x,k}}$, $\sigma^2_{\Delta d^{tilt}_{y,k}}$ are variance of incremental distance from tilt sensor in X and Y coordinate.

3.3 Sensor fusion between differential encoder as basic system of mobile robot and light intensity sensor system as relative complementary system with incremental translation

The analog light transmitter makes it possible to convert analog light values. The light value can be transmitted in parallel on 8 channels with sensors from Inex-Company, which are independently programmable, and can thereby be compared to 8 different threshold values.

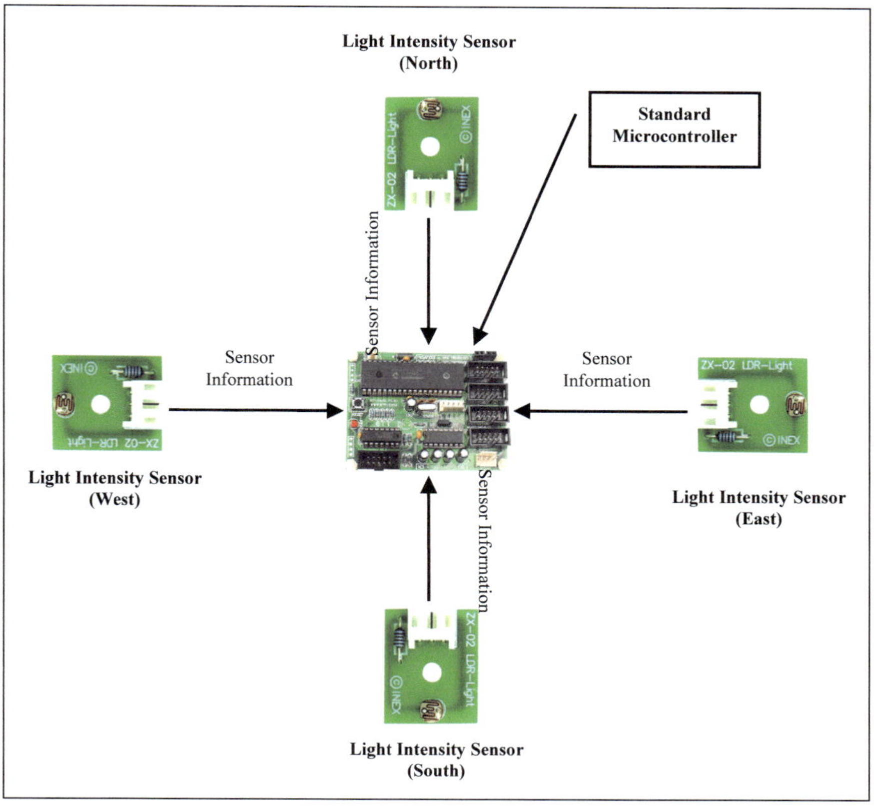

Fig. 7. Light intensity sensors system

The light intensity system is integrated on the mobile robot as shown in Fig.7. The light intensity system consists of four light intensity sensors and main controller. It is used to measure light intensity in north, south, west and east direction in environment (Surachai, P., et. al, 2009; Surachai, 2010b).

In a Cartesian coordinate system we can determine the robot's current position using the differences between opposite light intensity sensors shown in Fig.8. The incremental distance by light intensity sensors from position 1 to position 2 and can be calculated by $\Delta d_{x,k} \propto x_{2,k} - x_{1,k}$ and $\Delta d_{y,k} \propto y_{2,k} - y_{1,k}$, where the incremental distance $\Delta d_{x,k}$ in X and $\Delta d_{y,k}$ in Y coordinate are the robot's movement per time. The different information of the light intensity sensors $x_{2,k} - x_{1,k}$ and $y_{2,k} - y_{1,k}$ is in X and in Y coordinate as shown in Fig.9.

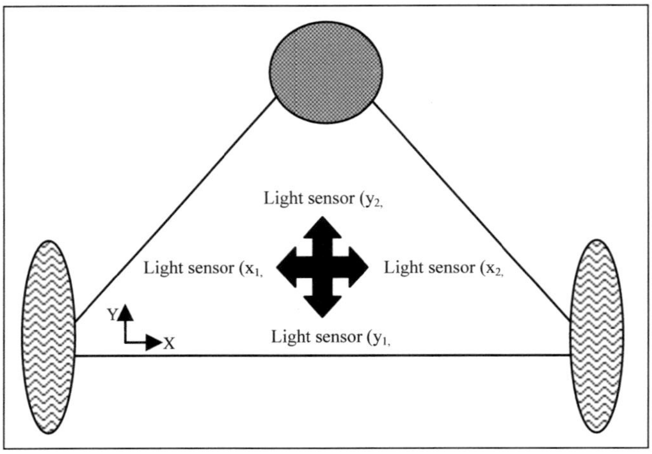

Fig. 8. Light intensity sensors System on mobile robot

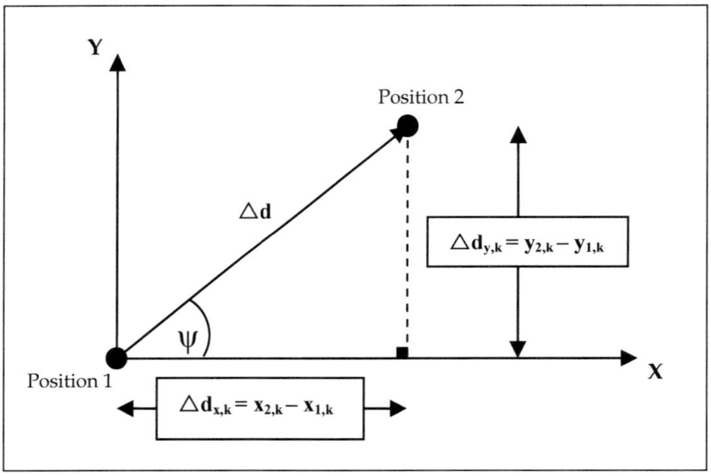

Fig. 9. Increment distance measured by light intensity sensors system

To eliminate the influence of changes in overall light intensity, we can normalize the readings for $\Delta d_{x,k}^{\text{light}}$ and $\Delta d_{y,k}^{\text{light}}$. The distance increment from light intensity sensors system is given as $\Delta d_{x,k}^{\text{light}}$ and $\Delta d_{y,k}^{\text{light}}$ in X and Y coordinate.

$$\Delta d_{x,k}^{\text{light}} = \frac{\Delta d_{x,k}}{\sqrt{\left(\Delta d_{x,k}\right)^2 + \left(\Delta d_{y,k}\right)^2}} \tag{13}$$

$$\Delta d_{y,k}^{\text{light}} = \frac{\Delta d_{y,k}}{\sqrt{\left(\Delta d_{x,k}\right)^2 + \left(\Delta d_{y,k}\right)^2}} \tag{14}$$

To satify equation (13) and (14), we assume that for all environments the overall intensity of light is constant, so that $\Delta d^{light}_{x,k}$ and $\Delta d^{light}_{y,k}$ are determined only by the robot's movement, not by the brightness of the environment.

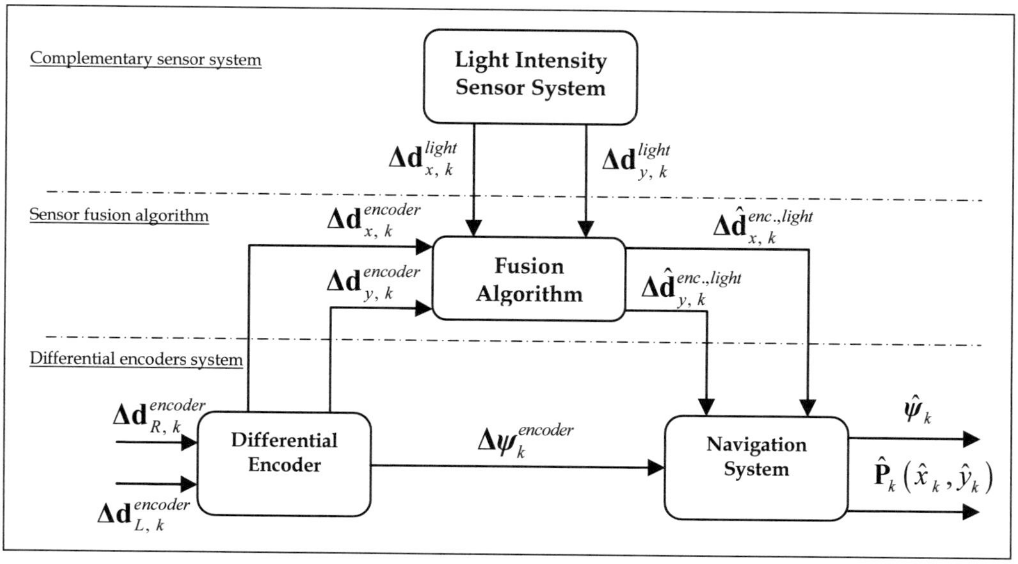

Fig. 10. Block diagram of the navigation system with the basic differential encoder compensated with light intensity sensor system.

The information from encoders and light intensity sensor can be combined by equations in X coordinate as

$$\Delta \hat{d}^{enc.,light}_{x,k} = \left(\frac{\sigma^2_{\Delta d^{light}_{x,k}}}{\sigma^2_{\Delta d^{encoder}_{x,k}} + \sigma^2_{\Delta d^{light}_{x,k}}} \right) * Z_{\Delta d^{encoder}_{x,k}} + \left(\frac{\sigma^2_{\Delta d^{encoder}_{x,k}}}{\sigma^2_{\Delta d^{encoder}_{x,k}} + \sigma^2_{\Delta d^{light}_{x,k}}} \right) * Z_{\Delta d^{light}_{x,k}} \quad (15)$$

and in Y coordinate as

$$\Delta \hat{d}^{enc.,light}_{y,k} = \left(\frac{\sigma^2_{\Delta d^{light}_{y,k}}}{\sigma^2_{\Delta d^{encoder}_{y,k}} + \sigma^2_{\Delta d^{light}_{y,k}}} \right) * Z_{\Delta d^{encoder}_{y,k}} + \left(\frac{\sigma^2_{\Delta d^{encoder}_{y,k}}}{\sigma^2_{\Delta d^{encoder}_{y,k}} + \sigma^2_{\Delta d^{light}_{y,k}}} \right) * Z_{\Delta d^{light}_{y,k}} \quad (16)$$

where

$\Delta \hat{d}^{enc.,light}_{x,k}$, $\Delta \hat{d}^{enc.,light}_{y,k}$ are estimated incremental distance from fusion algorithm in X and Y coordinate,

$Z_{\Delta d^{encoder}_{x,k}}$, $Z_{\Delta d^{encoder}_{y,k}}$ are the measurement of incremental distance from encoder in X and Y coordinate,

$Z_{\Delta d^{light}_{x,k}}$, $Z_{\Delta d^{light}_{y,k}}$ are incremental distance from light intensity sensor in X and Y coordinate,

$\sigma^2_{\Delta d^{encoder}_{x,k}}$, $\sigma^2_{\Delta d^{encoder}_{y,k}}$ are variance of incremental distance from encoder in X and Y coordinate and

$\sigma^2_{\Delta d^{light}_{x,k}}$, $\sigma^2_{\Delta d^{light}_{y,k}}$ are variance of incremental distance from light intensity sensor in X and Y coordinate.

3.4 Sensor fusion between differential encoder as basic system of mobile robot and compass as absolute complementary system of orientation

Absolute positioning systems use the geometry relation between the position and environment map (Byoung-Suk Choi, 2009). Some methods have been proposed for robot position reference. One of them has designed landmarks placed at specified positions with respect to the reference coordinate system. The current position and orientation of the robot can be obtained by geometry transformations between robot sensors and these landmarks. Kabuka and Arenas (1987) researched on absolute positioning by using barcodes. Kim (1996) described the experiments on orientation recovery of mobile robot by using compass. Ojeda, L. (2000) discussed on the use of electronic compasses in mobile robots with different error sources and solutions to correct these errors. From the geometry measurement the Cartesian coordinates (X, Y, θ) can be specified under certain assumption. If the position of the geometry reference is known, the position of mobile robot can be calculated usually with good accuracy. The compass can generate direct the output of absolute orientation $\psi_k^{compass}$ located at the mobile robot's center.

Fig. 11. The CMPS03 compass module.

In this chapter the CMPS03 compass module shown in Fig.11 is used, specifically designed for mobile robot as an aid to navigation with resolution ± 0.1° (Surachai, P., 2008).
The information from encoders and compass sensor can be combined by equations (17).

$$\hat{\psi}_k^{enc.,compass} = \left(\frac{\sigma^2_{\psi_k^{compass}}}{\sigma^2_{\psi_k^{encoder}} + \sigma^2_{\psi_k^{compass}}} \right) * Z_{\psi_k^{encoder}} + \left(\frac{\sigma^2_{\psi_k^{encoder}}}{\sigma^2_{\psi_k^{encoder}} + \sigma^2_{\psi_k^{compass}}} \right) * Z_{\psi_k^{compass}} \quad (17)$$

where

$\hat{\psi}_k^{enc.,compass}$ is the estimated orientation from fusion algorithm,

$Z_{\psi_k^{encoder}}$ is the measurement of orientation read from navigation system of differential encoder system,

$Z_{\psi_k^{compass}}$ is the measurement of orientation read from compass,

$\sigma^2_{\psi_k^{encoder}}$ is variance of the measurement of orientation read from navigation system of differential encoder system and

$\sigma^2_{\psi_k^{compass}}$ is variance of the measurement of orientation read from compass.

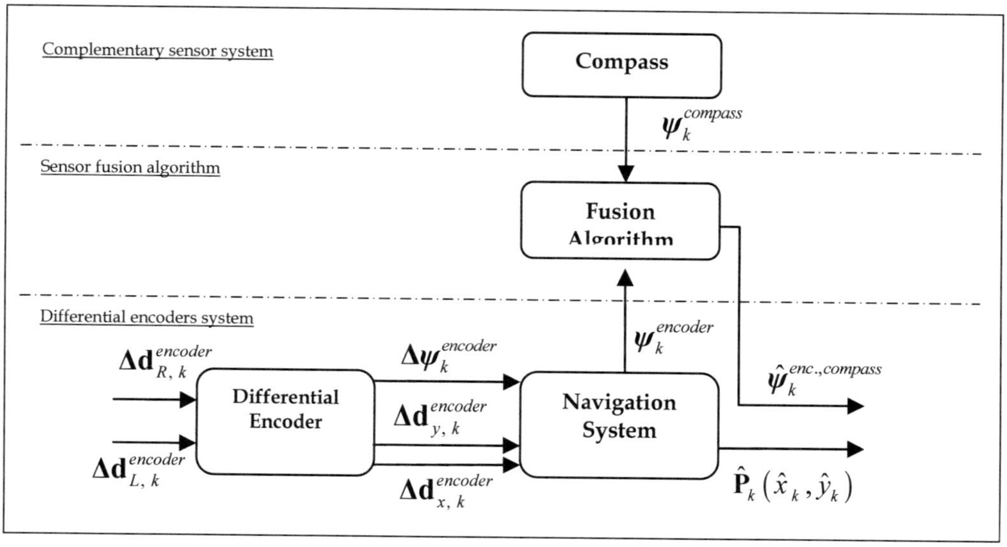

Fig. 12. Block diagram of the navigation system with the basic differential encoder combined with compass sensor

3.5 Sensor fusion between differential encoder as basic system of mobile robot and trilateration ultrasonic positioning system (TUPS) as absolute complementary system of position

The trilateration ultrasonic system is designed for mobile robot's absolute positioning. The three ultrasonic bases comprise one triangular area and each base measures the distance between itself and the robot, respectively. From the three distance measurements, the position of the mobile robot is determined as precisely as possible. The three circles are obtained by cutting three spheres whose centers are positions of bases **U₁**, **U₂** and **U₃** and radiuses are **d₁**, **d₂** and **d₃** as shown in Fig.13. If the measurements were ideally precise conducted, the mobile robot would be localized where the three circles intersect each other. However, the three circles do not intersect each other at one unique position since an actual measurement cannot be ideal determined. Therefore, the optimal point needs to be approximated, so the three points do not coincide with each other due to the measurement error, they actually comprise a triangle. The center of gravity of this triangle is used as an optimal point of mobile robot. It uses the principle of distance measurement that needs at least two landmarks with Cartesian's reference position. We can calculate the object's absolute position by using of the relation of distance measurement from ultrasonic bases (Surachai Panich and Nitin Afzulpurkar, 2010).

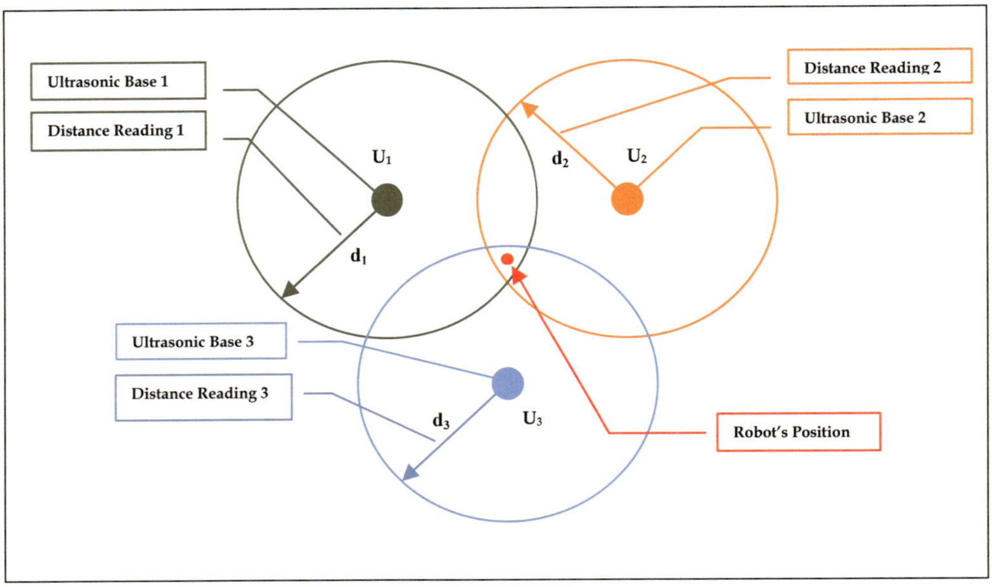

Fig. 13. Principle geometry of robot's work space

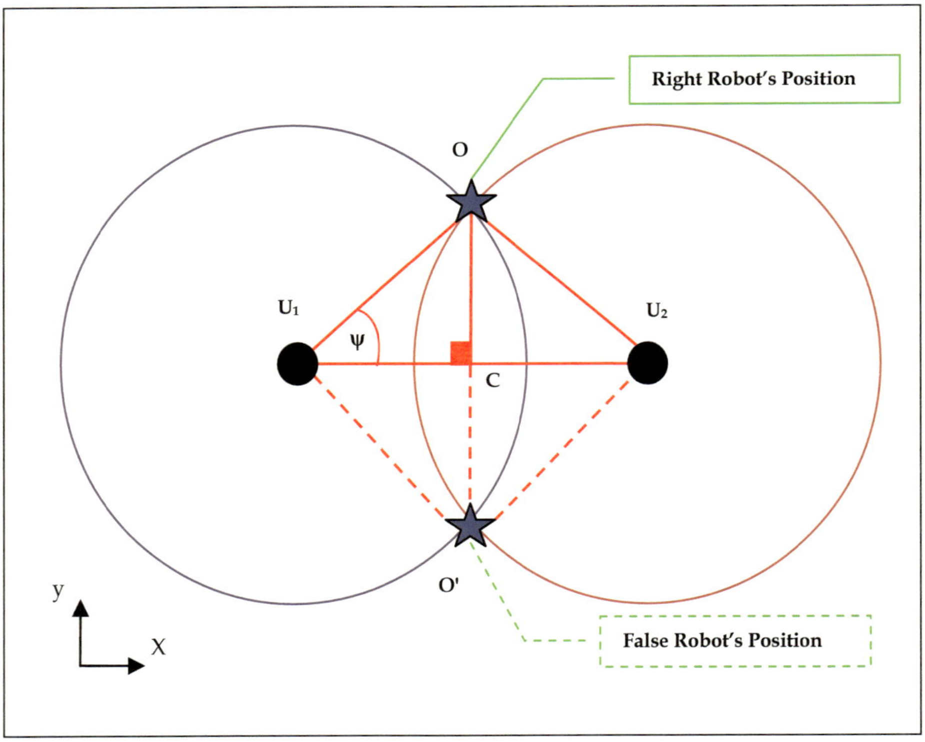

Fig. 14. Intersection points from two ultrasonic bases

With two ultrasonic distances from $\mathbf{U_1O}$ and $\mathbf{U_2O}$ and triangular relationship of $\Delta\mathbf{U_1U_2O}$ from Pythagorean Theorem, the robot's position can be calculated and detailed later. From Fig.14 two ultrasonic bases send two intersection points, which are possible to be right (**O**) and false (**O'**) solution, so the third ultrasonic play an important role to specify the right solution. In work space limited in half area only two ultrasonic bases are required to determine the right solution. The first ultrasonic base is located at point $\mathbf{U_1}$ $(\mathbf{X_{u1}}, \mathbf{Y_{u1}})$ and another ultrasonic base is placed at point $\mathbf{U_2}$ $(\mathbf{X_{u2}}, \mathbf{Y_{u2}})$. Before the object's position can be calculated, it must be considered that the intersection of two ultrasonic bases occurs or not.

If the distance between point $\mathbf{U_1}$ and point $\mathbf{U_2}$ is more than the summation of distance reading from ultrasonic base 1 and 2, the intersection of two ultrasonic does not occur.

Next, the robot's position (**O**) can be calculated from Pythagoras. From ultrasonic base 1, the relation is given as

$$(U_1C)^2 + (CO)^2 = (U_1O)^2 \tag{18}$$

and from ultrasonic base 2 is

$$(U_2C)^2 + (CO)^2 = (U_2O)^2 \tag{19}$$

From the subtraction of equation (18) and (19) the result is obtain as

$$(U_1C)^2 - (U_2C)^2 = (U_1O)^2 - (U_2O)^2 \tag{20}$$

, then

$$U_1C = \frac{(U_1O)^2 - (U_2O)^2 + (U_1U_2)^2}{2 * U_1U_2} \tag{21}$$

The angle (θ) is calculated as

$$\cos\psi = \frac{U_1C}{U_1O} \tag{22}$$

Finally, the robot's position is

$$X_k^{TUPS} = X_{U1} + (U_1O)\cos\psi \tag{23}$$

and

$$Y_k^{TUPS} = Y_{U1} + (U_1O)\sin\psi \tag{24}$$

The information from encoders and the trilateration ultrasonic system can be combined by equations (25).

The position with fusion algorithm in X coordinate is

$$\hat{X}_k^{enc.,TUPS} = \left(\frac{\sigma^2_{x_k^{TUPS}}}{\sigma^2_{x_k^{encoder}} + \sigma^2_{x_k^{TUPS}}} \right) * Z_{x_k^{encoder}} + \left(\frac{\sigma^2_{x_k^{encoder}}}{\sigma^2_{x_k^{encoder}} + \sigma^2_{x_k^{TUPS}}} \right) * Z_{x_k^{TUPS}} \tag{25}$$

and the position with fusion algorithm in Y coordinate is

$$\hat{Y}_k^{enc.,TUPS} = \left(\frac{\sigma^2_{y_k^{TUPS}}}{\sigma^2_{y_k^{encoder}} + \sigma^2_{y_k^{TUPS}}} \right) * Z_{y_k^{encoder}} + \left(\frac{\sigma^2_{y_k^{encoder}}}{\sigma^2_{y_k^{encoder}} + \sigma^2_{y_k^{TUPS}}} \right) * Z_{y_k^{TUPS}} \tag{26}$$

where

$\hat{X}_k^{enc.,TUPS}$, $\hat{Y}_k^{enc.,TUPS}$ are estimated position of mobile robot from fusion algorithm in X and Y coordinate,

$Z_{x_k^{encoder}}$, $Z_{y_k^{encoder}}$ are the measurement of mobile robot's position from navigation system with differential encoder system in X and Y coordinate,

$Z_{x_k^{TUPS}}$, $Z_{y_k^{TUPS}}$ are the measurement of mobile robot's position from the trilateration ultrasonic system in X and Y coordinate,

$\sigma^2_{x_k^{encoder}}$, $\sigma^2_{y_k^{encoder}}$ are variance of mobile robot's position from navigation system with differential encoder system in X and Y coordinate and

$\sigma^2_{x_k^{TUPS}}$, $\sigma^2_{y_k^{TUPS}}$ are variance of mobile robot's position from the trilateration ultrasonic system in X and Y coordinate.

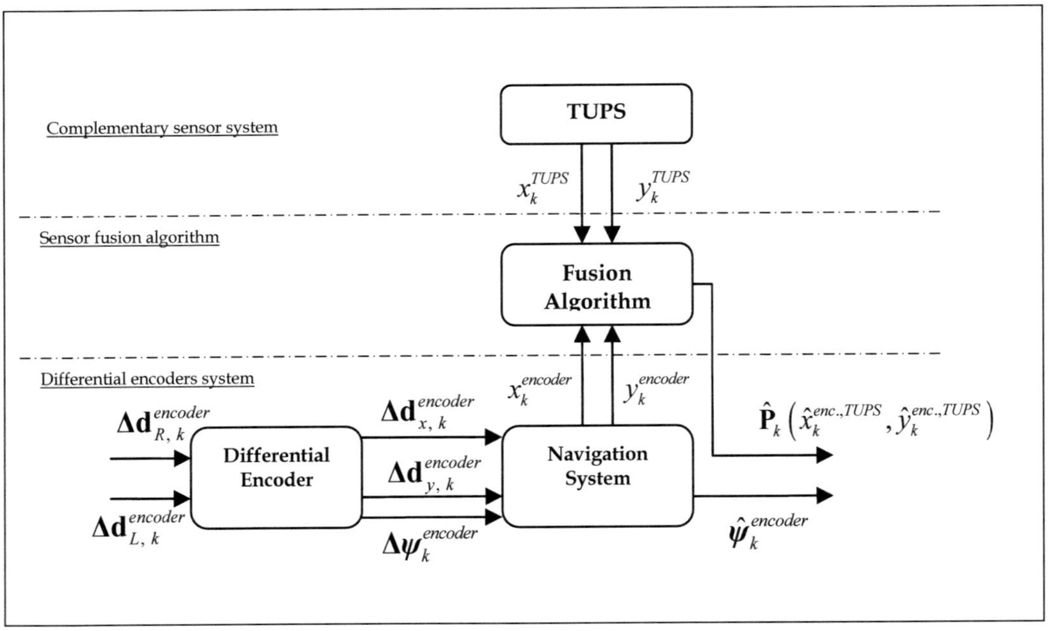

Fig. 15. Block diagram of the navigation system with the basic differential encoder compensated with the TUPS.

4. Simulation result

To explain the result of sensor fusion between differential encoder system and complementary sensor system, experiment result is simulated by Matlab software with real sensor information. For experiment, we select information from gyroscope to combine with differential encoder system for fusion of incremental rotaion. In experiment, the Pioneer-II mobile robot is used as the basic system driven by differential encoder system and gyroscope CRS03 is integrated on plate of Pioneer-II. The mobile robot is run in square shape and the sensor information both encoders and gyroscope are recorded and then we use the Matlab software to calculate error variance of encoder and gyroscope. From experiment we received good results, which can support our fusion algorithm as shown in Fig.16. Gyroscope information has better accuracy than differential encoder system as same physical parameter, so the accuracy of orientaion from differential encoder system integrated with gyroscope is better than from only differential encoder system.

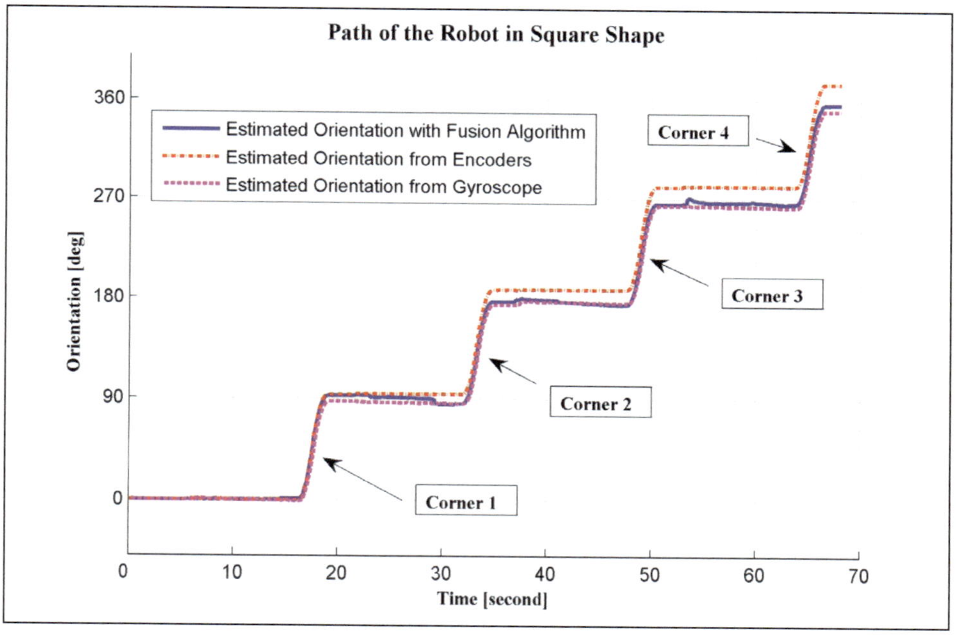

Fig. 16. Simulation result of estimated orientation from differential encoder system, gyroscope and differential encoder system integrated with gyroscope

5. Conclusion

In this chapter many complementary sensors for mobile robot driven with differential encoder system. Finally, we propose the idea for selection of complementary sensor system to combine sensor information of differential encoder system for increase of position accuracy. We can use the electric switches to control sensor information stream as shown in Fig.17 depended on task and the environment condition, which mobile robot has encounted. For example, if the mobile robot needs information from compass to combine with the differential encoder system, electric switch E will be enable. Then information from compass sensor will be only combined with differential encoder system as same physical parameter. For further work we can use more complicated algorithm, for example fuzzy logic, neural network or artificial intelligent system to fuse information between two sensors.

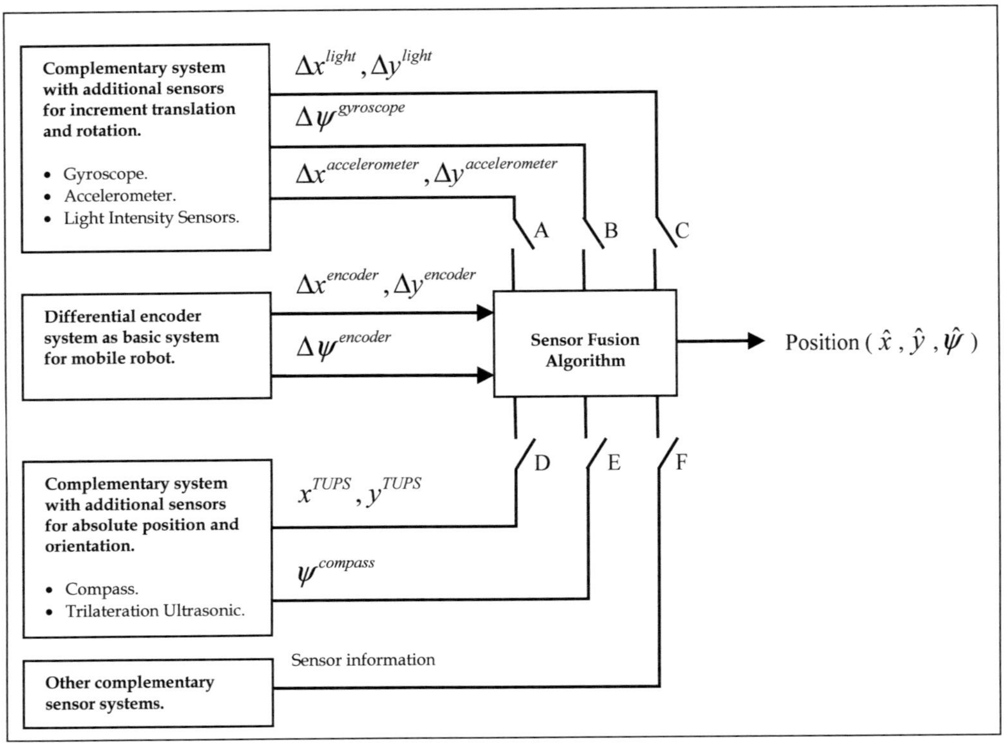

Fig. 17. Structure of sensor fusion between differential encoder system and complementary sensor systems

6. References

Hans-Joachim von der Hardt and Didier Wolf Rene Husson, 1996. The Dead Reckoning Localization System of the Wheeled Mobile Robot (ROMANE). Proceedings of the 1996 IEEE/SICE/RSJ International Conference on Multi-sensor Fusion and Integration for Intelligent Systems.

Hye Ri Park, et. al., 2009. A Dead Reckoning Sensor System and a Tracking Algorithm for Mobile Robot. ICROS-SICE International Joint Conference 2009 August 18-21, Fukuoka International Congress Center, Japan.

Cristina Tar in Sauer and Hermann Brugger, 2001. Odometry Error Correction by Sensor Fusion for Autonomous Mobile Robot Navigation. IEEE Instrumentation and Measurement Technology Conference Budapest, Hungary, May 21-23.

Houshangi, N. and Azizi, F., 2006. Mobile Robot Position Determination Using Data Integration of Odometry and Gyroscope. World Automation Congress, WAC '06 on 24-26 July.

Agostino Martinelli, 2002. The Odometry Error of a Mobile Robot with a Synchronous Drive System IEEE Transactions on Robotics and Automation, Vol. 18, No. 3, June.

Komoriya, K. and Oyama, E., 1994. Position estimation of a mobile robot using optical fiber gyroscope (OFG). Intelligent Robots and Systems, Advanced Robotic Systems and the Real World, IROS '94. Proceedings of the IEEE/RSJ/GI International Conference on 12-16 Sept. 1994.

Yikang Yang, Yong Lin and Yongxuan Huang, 2002. The Navigation Principle and Algorithm of Vehicular Dead Reckoning System with Differential Compass. Proceedings of the 4th World Congress an Intelligent Control and Automation, June 10-14, Shanghai, P.R.China.

Miao Yu and Li Shu-qin, 2010. A Method of Robot Navigation Based on the Multi-Sensor Fusion. Intelligent Systems and Applications (ISA), 2nd International Workshop on 22-23 May, DOI: 10.1109/IWISA.2010.5473355

Tao Guo and Silamu, W., 2009. A fuzzy logic information fusion algorithm for autonomous mobile robot avoidance based on multi-ultrasonic range sensors. Joint Conference of the 2009 Symposium on Piezoelectricity, Acoustic Waves, and Device Applications (SPAWDA) and Control Technology, DOI: 10.1109/SPAWDA.2009.5428888, Page(s): 84 - 84

Zhang, J., Wu, Y., Liu, W., Chen, X., 2010. Novel Approach to Position and Orientation Estimation in Vision-Based UAV Navigation. IEEE Transactions on Aerospace and Electronic Systems, Volume: 46 , Issue: 2 DOI: 10.1109/TAES.2010.5461649 , Page(s): 687 - 700

Chen, Yen-Sheng, Wang, Wei-Hsun, Juang, Jih-Gau, 2010. Application of intelligent computing to autonomous vehicle control. IEEE Congress on Evolutionary Computation (CEC). DOI: 10.1109/CEC.2010.5586475, Page(s): 1 - 8

Surachai, P.; Thiraporn, T.; Chaiyaporn, L.; Kittichai, T.; Tayawat, S., 2009. Sensor fusion for differential encoder integrated with light intensity sensors and accelerometer. 2009 IEEE International Symposium on Computational Intelligence in Robotics and Automation (CIRA), DOI: 10.1109/CIRA.2009.5423180, Page(s): 349 - 354

Gelb (1974). Applied Optimal Estimation. MIT Press, Massachusetts Institute of Technology Cambridge, Massachusetts 02142

Srinivasan, K. and Gu, J. Multiple Sensor Fusion in Mobile Robot Localization, Electrical and Computer Engineering, 2007. CCECE 2007 Canadian Conference on, 22-26 April 2007 Page(s): 1207 - 1210

Byoung-Suk Choi (2009). Mobile Robot Localization in Indoor Environment using RFID and Sonar Fusion System, *IEEE/RSJ International Conference on Intelligent Robots and Systems,* October 11-15, 2009 St. Louis, USA

Kakuba and Arenas (1987). Position Verification of a Mobile Robot Using Standard Pattern" *IEEE Journal of Robotics and Automation*, vol. RA-3, NO.6, December.

Ojeda, L. and Borenstein, J. (2000). Experimental Results with the KVH C-100 Fluxgate Compass in Mobile Robots, *Proceedings of the IASTED International Conference Robotics and Applications,* August 14-16, 2000 – Honolulu, Hawaii, USA.

Panich, S. (2008). A mobile robot with an inter-integrated circuit system, *10th International Conference on Control, Automation, Robotics and Vision,* Publication Year: 2008, Page(s): 2010 – 2014, Digital Object Identifier: 10.1109/ICARCV.2008.4795839

Surachai Panich (2010a). Mobile Robot Driven by Odometry System Integrated with Accelerometer. *Far East Journal of Electronics and Communications*, Volume 4, Issue 2 (June 2010) Page: 113 - 122

Surachai Panich (2010b). Dynamics, Control and Simulation of Robots with Differential Drive Integrated with Light Intensity Sensors. *Far East Journal of Dynamical Systems*, Volume 13, Issue 2 (July 2010) Page: 157 - 164

Surachai Panich and Nitin Afzulpurkar (2010). Absolute Positioning Instruments for Odometry System Integrated with Gyroscope by Using IKF, *Global Journal of Researches in Engineering*, Vol.10 Issue 4 (Ver 1.0) , September 2010, pp:63-74

7

Real-Time Fusion of Visual Images and Laser Data Images for Safe Navigation in Outdoor Environments

Maria C. Garcia-Alegre, David Martin,
D. Miguel Guinea and Domingo Guinea
Center for Automation and Robotics
Spanish Council for Scientific Research
Spain

1. Introduction

In recent years, two dimensional laser range finders mounted on vehicles is becoming a fruitful solution to achieve safety and environment recognition requirements (Keicher & Seufert, 2000), (Stentz et al., 2002), (DARPA, 2007). They provide real-time accurate range measurements in large angular fields at a fixed height above the ground plane, and enable robots and vehicles to perform more confidently a variety of tasks by fusing images from visual cameras with range data (Baltzakis et al., 2003). Lasers have normally been used in industrial surveillance applications to detect unexpected objects and persons in indoor environments. In the last decade, laser range finder are moving from indoor to outdoor rural and urban applications for 3D imaging (Yokota et al., 2004), vehicle guidance (Barawid et al., 2007), autonomous navigation (Garcia-Pérez et al., 2008), and objects recognition and classification (Lee & Ehsani, 2008), (Edan & Kondo, 2009), (Katz et al., 2010). Unlike industrial applications, which deal with simple, repetitive and well-defined objects, camera-laser systems on board off-road vehicles require advanced real-time techniques and algorithms to deal with dynamic unexpected objects. Natural environments are complex and loosely structured with great differences among consecutive scenes and scenarios. Vision systems still present severe drawbacks, caused by lighting variability that depends on unpredictable weather conditions. Camera-laser objects feature fusion and classification is still a challenge within the paradigm of artificial perception and mobile robotics in outdoor environments with the presence of dust, dirty, rain, and extreme temperature and humidity. Real time relevant objects perception, task driven, is a main issue for subsequent actions decision in safe unmanned navigation. In comparison with industrial automation systems, the precision required in objects location is usually low, as it is the speed of most rural vehicles that operate in bounded and low structured outdoor environments.
To this aim, current work is focused on the development of algorithms and strategies for fusing 2D laser data and visual images, to accomplish real-time detection and classification of unexpected objects close to the vehicle, to guarantee safe navigation. Next, class information can be integrated within the global navigation architecture, in control modules, such as, stop, obstacle avoidance, tracking or mapping.

Section 2 includes a description of the commercial vehicle, robot-tractor DEDALO and the vision systems on board. Section 3 addresses some drawbacks in outdoor perception. Section 4 analyses the proposed laser data and visual images fusion method, focused in the reduction of the visual image area to the region of interest wherein objects are detected by the laser. Two methods of segmentation are described in Section 5, to extract the shorter area of the visual image (ROI) resulting from the fusion process. Section 6 displays the colour based classification results of the largest segmented object in the region of interest. Some conclusions are outlined in Section 7, and acknowledgements and references are displayed in Section 8 and Section 9.

2. Experimental platform

Automation increases productivity of outdoor operations by enlarging the efficiency, reliability and precision, as well as reducing operator accidents and health risks. The automation has come to place human operators to the more comfortable position of supervisor and emergencies solver. On the other hand, wireless communication plays a fundamental role in the development of ever more autonomous systems. To this goal a commercial tractor has been adequately retrofitted to perform teleoperation and safe autonomous navigation.

2.1 Robot tractor DEDALO

The vehicle used as experimental platform is a hydraulic articulated commercial tractor (AGRIA S.A.), Figure 1. Safety, location and environment recognition sensors are installed on board, and mechanical adaptations have been undertaken to achieve automatic control of clutch, brake and steering. A 2D laser range finder and a visual camera are installed in the front of the tractor to ease the detection and recognition of close objects, Figure 2. The object features classification process entails control actions, such as a stop or a deviation under collision risk (Martin et al., 2009). Accounting for the objective to be accomplished, the on board sensors are classified as follows:

Fig. 1. Experimental vehicle: robot tractor DEDALO

Safety

- Bumper
- 2D Laser range finder
- Inclinometer

Location

- Odometer
- Digital compass
- GPS

Environment perception

- Camera
- 2D Laser range finder

Fig. 2. Detail of the laser range finder and the visual camera.

Three navigation modes have been developed and tested, in addition to the manual driving:

- Tele-operation
- Semi-autonomous navigation guided by natural objects in the scene
- Autonomous navigation guided by GPS

2.2 Visual camera

A compact and rugged digital camera was placed in the front of the vehicle, Figure 2. This location favours the correspondence process between pixel rows in the visual image and objects detected by the laser. The camera durable transparent cover provides excellent protection against dust, humidity and vibrations. Vision systems provide rich and meaningful information of the environment, where volume, shape and colour are the main cues, but lacks objects depth information. The main characteristics of the digital camera are summarized in Table1.

The image resolution has been set to 640x480 pixels, as higher resolutions imply greater computing time, incompatible with real time requirements.

Image sensor	1/3" Progressive Scan RGB CMOS 1.3 Mpixel
Lens	3.6 mm, F1.8, fixed iris, horizontal angle of view: 74°
Light sensitivity	3-10.000 lux
Shutter time	1/15.000 s to ¼ s
Camera angle adjustment	Pan +/- 10°, tilt 0-90°, rotation +/-10°
Pan/Tilt/Zoom	Digital PTZ, preset positions, guard tour
Resolution	160x90, 640x480, 1280x1024

Table 1. Some characteristics of the Axis 209 camera

2.3 Laser range finder

To obtain depth information from the environment, a 2D laser range finder (Sick LMS291) has been integrated in the vehicle, Figure 2. Main advantages of laser systems are a broad bandwidth and small beam divergence and footprint. They also offer a high immunity to atmospheric effects in opposition to the visual cameras. Laser range finder gives a sparse, but accurate map of the environment in a 2D plane. They use an infrared light beam of 905 nm, which receives directly the reflected signal from the objects, in polar coordinates. The laser operation mode is based on a time-of-flight (TOF) measurement principle: a single laser pulse is sent out and reflected by an object surface. The elapsed time between emission and reception allows the calculation of the distance between laser and object. The laser pulses sweep a radial range in front of the laser unit, via an integrated rotating mirror. Main laser characteristics are depicted in Table 2.

Range	Max. 80m
Angular resolution	0.25°/0.5°/1° (selectable)
Response time	53 ms/ 26 ms/ 13 ms
Measurement resolution	10 mm
Systems error (good visibility, T=23°)	Typ. +/- 35 mm, range 1.20 m

Table 2. Some characteristics of the 2D laser range finder

The laser range finder, located at 0.67 m height above ground level, has been set to a maximum distance of 8 m. with an angular resolution of 1°.

3. Major drawbacks in outdoor visual images

The outdoor visual image analysis has to face the sudden changes in the scene lighting, which depends on unpredictable weather conditions, as well as on seasonal and daily variations. High and unexpected illumination changes give rise to image in-homogeneities and require a dynamic processing to get optimum objects recognition . Sunny-days shadows make even more difficult the process of extracting relevant objects from an outdoor scene. Main disadvantages to image processing of natural scenes are:

- High lighting variation requiring continuous settings
- Large data volumes involved in the visual image
- Real-time detection of dynamic objects

Current work deals with visual images displaying high lighting variations and consequently requiring efficient techniques capable of self-adaptation. Three images of a semi-rural outdoor environment, acquired while the robot was navigating, are displayed in Figure 3. The region between the red lines in the laser data representation corresponds, to the 74° view angle of the camera. Thus, the targets detected by the camera come into view in the laser data representation (laser image) within the 53° to 127° interval. The small green rectangle located in the centre of the x axis, represents the robot-tractor. For each scene visualised in Figure 3, the left column displays the laser distance map in the 0-8 m interval. The right one,

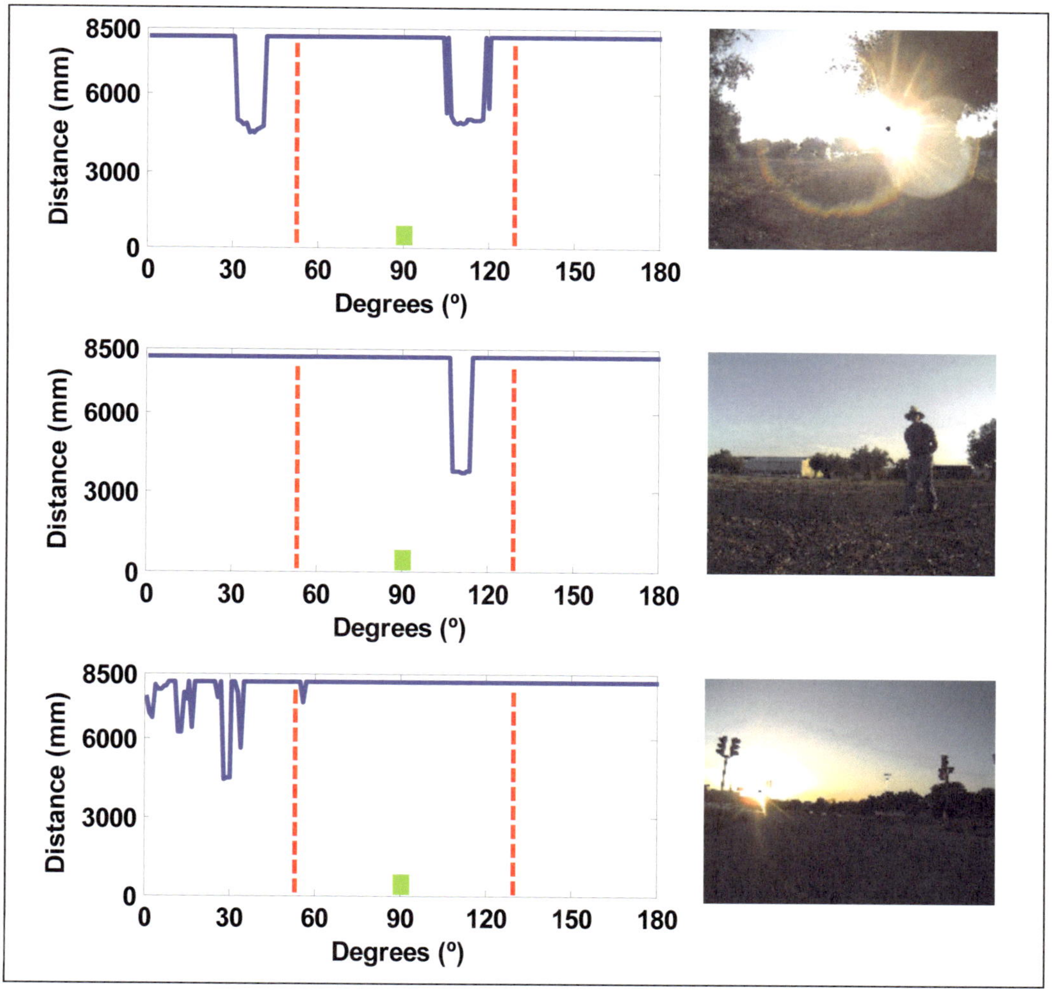

Fig. 3. Laser data and camera images of three semi-rural scenes acquired at nightfall in extreme conditions

displays the visual images acquired with the camera and field of view of 74°. The extreme lighting conditions of the first and third scenes points to the need of smart fusion methods to enhance the objects detection by combining features extracted from the information provided by heterogeneous sensors. The 2D laser representation of the first scene exhibit two obstacles, presumably trees, but from the visual image it is nearly impossible to obtain the same conclusion, due to the image in-homogeneities caused by direct incidence of the sunlight on the visual camera. In the third scene, the 2D laser representation of the environment outlines a thin mast, impossible to extract from the visual image which was acquired at the nightfall. Right mast appearing in the visual image is detected at a distance close to the 8 meter detection limit. The thin obstacles located in the 0°-35° interval on the 2D laser distance-angle representation, correspond to weeds rising on the left border of the asphalt road. Opposite, the second scene displays an operator in front of the vehicle, which is easily detected by the 2D laser range finder, but not so well by the camera because of the low illumination of the scene.

Thus, perception algorithms able to combine features extracted from heterogeneous sensor are essential for real-time interpretation of outdoor scenes.

4. Sensor fusion method

Laser range finders used in robotic systems provide accurate distance measurements of the objects in the environment. On the other hand, visual cameras are capable of providing dense 3D information of a scene, but lack depth information of the objects. The fusion method here proposed, combines the accuracy of laser distance measurements, with the complementary information obtained from a visual image that allows for the extraction of features such as, volume, shape, texture and colour. In the very recent years, works devoted to the accurate and automatic extrinsic calibration of a laser-camera rig (Zhang & Pless, 2004), (Li et al., 2007), (Kassir & Peynot, 2010) in static indoor environments, highlights the increasingly interest on this perception platform. Natural outdoor scenes are hard to interpret due to both illumination variability and quick variations from one scene to the other. The colour and texture of natural objects in a visual image change, not only by light incidence variability, but due to occlusion and shapes superposition effects. Moreover, the width and position of an object on a 2D laser distance map varies with its dynamic behaviour. The closer works, to current one, are described in both, (Katz et al., 2007) devoted to the self-supervised classification of dynamic obstacles but operating in urban environments, and (Naroditsky et al., 2011) illustrating the automatic alignment of a camera-LIDAR rig, with no comments on real time applications. The sensor fusion method here proposed, aims at the reduction of the drawbacks affecting each of the proposed sensor systems, enhancing real-time characterization of objects in outdoor environments. The laser data and visual images fusion process, pursuing close objects classification, initiates with the definition of the region of interest on the visual image, where objects detected by the laser in the 1 to 8 m interval, are mapped. This region, being a limited area of the global visual image, will greatly reduce the time-computing of the subsequent segmentation and classification processes requiring a real-time response. The analysis of the visual images and laser data has been accomplished with MATLAB (Martin et al., 2009), on account of the ease and fast development facilities provided by its multiple toolboxes and libraries, in the initial development stage of a vision application. The fusion method, here illustrated, enhances the real-time interpretation of a scene to generate the reactive motions required to reach the safety requirements.

4.1 Region of interest in the visual image

Despite the extensive research conducted in vision systems and robotics applications, camera-laser rig perception in outdoor environments still remains a challenge. The simulation of a rural environment has been accomplished with Google SketchUp platform, to visualize the effect of vertical parallax on the intersection of the laser plane on the visual image as a function of the objects distance. This is caused by the fact that the camera is located 6 cm below the laser, but in the same vertical axis. In the simulated world, composed of a vehicle, a warehouse, and several trees, two persons were located in front of the vehicle at 1 and 8 meter distance, Figure 4.

Fig. 4. Vehicle and operators locations in front of the vehicle in a simulated world.

The cut of the 2D laser plane on the visual image is highlighted in Figure 5. The laser plane cuts the operators, located at different distance, at distinct heights depending on the relative distance object-vehicle.

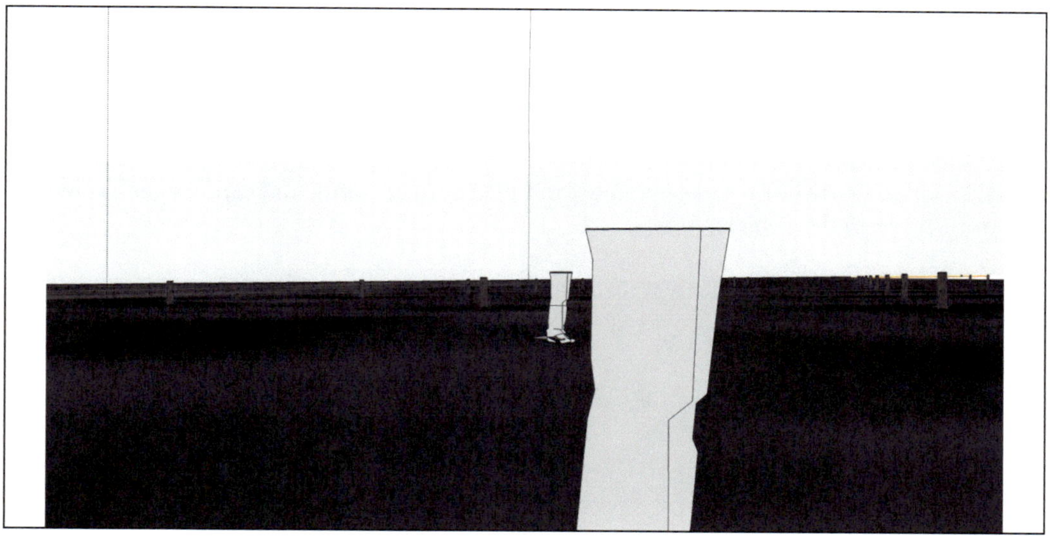

Fig. 5. Image acquired by the simulated camera displaying the operators cut by the 2D laser plane in the 1-8 m interval.

To calculate the limits of the laser plane displacement on the visual image, due to the vertical parallax effect exhibited in Figure 5, a calibration test has been accomplished in the real world. To this aim, a red card fixed at a stool, 0.67 m above the ground, has been used as a target. The stool was then placed at three different positions, 8, 4 and 1 meter distance from the vehicle. Then, laser angle-distance representations and visual camera images are acquired at each target position, as displayed in Figure 6 in the left and right columns respectively. In all three cases the red card was correctly detected by the laser and the camera. The greater the distance the lower is the card area in the visual image. The upward shift in the laser cut on the visual image (640 columns x 480 rows pixels) is visualised and calculated. Thus, for object distances in 1-8 meters interval, the target is always within the limited visual image area (640x80 pixels), parallax band, inscribed in the blue rectangle, Figure 7.

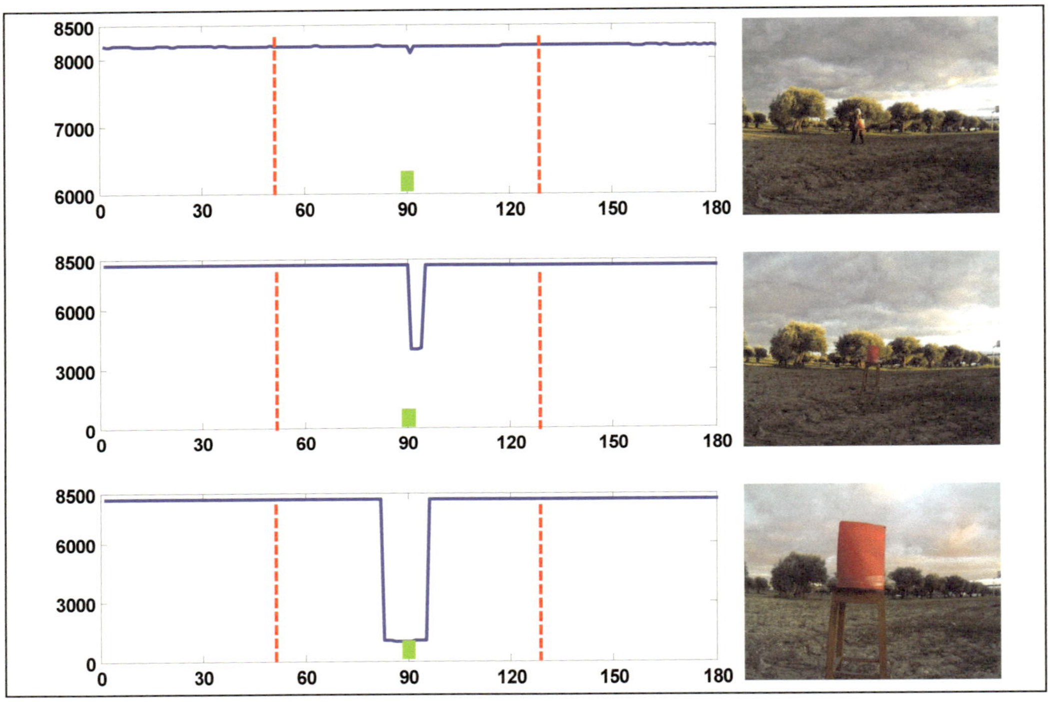

Fig. 6. Laser angle-distance representation and visual image of the red card target located at 8, 4 and 1 m. from the vehicle.

Fig. 7. Red card detected by the laser at 8, 4 and 1 m. distance from the vehicle, within the limited visual area inscribed in the blue frame.

Now, to experimentally determine the (640x80 pixels) parallax band location on the original visual image, the following test was carried on. This test determines the position of a calibration object, small yellow mast on top of the red sawhorse, as detected by the laser beam, on the corresponding visual image. The camera-laser rig is moved as a block vertically top-down to locate the obstacle on the laser image. Both visual image and laser data are synchronized to acquire data in unison. The test stops when the obstacle appearing at a well-known position on the visual image, emerges in the corresponding laser angle-distance representation.
The yellow low height mast, used as calibration obstacle is displayed about the centre of the Figure 8 (a).

(a) (b)

Fig. 8. (a) Visual image displaying the yellow mast on the red sawhorse, (b) Laser image displaying the mast distance.

The mast is detected by the laser plane at a distance of 5.1 m., Figure 8 (b), and appears on the 251 row of the original visual image.

4.2 Fusion of laser and camera images features

The calculation of the location (row 251) and height (+/- 40 pixels) of the visual region where objects detected by the laser are mapped, allows the processing of a shorter area visual image (ROI) that obviously will reduce time-computing. The horizontal correspondence between objects in laser and visual images is straightforward as the displacement between the sensors in this axis is very small. That is, the laser 53-127° range is matched within the -37° to +37° field of view of the camera. Therefore, a linear correspondence is performed between the 53-127° (74° width) range, in the laser angle-distance representation, and the 640 pixel width of the visual image. Consequently, a region centred in the image row 251 with a total height of 80 rows (pixel), 15% of the global visual image, is selected for the segmentation and classification processing. Thus, sensor fusion is accomplished from the determination of the visual image masks (parts of the ROI) of the different objects detected by the laser in an outdoor scene. An example of the sensor fusion method, is illustrated in Figure 9. The two black masks, Figure 9(c), correspond to objects detected by the laser at about 3 and 7 m. distance, Figure 9(b). The centre of the mask is located in the row 251 of a visual image frame, with a fixed vertical height of 80 rows. The horizontal length of each mask is proportional to the obstacle width (in degrees) as detected by the laser range finder.

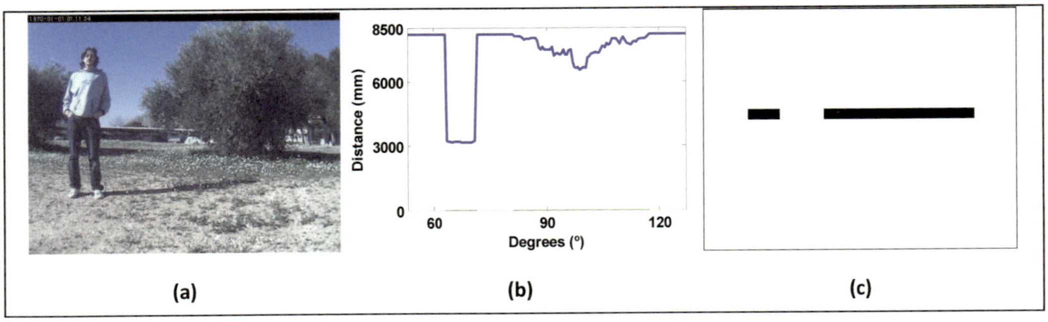

Fig. 9. (a) Visual image of an outdoor scene, (b) laser angle-distance representation, (c) visual masks of the two objects detected by the laser in the 1-8 m. range, mapped on a white visual frame.

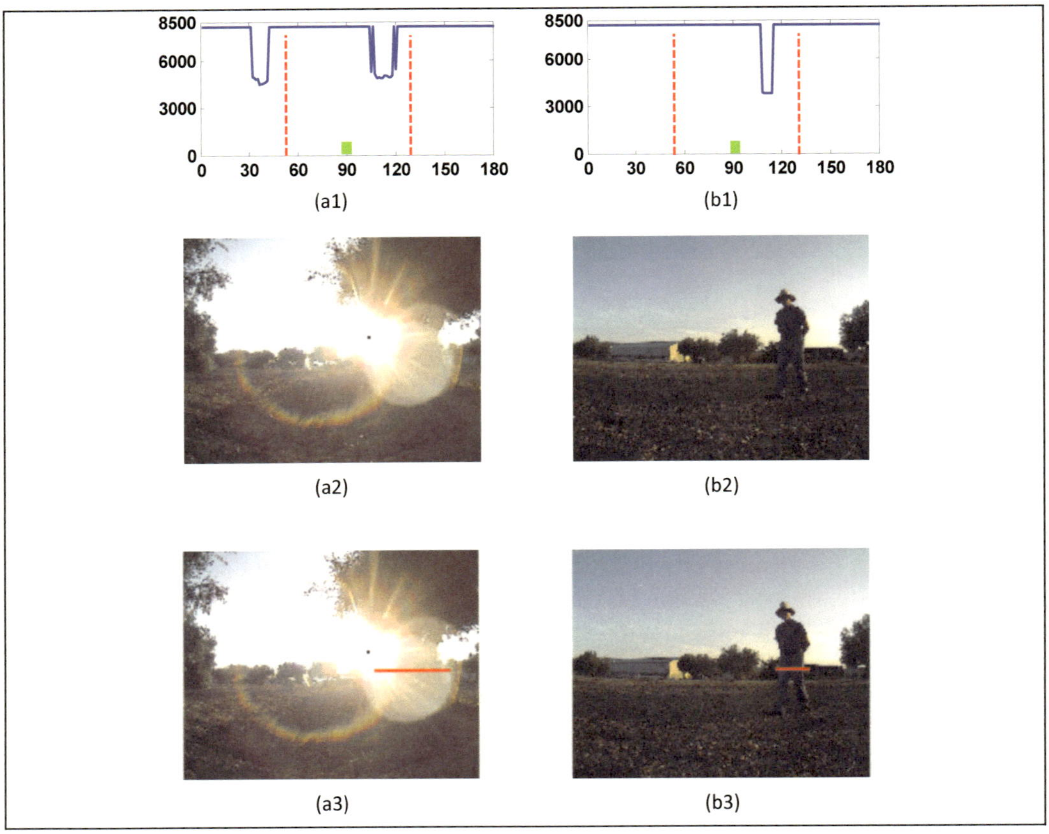

Fig. 10. (a1) and (b1) are the laser data representations; (a2) and (b2) the visual images and (a3) and (b3) are the results of the sensor fusion process.

The second step of the sensor fusion process is concerned with the determination of the visual image regions associated to the two masks. Only in the selected visual region, the segmentation and classification processes will be executed. Two scenes are displayed in Figure 10 (a) and (b), to illustrate the steps of the fusion process, where the extraction of objects in the visual image is difficult caused by light incidence angle on the camera and low

lighting in the scenes. The first row displays the 2D laser angle-distance representation, the second row corresponds to the acquired visual images and the third one exhibits the performed fusion process. The masks are located on the visual image regions corresponding to the objects detected by the 2D laser range finder, Figures 10 (a1) and (b1).
The next step, after the sensor fusion process, is the segmentation of the visual image regions (ROI) that intersects the visual masks, followed by the classification process of the segmented images.

5. Image segmentation

Image segmentation is one of the most widely used steps in the process of reducing images to information. It is accomplished by dividing the image into regions that correspond to the different structures present in the scene. This process is often described, in analogy to the human visual processes, as a foreground/background separation, focused in the recognition of one o several structures of interest in the scene, disregarding the rest. That means focusing the attention in the objects defined as relevant for the task to be performed. One of the problems of the segmentation algorithms is their difficulty to meet the real-time requirement (Sun et al., 2008).

5.1 Region growing algorithm

Region growing algorithm, based on active contours, is a simple region-based image segmentation method (Chan et al., 2001). It aims to group pixels together into regions of similarity, starting from an initial set of pixels. It is classified as a pixel-based image segmentation method because it involves the selection of initial seed points.
The method initiates defining an arbitrary pixel that acts like a seed and is compared to neighbours to determine its aggregation to the region. The method works iteratively for growing the initial regions of the visual image mask, Figure 11(a). The seeds are selected from the regions of former initialization image mask that allows applying the active contours method only where obstacles were detected by the laser range finder. The algorithm stops when 120 iterations are performed.

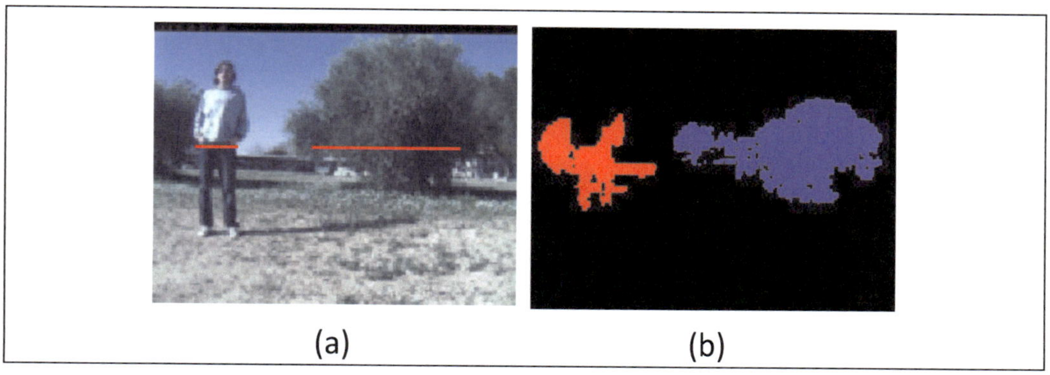

Fig. 11. (a) camera-laser fusion visual image, (b) visual segmented image by Region growing algorithm (objects distance is represented by false colour).

The false colour palette displayed in Figure 12, allows a quick interpretation of the outdoor scene objects. The precise value of the distance vehicle-object12 is provided by the 2D laser, Figure 11(b).

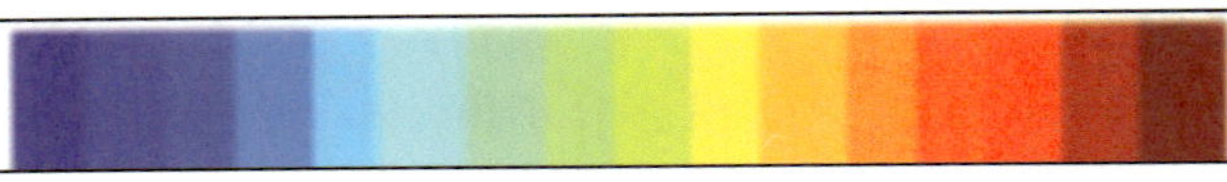

Fig. 12. Colour palette for a qualitative representation of object-vehicle distance

As it becomes evident,, the segmentation process based on Region growing algorithm is imprecise, Figure 11(b). The shape of the person in the segmented image lacks in exactness caused by an inadequate contour calculated by the algorithm. To improve this process, the Mean-shift segmentation algorithm is selected.

5.2 Mean shift algorithm: global image

The Mean shift algorithm is a nonparametric clustering technique which neither requires prior knowledge on the number of clusters, nor any constraint in the shape of the cluster. It is specially addressed for outdoor environments (Comaniciu and Meer, 2002). This unsupervised clustering algorithm and the segmentation based on colour, aids to handle real and dynamic outdoor environment characterization. The algorithm has been chosen opposite to other methods that require either a priori information about the number and shape of groups, or that lack the use of colour in the feature analysis. The Mean-shift algorithm was successfully used in the tracking of objects in image sequences with complex backgrounds (She et al., 2004). The algorithm requires high computing time, even with standard (640 * 480 pixel) images. Another drawback of the segmentation of global images is related to both the appearance of shadows and the region combination. The Mean-shift algorithm applied to the segmentation of an outdoor global image with a shadow on the ground is illustrated in Figure 13 (a) and (b). In the first scene the shadow is segmented as an obstacle. In the second scene, the blue trouser of a person in the global image, Figure 13 (c), is joined with a blue region of the sky Figure 13 (d).

Fig. 13. (a) and (c) original images, (b) and (d) visual images segmented by the Mean-Shift algorithm

For these reasons, the segmentation process is only performed on the visual image region detected with the fusion process (ROI) associated with the closest object.

5.3 Mean-shift algorithm: region of interest

The segmentation of the selected region on the visual image (ROI), obtained by the sensor fusion method with the Mean-shift algorithm, is now illustrated. The algorithm configuration parameters are defined to obtain large segmented regions on the image. The Matlab interface used (Bagon, 2010), is based on the EDISON program that implements the Mean-shift algorithm. First parameter is the "Minimum Region Area", that is calculated accounting for the extension, in pixels, of the detected obstacle times a constant, and varies according to the obstacle extension detected by the laser range finder. This parameter is initially set to 10. On the other hand, the height of the region of interest (ROI) on the visual image has been empirically determined to 80 pixels, as a relevant value for the characterization of the objects of interest in this domain. The values of both "Spatial Band Width" and "Range Band Width " parameters have been fixed to 2, so as to highlight the more relevant segmented regions, in this case large areas. Finally, the largest segmented region is classified through its colour feature by using expert knowledge. The segmentation process as performed only on the region of interest (ROI) reduces the computing time of the Mean-shift algorithm to less than 1 second per scene. The classification results are better than those obtained from the segmentation of the global image. Thus, the Mean-shift method is appropriated for the segmentation when applied to a part of the global image, selected by means of the sensor fusion algorithm here proposed. The results obtained from the application of both the sensor fusion method and the Mean-shift segmentation algorithm for the classification of the closest object, are presented in the next section.

6. Real-time classification results

Three types of objects have been detected and classified in a specific semi-rural scenario consisting of trees, human operators and vehicles. The fusion and segmentation method here proposed is invariant to both brightness variations on the image and position or orientation of the static or dynamic objects. The ultimate aim of the characterization of the objects is their classification through a set of descriptors: extension, location, distance and colour. These descriptors conveniently integrated with domain knowledge result in smart strategies for safe reactive piloting either in rural or rural-urban domains. The first result concerns the characterization of the object "tree". Six scenes, with large illumination variations, displaying trees in front of the robot that navigates along a trajectory, are displayed in Figure 14. The first column exhibits the 2D laser angle-distance representation, reflecting the distance and angular extension of the detected objects, second column exhibits the visual images, third one the results of the fusion method, and fourth the segmentation of the region of interest (ROI), where clusters are represented in smooth colour (close to real colour). The largest pixel area corresponds to the closest object detected by the laser.

The first scene displays three clusters and the largest one is the most representative. Colour expert knowledge, domain dependent, is integrated to automatically classify the type of obstacle among the three possible classes: tree (green or brown), human operator (blue) and vehicle (white). Thus, the pixels of the most representative cluster are matched with the corresponding pixels on the original global image, and the mean colour is calculated to accurately determine the real colour of the obstacle. In current domain, expert knowledge states that objects displaying green or brown colours are "tree".

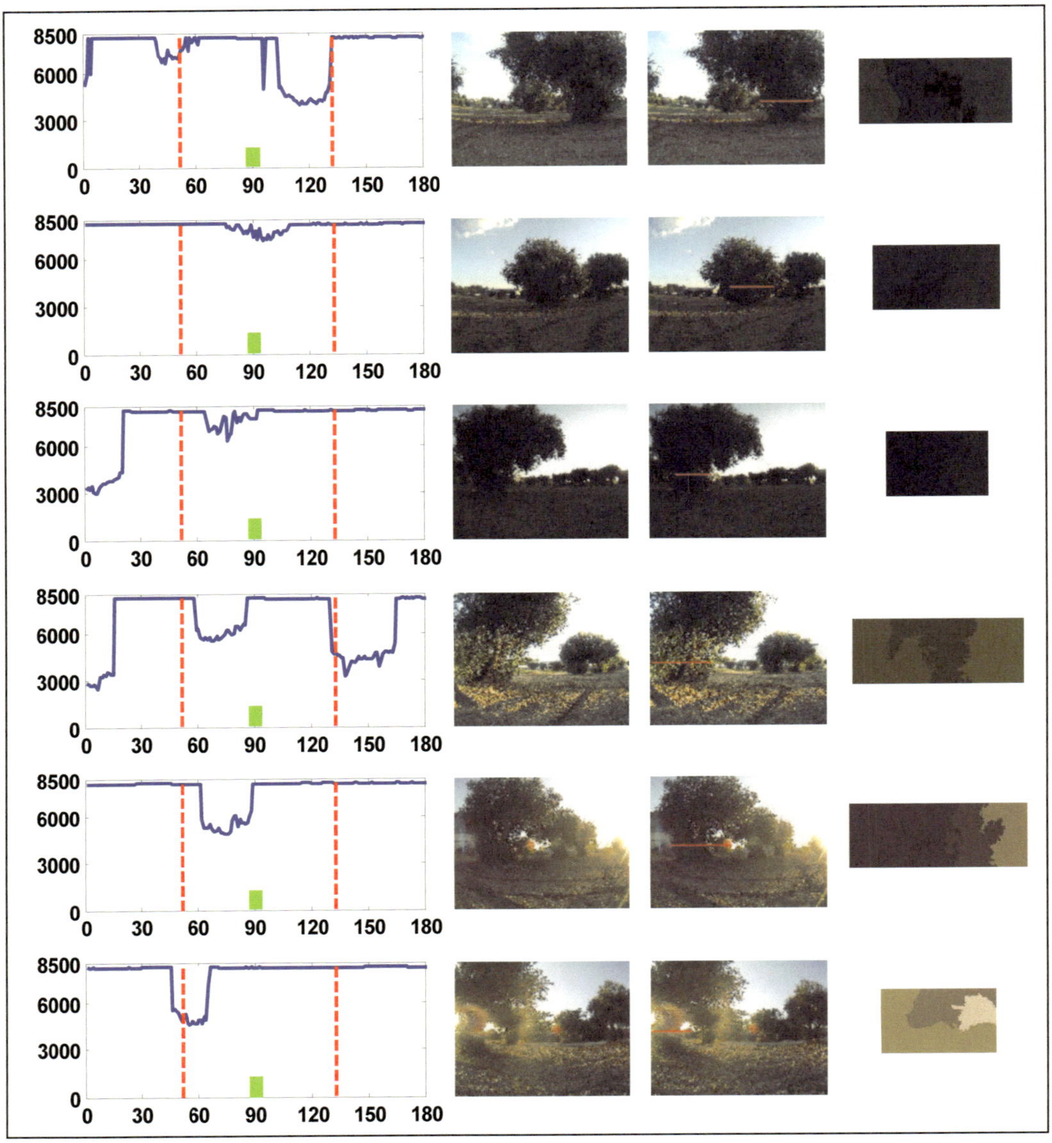

Fig. 14. Six rural scenes at different illumination conditions. First and second columns correspond to 2D laser angle-distance representations and visual images. Third and fourth to the visual image resulting from the sensor fusion method and the segmented regions of the ROI visual image.

The first to third rows, in Figure 14, show three cases with low illumination, and fourth to sixth rows three scenes with high illumination. The results of sensor fusion and segmentation of the region of interest (ROI) demonstrate that the proposed methods are robust to illumination changes. Both, the first and second rows display an obstacle at about four and eighth meters distance, respectively. The sensor fusion method locates the visual region to be segmented, and the segmentation algorithm gives rise to three and two clusters, respectively. The search for the real colour, of the largest cluster, in its original visual image confirms green colour in both cases and the objects are assigned to the class "tree". The fourth row scene is difficult to segment due to shadows on the ground. The fusion and segmentation algorithms highlight three clusters and the classification of the largest one is, "tree". The fifth and sixth rows show scenes which are difficult to analyse because of the visual image in-homogeneities. Objects are detected at 4 and 5 meters, in the front of the vehicle, respectively, and are automatically classified as "tree".

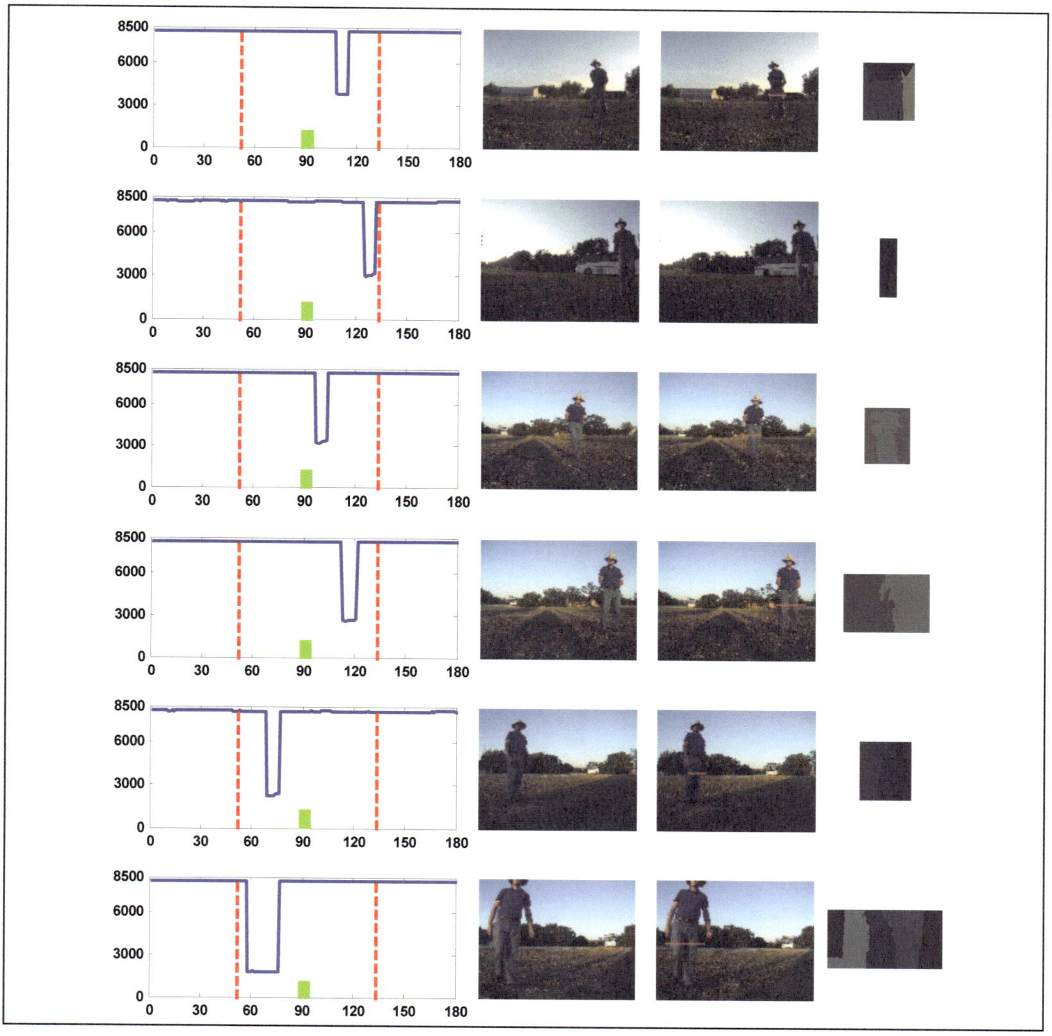

Fig. 15. A sequence of six images displaying a human operator approaching the vehicle.

The Figure 15 displays six rural scenes where a human operator is moving in front of the robot from right to left. As in former scenes, the largest cluster (smooth colour) in each of the six images, is matched with the same region in the original image to determine the real colour, which in this case corresponds to blue. Thus, the detected object is classified as "operator". The first row shows a scene with low illumination where the laser image accurately detects the human operator at 4 m. distance. However, the automatic recognition, based only in the visual image analysis, is difficult due to the colour similarity between human operator and ground. The second row shows the human operator and a white vehicle at the back at a distance greater than 8 m., so it is not detected by the laser. Only one cluster is obtained, that is classified as "operator". The third and fourth scenes present shadows in the ground, however the sensor fusion and the segmentation methods, classify the largest cluster as "operator". In the fifth and sixth scenes, the operator is close to the robot and the colour of his trouser as well as its vertical pattern are correctly classified.

Finally, six rural-urban scenes showing two vehicles parked on the right side of the robot-tractor trajectory are illustrated in Figure 16. The first and second columns introduce the 2D laser angle-distance representation and the visual images where the robot-tractor is close to the bus. The first to third rows corresponds to images where the robot is approaching a first bus parked almost in parallel to the asphalt road. The segmented image (fourth column) displays six clusters with vertical and horizontal patterns. The vertical patterns correspond to the doors and the horizontal pattern to the front. The second and third row images, Figure 16, correspond to the detection of the lateral sides of the first bus. Third and fourth

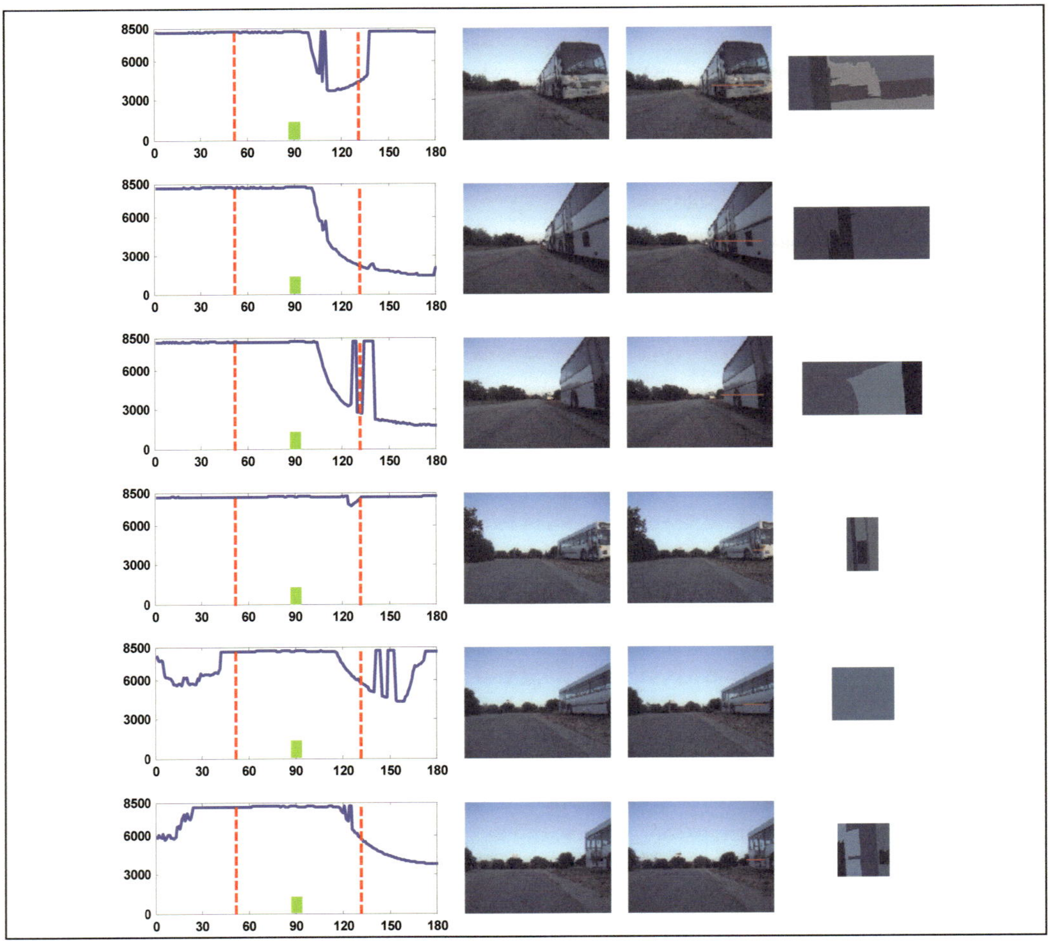

Fig. 16. Six rural-urban scenes with two large vehicle and several trees.

columns present the images resulting from the sensor fusion and the segmented areas of the region of interest (ROI), respectively. In both cases the largest cluster corresponds to white colour in the original image, so as it is assigned to the class "vehicle". Fourth to sixth rows show images acquired while the robot-tractor is navigating close to a second bus. The fourth row images show the detection of the front corner of this bus. The vertical pattern emerging in the segmented image corresponds to the front door of the bus. The fifth and sixth rows of Figure 16, show in the segmented images the lateral and the back of the second bus. The region of interest is larger for the first bus than for the second one, as it is detected closer to the robot.The proposed methodology has been verified by extensive experimental tests, with a 98% success in the classification of the types of objects defined by their colour feature, in this scenario.

7. Conclusion

The fusion method proposed in current work is based on the combination of visual image and 2D laser range finder data. The method improves the recognition of outdoor objects in extreme illumination conditions by integrating objects features from both representations.
The fusion method succeeds in the most extreme and variable weather conditions in dynamic outdoor environments, combining the rich and colourful representation provided by the visual images with the precise depth planar picture supplied by the 2D laser range finder.
The Mean-shift algorithm used for the segmentation of a region of the image (ROI) resulting from the fusion of both laser and visual image features, and the integration of expert knowledge on the specific domain, permits the classification of some common objects: vehicles, trees and human operators. Experiments show 98% classification accuracy of dynamic objects.
The real-time detection and classification of unexpected objects, guarantees safety and permits the instantaneous generation of reflex actions to either deviate or stop the vehicle. Computing time of the whole perception and classification modules is about 1 second, but it could be greatly reduced by integrating real time processor systems.

8. Acknowledgments

Present work has been supported through projects: CICYT- DPI-2006-14497 by the Science and Innovation Ministry, ROBOCITY2030 I y II: Service Robots-PRICIT-CAM-P-DPI-000176-0505, and SEGVAUTO: Vehicle Safety-PRICIT-CAM-S2009-DPI-1509 by Madrid State Government.

9. References

Bagon, S. (2010) Matlab interface for EDISON. The Weizmann Institute of Science, Faculty of Mathematics and Computer Science, *Computer Vision Lab, Rehovot 76100,* Israel.

Baltzakis, H.; Argyros, A. & Trahanias, P. (2003) Fusion of laser and visual data for robot motion planning and collision avoidance. *Machine Vision and Applications,* Vol.12, pp.431-441.

Barawid, O.C.; Mizushima, A.; Ishii, K. & Noguchi, N. (2007). Development of an Autonomous Navigation System using a Two-dimensional Laser Scanner in an Orchard Application. *Biosystems Engineering,* Vol.96, No.2, pp.139-149.

Comaniciu, D. & Meer, P. (2002) Mean-Shift a Robust Approach toward Feature Space Analysis. *IEEE Transactions on Pattern Analysis and Machine Intelligence,* Vol.24, pp.603-609.

Chan, T.F. & Vese, L.A. (2001). Active Contours Without Edges. *IEEE Transactions on Image Processing,* Vol. 10, pp. 266–277.

DARPA (2007). DARPA Urban Challenge. http://www.darpa.mil/grandchallenge/.

Edan, Y.; Han, S. & Kondo, N. (2009). Automation in Agriculture, In: *Handbook of Automation*, pp.1095-1128. ISBN 978-3-540-78830-0. doi 10.1007/978-3-540-78831-7

Garcia-Lopez, L.; Garcia-Alegre, M.C.; Ribeiro, A. & Guinea, D. (2008). An agent of Behaviour Architecture for Unmanned Control of a Farming Vehicle. *Computers and Electronics in Agriculture*, Vol. 60, pp.39-48.

Kassir, A. & Peynot, T. (2010) Reliable Automatic Camera-Laser Calibration. *Proc of the 2010 Australasian Conference on Robotics & Automation (ACRA2010) (Wyeth.G. & Upcroft B. Eds)* Brisbane, Australia. ISBN: [978-0-9807404-1-7].

Katz, R.; Nieto, J.; Nebot, E. & Douillard, B. (2010) Track-based Self-supervised Classification of Dynamic Obstacles. *Autonomous Robots*, Vol.29, pp.219-233. DOI 10.1007/s10514-010-9193-0.

Keicher, R. & Seufert, H. (2000). Automatic guidance for Agricultural Vehicles in Europe. *Computers and Electronics in Agriculture,*Vol.25, pp. 169-194.

Lee, K.H. & Ehsani, R. (2008). Comparison of two 2D Laser Scanners for Sensing Object Distances, Shapes and Surface patterns. *Computers and Electronics in Agriculture,* Vol.60, pp. 250-262.

Li, G.; Liu, Y.; Dong, L; Cai, X. And Zhou, D. (2007). An Algorithm for Extrinsic Parameters Calibration of a Camera and a Laser Range Finder using Line Features. *Proc of IEEE/RSJ Intern Conf on Intelligent Robots and Systems*, pp. 3854-3859, San Diego, CA.

Martin, D; Garcia-Alegre, M.C.; Piva, A. & Guinea, D. (2009) Camera and Laser Vision Fusion for Safe Navigation of a Commercial Tractor. *Proc. IEEE Intern. Conf. on Robotics and Appl. (ICRA2009)* Special Sessions, Kobe, Japan.

Naroditsky, O; Patterson, A. & Daniilidis, K. (2011) Automatic Alignment of a Camera with a Line Scan LIDAR System, *IEEE Inter. Conf on Robotics and Automation (ICRA2011)*, 9-13 May, Shangai, China.

She, K.; Bebis, G.; Gu, H. & Miller, R. (2004). Vehicle Tracking using On-line Fusion of Colour and Shape Features. *IEEE Intelligent Transportation Systems Conference,* Washington, D.C., USA, October 3-6.

Stentz, A.; Dima, C.; Wellington, C.; Herman, H. & Stager, D. (2002). A System for Semi-autonomous Tractor Operations, *Autonomous Robots*, Vol.13, pp. 83-104.

Sun, Z.; Zhu, S.-A. & Zhang, D. (2008). Real-time and Automatic Segmentation Technique for Multiple moving Objects in Video sequence. *IEEE Intern. Conf on Control and Automatic (ICCA)*, Article number 4376472, pp.825-829.

Yokota, M.; Mizushima, A.; Ishii, K. & Noguchi, N. (2004) 3-D Map Generation for a Robot Tractor Equipped with a Laser Range Finder, *Proceedings of ASAE, Automation Technology for Off-Road Equipment*, pp.374-379. ISBN 1-892769-45-X, Kyoto, Japan, 7-8 October 2004.

Zhang, Q.L. & Pless, R. (2004). Extrinsic Calibration of a Camera and Laser Range Finder (Improves Camera Calibration). *Proc. IEEE/RSJ Intern. Conf. on Intelligent Robots and Systems,* Vol.3, pp.2301-2306. Sendai. Japan.

8

Detecting, Tracking, and Identifying Airborne Threats with Netted Sensor Fence

Weiqun Shi, Gus Arabadjis, Brett Bishop, Peter Hill, Rich Plasse and John Yoder
The MITRE Corporation
Bedford, Massachusetts
U.S.A

1. Introduction

Today's technological advances allow for the development of unmanned aerial systems, fixed-wing aircraft, that are small enough and fly low enough to elude conventional radar detection. Such aircraft could carry out chemical, biological, or nuclear attacks, or they could be employed to smuggle drugs or illegal immigrants across the border.

This chapter describes a low cost, low power (potentially disposable) methodology for performing key 24/7 sentry functions to protect critical civilian and military infrastructure from airborne threats. The methodology is based on joint multi-sensor exploitation technology by designing and developing a forward-based fence that contains a mix of various low cost, low power, netted sensors including a simple radar, acoustic microphones and optical (Infrared and visible) cameras to detect, track and discriminate potential airborne targets. An in-depth understanding of candidate target signature phenomenologies is developed through theoretical, numerical assessments and proof-of-concept field experiments. An integrated (over sensor modality) detection, tracking and discrimination process is developed which forms the basis of the fence's friend/foe sentry capability and ability to provide accurate/timely intercept information. An experimental prototype end-to-end proof of concept system with deployable software, hardware and connectivity has also been developed to perform the field demonstration.

2. System concept and design

The primary detection component in the system is a radar fence. The radar fence is designed to detect approaching targets and provide a cue to the acoustic and infrared sensors that perform the discrimination task. The radar fence consists of multiple, low power (10 Watts), non-scanning (for low cost and complexity), UHF, pulse-Doppler radars (to estimate target speed, range and eliminate birds and ground clutter), with a radar-to-radar separation of approximately 5 km (Figure 1). Each radar operates with a different carrier frequency (to avoid crosstalk between radars) and has a beamwidth that is broad in both azimuth (so that the number of radars can be kept small) and elevation (to detect both high and low-flying targets). The radars measure target range and radial speed five times per second and report these values to a central processing station that cues the acoustic and infrared sensors (if a target report is issued), and then fuses the reports from all sensors (radars, acoustic, etc.) to form a target track and alert rear-area weapons systems or potential interceptors so that defensive action can be taken. A complete description of the radar parameters and detection characteristics is contained in Tables 1.

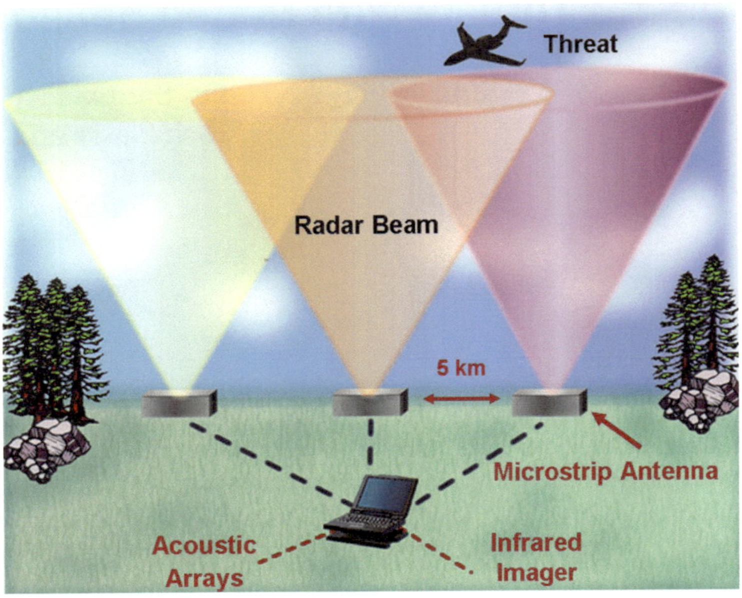

Fig. 1. A conceptual illustration of the netted sensor fence

Acoustic microphone arrays are used as the second sensor modality in the system to detect broadband acoustic emissions from approaching targets. Acoustic sensors are non-line-of-sight, passive, low-cost, and portable sensors that can be effectively deployed in wide areas. Primary objectives of acoustic sensors in this sensor fence system are: a) to provide target direction of arrival (DOA) estimates that will then be fused with radar measurements to form a target track; b) to provide a means for target identification and classification; and c) to mitigate false alarms. The system is designed to contain several equally spaced, diagonally-arranged microphone arrays.

The third sensor modality in the fence is an optical system which is cued by the radar and/or acoustic sensors and slews in angle to acquire track and identify the potential airborne threat. The system is designed to contain an uncooled infrared detector sensitive to the 8-12 um waveband to provide day and night time operation. The uncooled IR detector array uses only several Watts of power. A boresighted visible camera is also used for improved target resolution during the daytime. Visible cameras are inexpensive and have improved resolution compared to the infrared detector array.

3. Detection

3.1 Radar detection

Assessments of radar detectability and detection range are accomplished via numerical simulation. In Table 1, the range at which the probability of detection equals 0.9 was calculated assuming a non-fluctuating target (this is why the UHF frequency band was chosen), with the signal-to-noise ratio at target range R calculated from,

$$\frac{E}{N_o} = \frac{PG^2\sigma\lambda^2\tau}{(4\pi)^3 R^4 kT_o FL} \tag{1}$$

for ranges where the target is completely uneclipsed (eclipsing occurs if part or all of the return from the target is received while the transmit pulse is still on), which implies R > cT/2, where c = speed of light and T = duration of the uncompressed radar pulse. Also, P = average transmitted power, G = antenna gain at the target (this varies with target location), σ = target radar cross section, τ = coherent integration time, λ = wavelength, k = Boltzmann's constant, T0 = 290°K, L = loss, and F = receiver noise figure. When the target is partially eclipsed, the signal-to-noise ratio in Equation (1) is reduced by (2R/cT). The pulse repetition frequency was chosen so that the radar is unambiguous (unambiguous range = c/2·PRF, unambiguous speed = λ·PRF/2) in both target range and speed for all targets of interest. Plots of E/N0 (including antenna patterns) for a target with a 3 m^2 radar cross section flying at 100 m altitude and crossing the fence directly above radar (X = 0) and at 2 km from radar are shown in Figure 2. These values were used with the standard curves (Skolnik, 1990) of probability of detection per look versus signal-to-noise ratio to calculate the probability, Pd(n), that the target is detected at the end of the coherent integration interval τ at a time when the target is at a range Rn. Finally, the cumulative probability that the target has been detected by the time it reaches range Rm is

$$P_c(m) = 1 - \prod_{n=0}^{m} [1 - P_d(m)] \tag{2}$$

where R_0 (n = 0) corresponds to the range at which the target comes over the horizon. The range Rm where Pc(m) = 0.9, for the case when X = 2 km, is the value cited in Table 1.

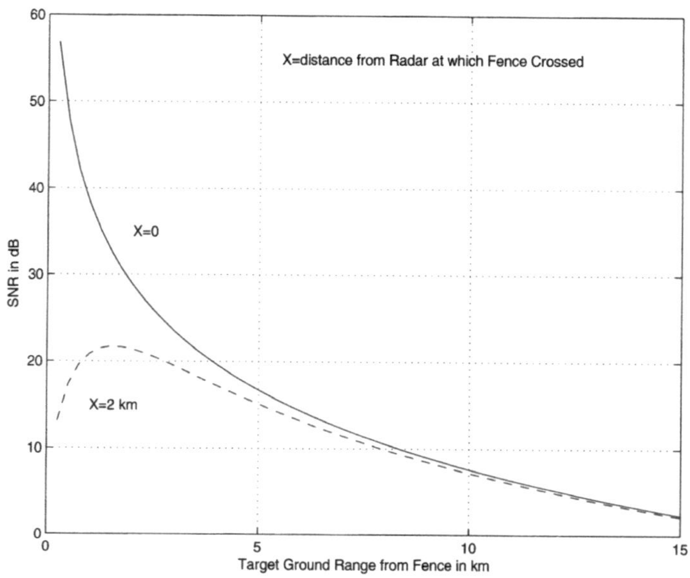

Fig. 2. SNR for Low-Flying Target with a Radar Cross Section = 3 m^2

Type	Pulse/Doppler
Frequency	UHF
Scan	None (Radar Stares Forward)
Antenna	Broadbeam, Gain ~ 3 dB
Polarization	Circular
Average Power	10 W
Duty Factor	10%
Pulse Repetition Frequency	10 kHz
Coherent Integration Time	0.2 sec. (5 Hz Update Rate)
Unambiguous Range	75 km
Unambiguous Speed	750 m/sec
Range Resolution	150 m
Speed Resolution	1.9 m/sec
Minimum Detection Range	25 km
Range for Probability Detection = 0.9	7 km (Small Target , 15 km (Large Aircraft)
Noise-Induced False Alarm Rate	1 Per Month
Clutter/Bird Cancellation Approach	Blank Lowest few Doppler Bins
Quantities Measured	Target Range and Speed (Azimuth and Height Estimated)
Target Discrimination	Based on Speed Only. Need Other Sensor Types

Table 1. Individual Radar Properties

3.2 Acoustic detection

Acoustic detection assessments are primarily accomplished by performing several field measurements of different types of aircraft to obtain information on target acoustic detection and signature characteristics. Although many aircraft acoustic data are available in literature (Ferguson & Lo 2000; Pham & Srour 2004), for the purpose of developing and testing multi-modal sensor fence detection, tracking and classification algorithms, it is critical to simultaneously obtain data measurements from all the fence sensors at the same time. A typical experiment layout and the 8-element acoustic microphone array (with equal element spacing of 0.5m) are shown in Figure 3. The sensor suite which includes acoustic array and IR/visible cameras are positioned near the end of an airport runway. The test aircraft are flying at a flight test matrix with multiple combinations of altitude and engine RPM. GPS data recording systems are mounted on the aircraft so the ground truth information can be retrieved and later can be used for target validation after the flight.

Figure 4 shows a spectrogram of a measured acoustic data showing a rich array of features corresponding to two crop dusters flying above the sensor array. Strong Doppler shifted harmonic structures caused by aircraft' engine noises are observed. The parabolic energy distortion (in the time period between 60-80 seconds) in the spectrum is caused by the multipath ground reflection interference when the target is passing directly above the sensor array. The spreading width of the parabolas is directly related to the speed of the targets.

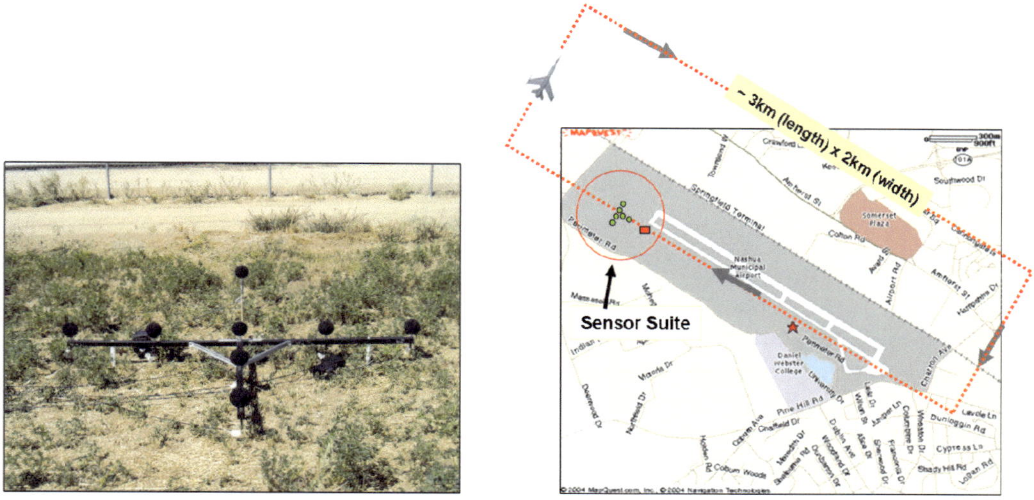

Fig. 3. Field experiment set up (right) and the acoustic microphone array (left)

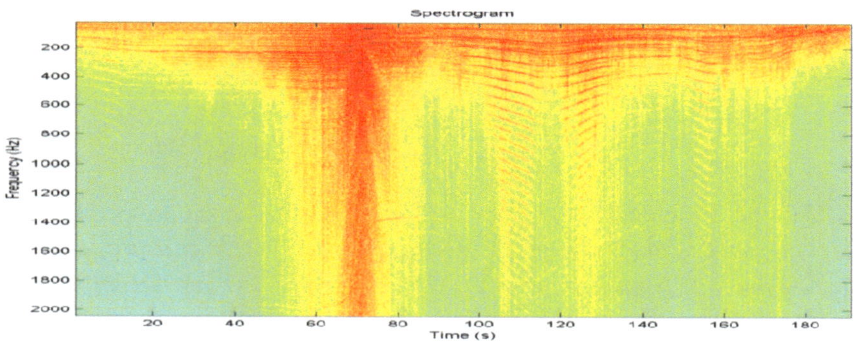

Fig. 4. Spectrogram of two crop dusters flying above the sensor array

3.3 IR detection

The development of the optical subsystem started with an evaluation of IR detectors, wavebands and target signatures in order to calculate the effective range of these IR systems, design and optimize the optical sensing system performance. This work is followed by designing an experimental set-up and experiments to collect images of airborne targets.

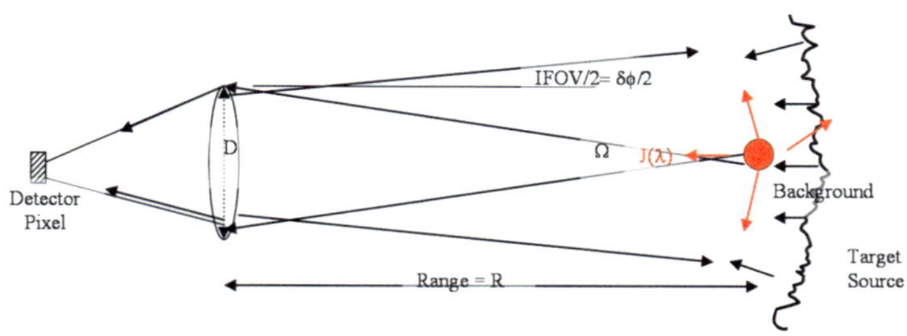

Fig. 5. Diagram of the Optical System and the Target Environment

Detector received power for an optical system (Figure 5) whose diameter is D, pixel angular field-of-view is δϕ, target range is R and target spectral radiance is J(λ) W/Sr/μm is calculated below. The backgrounds are clouds and atmospheric aerosol scattering. The magnitude of these backgrounds can be calculated by MODTRAN code runs for any specific defined atmospheric condition

The signal power received from the source by the detector is given by

$$P_S = \int_{\lambda_1}^{\lambda_2} T_{optics} T_{atmosphere}(\lambda) J(\lambda) * \frac{\pi D^2}{4R^2} d\lambda = \frac{\pi D^2}{4R^2} T_{optics} \int_{\lambda_1}^{\lambda_2} T_{atmosphere}(\lambda) J(\lambda) * d\lambda \tag{3}$$

Where T_{optics} is the transmittance (or transmission coefficient) of the optical system and $T_{atmosphere}$ is the transmission of the atmosphere along the path from the source to the sensor. Generally, with proper optics design, the transmission of the optics is a constant independent of wavelength, but the transmission of the atmosphere is a function of wavelength. The integration is performed over the spectral bandpass of the optical system. This derivation assumes that the source is incident on a single pixel. The expression can be approximately corrected for multiple pixels on the target by dividing by the number of pixels on the target, n.

The power on each detector pixel from the background scene in the field-of-view of each pixel (see figure) is given by

$$P_B = (IFOV)^2 \frac{\pi D^2}{4} T_{optics} \int_{\lambda_1}^{\lambda_2} T_{atmosphere}(\lambda) N_{background}(\lambda) d\lambda \tag{4}$$

where, IFOV is the instantaneous angular field of view of each square pixel (radians) and $T_{atmosphere} * N_{background}$ is the background spectral radiance W/(Sr cm2 μm) at the aperture. The term $T_{atmosphere} * N_{background}$ can be obtained from the MODTRAN atmospheric modeling code and already contains the effects of atmospheric transmission between the noise sources (background scene and/or aerosol scattering) and the receiver aperture. Thus optic transmission term can be dropped from the expression. Note: R and D have units of centimeters, and λ has units of μm.

The circuit power is proportional to the square of the current so, the power signal to noise ratio is given by [(Kingdton, 1978)

$$\left(\frac{S}{N}\right)_P = \frac{i_S^2}{\left(i_N^2\right)_{Mean}} = \frac{\eta^2 e^2 P_S^2}{(h\nu)^2} * \frac{h\nu}{2\eta e^2 (P_S + P_{B'})} = \frac{\eta P_S^2}{2h\nu B (P_S + P_{B'})} \tag{5}$$

For the typical infrared case of the signal power being much less than the background power $P_s << P_{B'}$, (background limited detection) then

$$\left(\frac{S}{N}\right)_P = \frac{\eta P_S^2}{2h\nu B P_{B'}} \tag{6}$$

[Note: the background power, P_B', is due to all non-signal sources, including (a) nearby detector cryogenic Dewar radiation, (b) optics radiation, (c) radiation from the field-of-view limiting shrouds and (d) other nearby sources]. We can invoke the "pure detection" criterion (Howe, 1996) that if the signal to noise ratio with the target in a pixel minus the signal to noise ratio with the target not in the pixel (SNR= SNR_{target}-$SNR_{no-target}$)\i) is greater than 5, the target will be detected against a cluttered background. The range of detection of an airborne target at various altitudes can then be calculated and the results are shown in Figure 6.

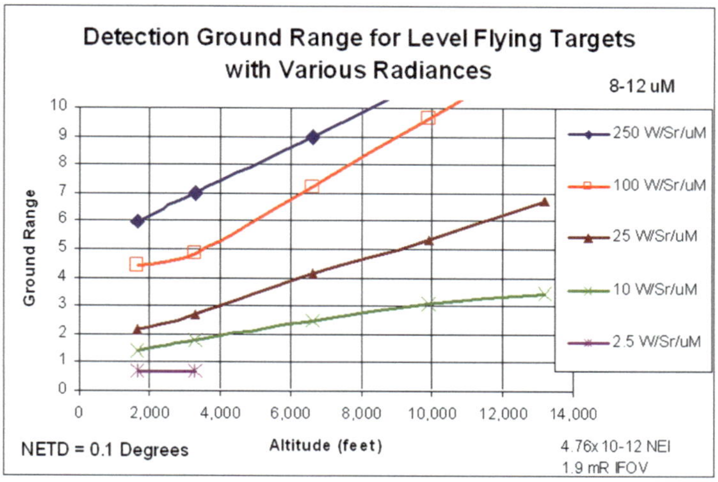

Fig. 6. Attitude and ground range for the detection of airborne targets of various values of target radiance

This calculation is for an uncooled 8-12 μm detector with a noise equivalent temperature difference (NETD) of 0.1 degrees and shows the altitude and ground range for the detection of airborne targets of various values of target radiance. Typical aircraft have spectral radiance of about 100 W/Sr/μm so targets can be detected at ranges in excess of 5 km even if they are at low altitudes. (Note a NEI of 0.1 degrees corresponds to a NEI of about 4.76x10-12 W/cm2 for a 1.9 milliradian IFOV.)

The experimental set-up to collect visible and IR field data is shown in Figure 7. The visible and IR cameras are boresighted with each other to take simultaneous visible and IR images of the target. A frame grabber and computer is used to collect the data and display the images side-by-side. The IR camera is a BAE 320x280 micro-bolometer array which is sensitive over the wavelength range of 7 to 14 μm. The FOV of the array is 7.7x10.3 degrees and the IFOV of each pixel is about 0.56 milliradians. It has a NETD of aboutabut 0.08 Celsius and consumes only 6 Watts of power.

Typical images at short range are shown in Figure 8 at the lower left for a Cessna and the middle for a crop duster. The detection of a twin engine Cessna is at 6.1 km shown at the lower right.

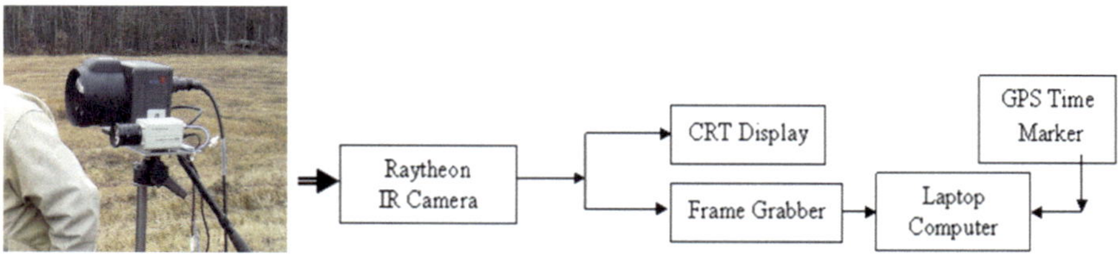

Fig. 7. IR/visible data collection set up.

Fig. 8. The detection of single engine Cessna (left), a crop duster (middle) and a twin engine Cessna at 6.1 km (right)

4. Tracking

A major challenge of the project has been the design and development of sensor fusion techniques which combine data from multi-modality and multi-sensor nodes to achieve improved accuracy in target detection, tracking, and classification. The system employs a Kalman filter (Anderson & Moore, 1979; Grewal & Andrews, 2001) to establish multi-target tracks using radar and acoustic measurements as input. The tracker is designed to handle multiple targets and false reports. It is also designed to have data input flexibilities such as allowing input data measurements from each sensor (radar and acoustic) that are not time coordinated. The tracker must allow that some tracks it creates may be based on false reports, and therefore these tracks must be dropped if they behave erratically or do not associate with further detections at later times. Tracks can be promoted or demoted by evaluating cumulative properties of a score that was originally assigned to the initial tracks. Constant velocity with additive white noise acceleration is introduced in the plant noise assumption. Since the range of radar measurements is on the order of ~10 km, it is adequate to use a flat earth model when calculating tracker updates. The objective of the tracker is to fuse asynchronous radar and acoustic data to predict kinematic properties such as the location, the speed, the heading, and the flight trajectory of the target. This prediction is then used to automatically aim a camera to the predicated point and photograph the target.

4.1 Tracker requirements, functional capability, and restrictions

For an operational system the tracker can be designed to sit at a central location (a/k central node). The remote nodes transmit detections of an aircraft target (time, range, azimuth, elevation) to the central node. Each transmitted signal packet contains all the detections (there may be none) accumulated since the previous transmission. The data is time-stamped at each remote site with the time at which the signal arrives to the sensor. For such multi-modal sensor system the tracker must be designed to accommodate asynchronous data streams from multiple remote sites. The tracker is also required to allow for an arbitrary number of remote sites, remote site dropout during a run, false reports in the data, the correction for propagation time delay of the acoustic signals, and the prediction of the target location in the future. The tracker must also be operating in real time.

These requirements in turn imply that the tracker must have the following functional capabilities:

- Track initialization, i.e., the capability to start a track from the data
- Rejection gates to eliminate false reports
- Data association, i.e., the capability to associate a new report with a track for track updating

- Kalman filter remote site data fusion for minimum variance state estimate
- Dynamic status tables identifying which sites are active or inactive
- Time management and coordination logic to allow for asynchronous data streams and propagation time delays of acoustic data.

The development of such a complex tracker involves considerable effort. In order to keep the tracker development effort within manageable bounds, several key restrictions and simplifications were imposed on the tracker capability. These include a single target track, and a flat earth model assumption

4.2 Tracker processing and time delay correction

The multi-modal kinematic tracker employs a Kalman filter to update the data measurement. At each cycle, the tracker corrects the acoustic data timestamp for propagation delay, attempts to initialize a track if none exists, performs the association function, discards false reports, and performs the Kalman filter update of the time sequenced data.

Assume at time t the state vector of the track is a six dimensional vector of target position and velocity $(\dot{x}(t), \dot{y}(t), \dot{z}(t))$ in Cartesian coordinates relative to an east-north-up topocentric coordinate system with origin at the central node, $\mathbf{x}(t) = [x(t) \ \ y(t) \ \ z(t) \ \ \dot{x}(t) \ \ \dot{y}(t) \ \ \dot{z}(t)]^{\mathrm{T}}$. Let $Z_i(t)$ be the measurement vector from remote node i at time t and $\mathbf{R}_i$ the corresponding covariance matrix. For the radar range measurement, $Z_i(t) = [r_i(t)]$, $\mathbf{R}_i = [\sigma_R^2]$, where

$r_i(t)$ is the range measurement value from remote node i, and σ_R is the standard deviation of the covariance matrix. In parallel, for the acoustic angle measurement $Z_i(t) = \begin{bmatrix} Az_i(t) \\ El_i(t) \end{bmatrix}$, $\mathbf{R}_i = \begin{bmatrix} \sigma_{Az}^2 & 0 \\ 0 & \sigma_{El}^0 \end{bmatrix}$. Where $Az_i(t)$ is the azimuth angle and $El_i(t)$ is the elevation angle. Due to significant propagation time differences between the radar and the acoustic data (e.g. at a range of 5 km the acoustic sensor data corresponds to a point on the flight path that is 5000/340 = 14.7 sec in the past, while the radar data is virtually instantaneous), it is necessary to correct the timestamp of the acoustic sensor data to correspond to the time the signal left the target.

Assume at time *t*, $\mathbf{r} = [x \ \ y \ \ z]$ is the position and $\dot{\mathbf{r}} = [\dot{x} \ \ \dot{y} \ \ \dot{z}]$ is the velocity of the target. The acoustic measurement received at time t should correspond to an earlier time t_0 at which the position of the target is $\mathbf{r}_0$, as shown in Figure 9 . For a target traveling at a constant speed v ($0<v<c$, where c isthe spend of sound in the air) between locations of $\mathbf{r}_0$ and $\mathbf{r}$, through simple geometrical derivations, it is easy to obtain,

$$r_0 = \frac{c\sqrt{a^2 + (c^2 - v^2)r^2} - ac}{c^2 - v^2}, \quad 0 < v < c \tag{7}$$

Where $v = |\dot{\mathbf{r}}| = \sqrt{\dot{\mathbf{r}}\dot{\mathbf{r}}^{\mathrm{T}}} = \sqrt{\dot{x}^2 + \dot{y}^2 + \dot{z}^2} = |\dot{\mathbf{r}}_0| = v_0$ and $a = \mathbf{r}\dot{\mathbf{r}}^{\mathrm{T}} = x\dot{x} + y\dot{y} + z\dot{z}$

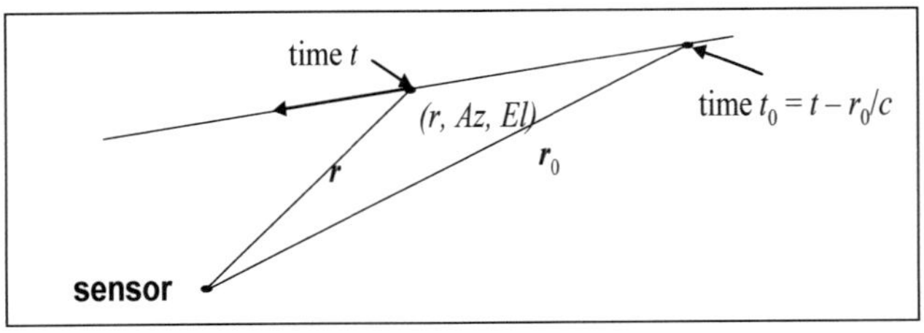

Fig. 9. Diagram of acoustic travel time correction

4.3 Tracker initialization

The Kalman filter is a recursive algorithm which starts with an estimated state vector based on past data and updates it with new data. In order to start the algorithm it is necessary to have an initial estimate of the state. This is done by generating an initial guess of a constant-velocity path via a least-squares fit to the batch of collected data. This method allows us to separate a reasonable collection of false alarms from real data without relying on prior knowledge of the target's position. The least-square fit minimizes the following objective function for a measurement Z_i that has an expected value of E_i,

$$Q = \sum_i \frac{(E_i - Z_i)^2}{\sigma_i^2} \tag{8}$$

This formula is modified slightly for acoustic measurements in that it combines azimuth and elevation measurements into a single term using the law of cosines to calculate the great circle distance on a unit circle:

$$E_i - Z_i = \arccos(\sin(El_i)\sin(E_{El}) + \cos(El_i)\cos(E_{El})\cos(E_{Az} - Az_i)) \tag{9}$$

While minimizing Q is the core function of the initialization tracker, it is not nearly adequate for consistently producing reliable tracks. Steps must be taken to both minimize the number of false tracks and maximize the chances of generating acceptable tracks when real targets are present. These steps include windowing, setting a minimum number of data points, placing bounds on target speed for acceptable tracks, discarding outliers, selecting the best of multiple independently generated tracks, and setting maximum values for Q such that a track is still valid.

4.4 Data association

Data arrives to the tracker from various remote nodes. The data from each node consists of either radar range data alone, acoustic sensor azimuth and elevation data alone, or both radar and acoustic sensor data. At each time point the acoustic sensor outputs only one detection (or possibly none) and this probably corresponds to the loudest source. By contrast the radar can output any number of detections at each time point, as many as cross the detection threshold. The tracker must allow that many of these detections could be false alarms, arising from random noise and clutter. They may also be detections of other real targets within the range of the sensors. The association process is an attempt to weed out the irrelevant detections so as not to corrupt the updating of the track.

This association is performed by comparing each new measurement to some previously generated expectation of the target's location. This previous expectation is generally the result of the most recent tracker output. The comparison between the measurement and the expected target location begins by creating an expected state vector for time t where t is the time of the measurement, and then convert the expected state vector to the measurement vector. The differences between the measurement and the expected value is denoted as, $\boldsymbol{\varepsilon} = Z(t) - \bar{Z}(t)$, where $Z(t)$ is the measurement at time t, and $\bar{Z}(t)$ is the expected measurement converted from the expected state vector. A set of typical range, azimuth and elevation gate G_R, G_{ang} can be defined based on sensor properties. The error ε must be within those gates in order for a data measurement to associate.

4.5 Kalman filter data fusion

Kalman filter is employed for data update and predictions. It is assumed that the flight path is a constant velocity plus a Gaussian white noise acceleration term (plant noise). This implies that the state obeys the linear difference equation

$$\mathbf{x}(t+\tau) = \mathbf{\Phi}(\tau)\mathbf{x}(t) + \mathbf{w}(t,t+\tau) \; 10)$$

Where $\mathbf{\Phi}(\tau)$ is the transition matrix, $\mathbf{\Phi}(\tau) = \begin{bmatrix} 1 & 0 & 0 & \tau & 0 & 0 \\ 0 & 1 & 0 & 0 & \tau & 0 \\ 0 & 0 & 1 & 0 & 0 & \tau \\ 0 & 0 & 0 & 1 & 0 & 0 \\ 0 & 0 & 0 & 0 & 1 & 0 \\ 0 & 0 & 0 & 0 & 0 & 1 \end{bmatrix}$ (11)

and $w(t,t+\tau)$ is a random zero mean Gaussian plant noise process with covariance matrix,

$$\mathbf{Q}(\tau) = \operatorname{cov} \mathbf{w}(t,t+\tau) = \begin{bmatrix} q_x^2\tau^3/3 & 0 & 0 & q_x^2\tau^2/2 & 0 & 0 \\ 0 & q_y^2\tau^3/3 & 0 & 0 & q_y^2\tau^2/2 & 0 \\ 0 & 0 & q_z^2\tau^3/3 & 0 & 0 & q_z^2\tau^2/2 \\ q_x^2\tau^2/2 & 0 & 0 & q_x^2\tau & 0 & 0 \\ 0 & q_y^2\tau^2/2 & 0 & 0 & q_y^2\tau & 0 \\ 0 & 0 & q_z^2\tau^2/2 & 0 & 0 & q_z^2\tau \end{bmatrix} \quad (12)$$

Where q_x, q_y, q_z are plant noise intensities in x, y, z directions.

Let $\hat{\mathbf{x}}(t|t_0)$ be the Kalman filter optimum estimate of the state $\mathbf{x}(t)$ at time t based on data taken up to and including time t_0, $\mathbf{P}(t|t_0)$ is the covariance matrix of the errors in this estimate of the state, the recursive Kalman filter estimate of the updated tracker from time t_0 to time $t = t_0 + \tau$ is thus written,

$$\begin{aligned} \hat{\mathbf{x}}(t|t_0) &= \mathbf{\Phi}(t-t_0)\hat{\mathbf{x}}(t_0) \\ \mathbf{P}(t|t_0) &= \mathbf{\Phi}(t-t_0)\mathbf{P}(t_0)\mathbf{\Phi}^{\mathsf{T}}(t-t_0) + \mathbf{Q}(t-t_0) \end{aligned} \quad (13)$$

Transform the predicted track $\hat{\mathbf{x}}(t|t_0)$ to the measurement the measurement variables $\hat{Z}_i(t|t_0)$, Form the innovation, which is the difference between the actual measurement $Z_i(t)$ and the predicted measurement $\hat{Z}_i(t|t_0)$, $\tilde{Z}_i(t|t_0) = Z_i(t) - \hat{Z}_i(t|t_0)$.
The covariance matrix of the innovation is

$$\mathbf{S}_i(t|t_0) = \mathbf{H}_i(t)\mathbf{P}(t|t_0)\mathbf{H}_i^{\mathrm{T}}(t) + \mathbf{R}_i \tag{14}$$

Where $\mathbf{H}_i(t)$ is the matrix of partial derivatives of the measurement variables of node i with respect to the state variables at time t. Thus for the radar measurement

$$\mathbf{H}_i(t) = \begin{bmatrix} \frac{\partial R_i(t)}{\partial x} & \frac{\partial R_i(t)}{\partial y} & \frac{\partial R_i(t)}{\partial z} & 0 & 0 & 0 \end{bmatrix} \tag{15}$$

and for the acoustic measurement

$$\mathbf{H}_i(t) = \begin{bmatrix} \frac{\partial Az_i(t)}{\partial x} & \frac{\partial Az_i(t)}{\partial y} & \frac{\partial Az_i(t)}{\partial z} & 0 & 0 & 0 \\ \frac{\partial El_i(t)}{\partial x} & \frac{\partial El_i(t)}{\partial y} & \frac{\partial El_i(t)}{\partial z} & 0 & 0 & 0 \end{bmatrix} \tag{16}$$

At this point the tracker makes a test to assure that the innovation is consistent with its covariance matrix. The Mahalonobis distance d between the measurement and its predicted value is

$$d = \tilde{Z}_i^{\mathrm{T}}(t|t_0)\mathbf{S}_i^{-1}(t|t_0)\tilde{Z}_i(t|t_0) \tag{17}$$

Where d is a chi-squared distributed variable with $n = 1$ (radar) or $n = 2$ (acoustic sensor) degrees of freedom. If $d \le T$, for a given threshold T, then the data is accepted. If $d > T$ then the Mahalonobis distance is too large and the data is rejected. The threshold has been chosen at the 10% level, i.e., the probability that d is larger than T is one in ten. The Kalman filter then updates the state and the associated covariance matrix,

$$\begin{aligned} \hat{\mathbf{x}}(t) &= \hat{\mathbf{x}}(t|t_0) + \mathbf{K}_i(t|t_0)\tilde{Z}_i(t|t_0) \\ \mathbf{P}(t) &= (\mathbf{I} - \mathbf{K}_i(t|t_0)\mathbf{H}_i(t))\mathbf{P}(t|t_0) \end{aligned} \tag{18}$$

Where $\mathbf{K}_i(t|t_0)$ is the Kalman gain matrix given by,

$$\mathbf{K}_i(t|t_0) = \mathbf{P}(t|t_0)\mathbf{H}_i^{\mathrm{T}}(t)\mathbf{S}_i^{-1}(t|t_0) \tag{19}$$

5. Classification

Target classification is performed as part of the sensor fusion. Once the target track is established from fusing the radar range detections and the acoustic angle detections, the kinematic properties of the approaching targets such as target velocity, range and location can be extracted from the tracker to give an initial classification of target types. This tracker result is also used to automatically aim a camera to the predicated point and photograph the target. Acoustic measurements can be used to further divide the target groups based on Harmonic Line Association (HLA) method by extracting a set of feature vectors from acoustic spectrograms and comparing them against the acoustic target database. Therefore targets such as typical false alarms (e.g., birds, ducks, etc.), propeller driven aircraft (civilian small aircraft), helicopters, and jets can be classified.

5.1 Acoustic feature extraction and classification

As mentioned earlier, the primary targets of interest in this study are small, low-flying aircraft. Such small aircraft tend to emit strong harmonic lines produced by propeller or profane noise. This suggests that a target classification algorithm can be developed based on the Harmonic Line Association (HLA) method.

Given an acoustic time sequence and the corresponding sampling rate, an FFT spectrum is computed at each buffered data frame. A noise spectrum is calculated using a two-pass notched moving average approach with a single-sided window width and a given detection threshold estimated from past experimental data. Spectra peaks, defined as a sequence of 3 FFT bins where a local max occurs are then detected, and the frequencies at which the peaks are detected are accurately determined by doing a parabolic curve fitting to the peak profiles. Using the most significant peak as an anchor, those harmonically related frequency peaks are grouped together to form a hypothetical harmonic feature vector set. This process is then repeated until all the harmonic feature vector sets are extracted for each data frame from all the frames available. It has been found from the field experiments that the most informative aircraft harmonic signatures for small civilian aircraft usually exist within the frequency range from 20-2000 Hz. Given a typical fundamental frequency of small civilian aircraft that are on the order of 50 Hz during normal flight, the first 40 harmonics are selected to form a 40-component feature vector which will be used for the classification. In order to minimize the sound propagation effect and make the feature vector essentially distance invariant, the magnitude of each component is normalized relative to the sum of the magnitudes of the two highest harmonics in the set. Finally the derived feature vectors from each data frame are statistically averaged to form a feature vector template which distinctively represents the aircraft target. The above workflow is summarized in Figure 10. The final classification is performed using a Nearest Neighbor classifier

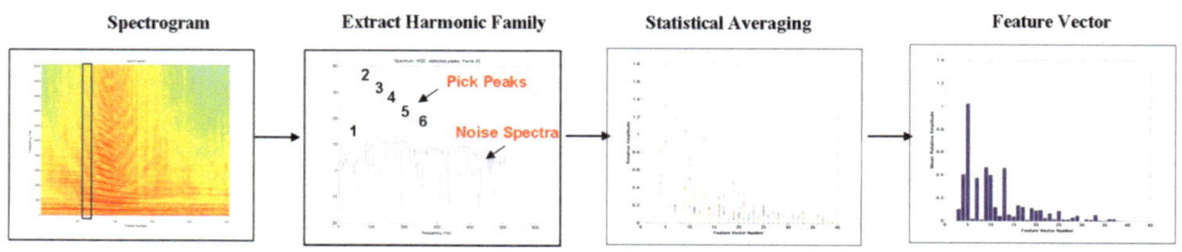

Fig. 10. Acoustic feature extraction processes

5.2 IR feature extraction and recognition

Aircraft recognition from IR images is done based on the Moment Invariants method. The Moment Invariants method has been frequently used as a feature extraction technique for image processing, remote sensing, shape recognition and classification (Keyes &

Winstanley, 2000) The method extracts a set of numerical attributes - the moment feature vectors which uniquely characterize the shape of an object and yet have the desired property of invariance under image translation and rotation. The method was first applied to aircraft shape identification from binary television images by Dudani, etc. (Dudani, etc. 1977) and was shown to be quick and reliable.
The mathematical foundation of Moment Invariants for two-dimensional shape recognition was first introduced by Hu (Hu 1962) in which a set of shape descriptor values were computed from central moments through order three that are independent to object translation, scale and orientation. Translation invariance is achieved by computing moments that are normalized with respect to the centre of gravity so that the centre of mass of the distribution is at the origin (central moments). Size invariant moments are derived from introducing a simple size normalization factor. From the second and third order values of the normalized central moments a set of invariant moments can be computed which are independent of rotation.
In this paper six invariant moment functions that appear to be suitable for the present problem are selected with their mathematical expressions given below,

$$\begin{aligned}
M_1 &= ((\mu_{20}-\mu_{02})^2 + 4\mu_{11}{}^2)/r^4 \\
M_2 &= ((\mu_{30}-3\mu_{12})^2 + (3\mu_{21}-\mu_{03})^2)/r^6 \\
M_3 &= ((\mu_{30}+\mu_{12})^2 + (\mu_{21}+\mu_{03})^2)/r^6 \\
M_4 &= ((\mu_{30}-\mu u_{12})(\mu_{30}+\mu_{12})\cdot[(\mu_{30}+\mu_{12})^2 - 3(\mu_{21}+\mu_{03})^2] \\
&\quad +(3\mu_{21}-\mu_{03})(\mu_{21}+\mu_{03})\cdot[3(\mu_{30}+\mu_{12})^2 - (\mu_{21}+\mu_{03})^2])/r^{12} \\
M_5 &= ((\mu_{20}-\mu_{02})\cdot[(\mu_{30}+\mu_{12})^2 - (\mu_{21}+\mu_{03})^2] + 4\mu_{11}(\mu_{30}+\mu_{12})(\mu_{21}+\mu_{03}))/r^8 \\
M_6 &= ((3\mu_{21}-u_{03})(\mu_{30}+\mu_{12})\cdot[(\mu_{30}+\mu_{12})^2 - 3(\mu_{21}+\mu_{03})^2] \\
&\quad +(\mu_{301}-3\mu_{12})(\mu_{21}+\mu_{03})\cdot[3(\mu_{30}+\mu_{12})^2 - (\mu_{21}+\mu_{03})^2])/r^{12}
\end{aligned} \tag{20}$$

Where $\mu_{pq} = \frac{1}{N}\sum_{i=1}^{N}(u_i-\bar{u})^p(v_i-\bar{v})^q$ are the central moments, u and v are the image coordinates, and $r = \sqrt{(\mu_{20}+\mu_{02})}$ the gyration factor which is used to normalize the moment functions in order to obtain the desired size invariance.
A preprocessing of IR images is performed before the final recognition process. After detection, the area that contains the potential target is first cropped from the original image. Then a binary image is formed by a simple threshold circuit. The aircraft silhouette is next extracted from the resulting binary image and its coordinates are used for the invariance moments feature vector exaction. Figure (11) illustrates the above workflow.

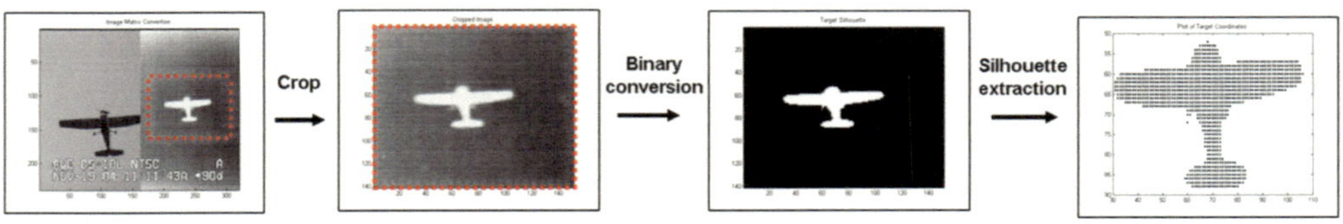

Fig. 11. IR image preprocessing workflow for target classification

To perform an initial classification test, a collection of numerically generated three-dimensional models representing classes of targets of interest (small civilian aircraft, military helicopters, large jets, missile) were chosen to be compared against the IR image collections from the field test. The three-dimensional models, as shown in Figure (12), consist of a Cessna 172, a Black Hawk helicopter, a Lear jet 35, and a missile . These models are constructed based on scaled drawings of the geometric models of each type.

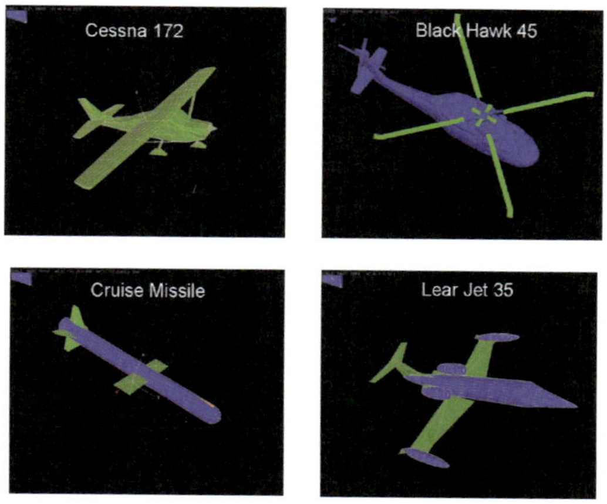

Fig. 12. Three-dimensional models used in IR image classification

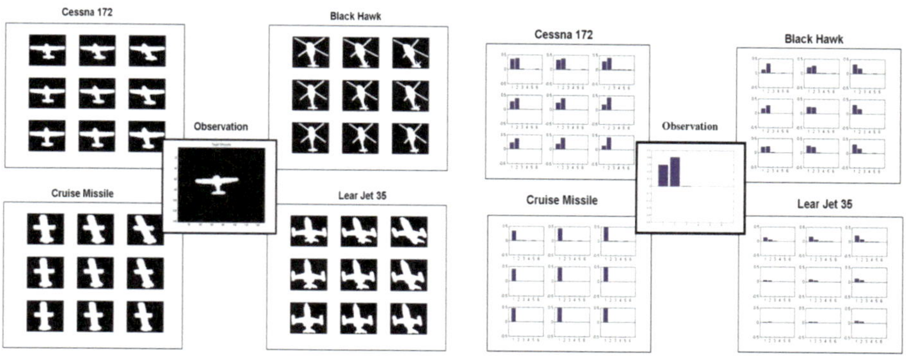

Fig. 13. A measured IR image (center) and the projected 2D images of the selected aircraft (left plot), and the extracted invariant moment vectors (right plot)

In order to compare the candidate target templates to the observed IR images, the corresponding three-dimensional models must be projected into a two-dimensional image plane with the appropriate azimuth ϕ and elevation θ angles corresponding to the camera viewing angles. Taking advantage of the multi-modal sensor character, the azimuth and the elevation angles can be effectively determined from the output of the kinematic tracker derived from the corresponding radar and acoustic measurements. Since the tracker provides the kinematic parameters of the target which include the range, the angles of arrival, and the flight trajectory, the perspective azimuth and elevation angles (ϕ, θ) of the target can be derived from a simple geometrical translation. Figure (13) shows an example of a measured IR image and the projected 2D images centered on the observed azimuth and elevation (ϕ=110, θ=60)) with variations of ±10 degrees on both angles. The corresponded invariant moments are extracted from these images and the results are also shown in the figure.

6. Prototype system and field demonstration

An experimental prototype consisting of three remote sensor nodes and a central processing node has been developed and built using COTs components. Figure 14 show the system hardware diagram, respectively. Each remote sensor node contains a low cost, low-power range only radar sensor (NobelTec IR2 X-band marine radar); an equally spaced, 4-element rectangularly-arranged acoustic array (B&K microphones), and a mini-computer (Slim Pro PC) that performs target range and angle detection and reports those results to the central node. A first order classification based on target acoustic signature is also performed at each remote node and the result is reported back to the central node. The central node contains an IR camera (uncooled BAE Micro IR sensitive to the 8-12 um waveband), a Pelco pan and tilt controller device mounted on a small tower, and the central computer that performs data fusion and final target classification. The connectivity is provided through a simple point-to-point 802.11 wireless communications network consisting of signal boosters and omni-directional antenna located at the remote node and a Yagi-type directional antenna located at the central node. This modular and compact system allows for rapid and inexpensive production of nodes and rapid deployment of the netted sensor fence system.

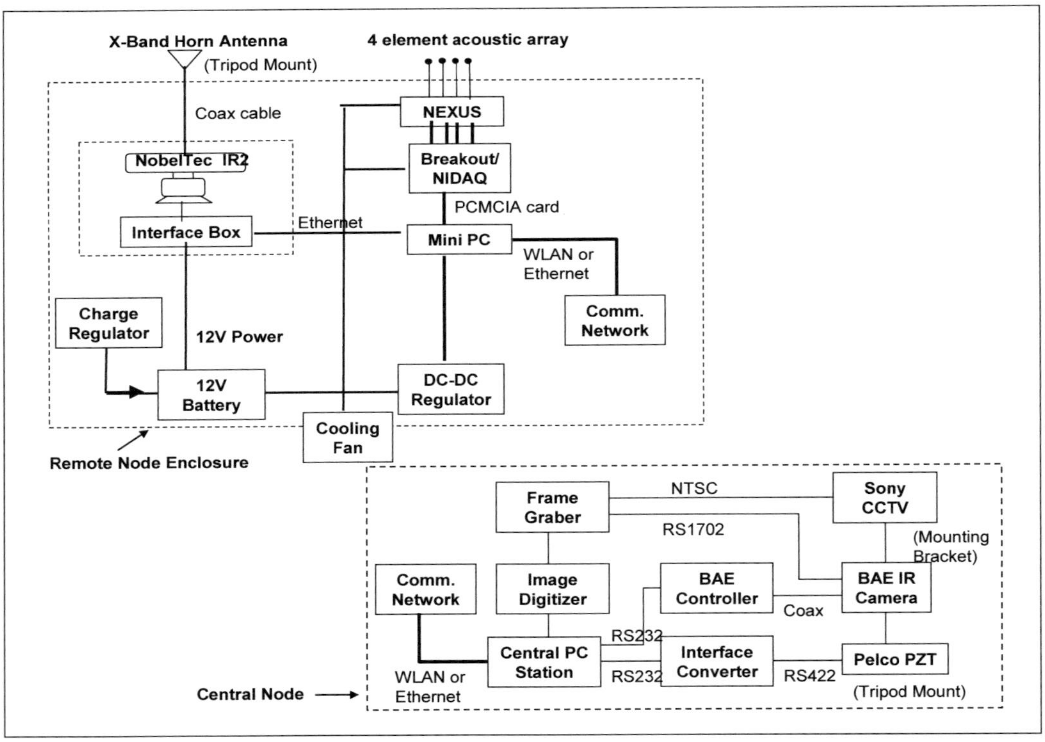

Fig. 14. System hardware diagram

Field tests of the netted sensor fence have been conducted at Nashua Municipal Airport, Nashua, New Hampshire. Typical experiment layout and sensor array positions are shown in Figure 15. The sensor suite is positioned near the end of the runway. The test aircraft are flying at a flight test matrix with multiple combinations of altitude and engine RPM. GPS data recording systems are mounted on the aircraft so the ground truth information can be transmitted in real time to the central node for target validation. The target aircraft used in one of the most recent tests was a Beech BE-76 Duchess. The remote sensor nodes node 1 & 2 were placed at the end of the runway so that planes taking off and those flying parallel to the runway would cross the fence. Due to space constraints, the remote nodes were placed in a T configuration with spacing of approximately 200 meters. The central node was collocated with one of the remote nodes (remote node 3) at the base of the T. Figure 16 shows plots depicting the tracker performance via comparisons of the tracker results with the ground truth recorded by an on board GPS. In general the tracker results show good agreement with the GPS ground truth data.

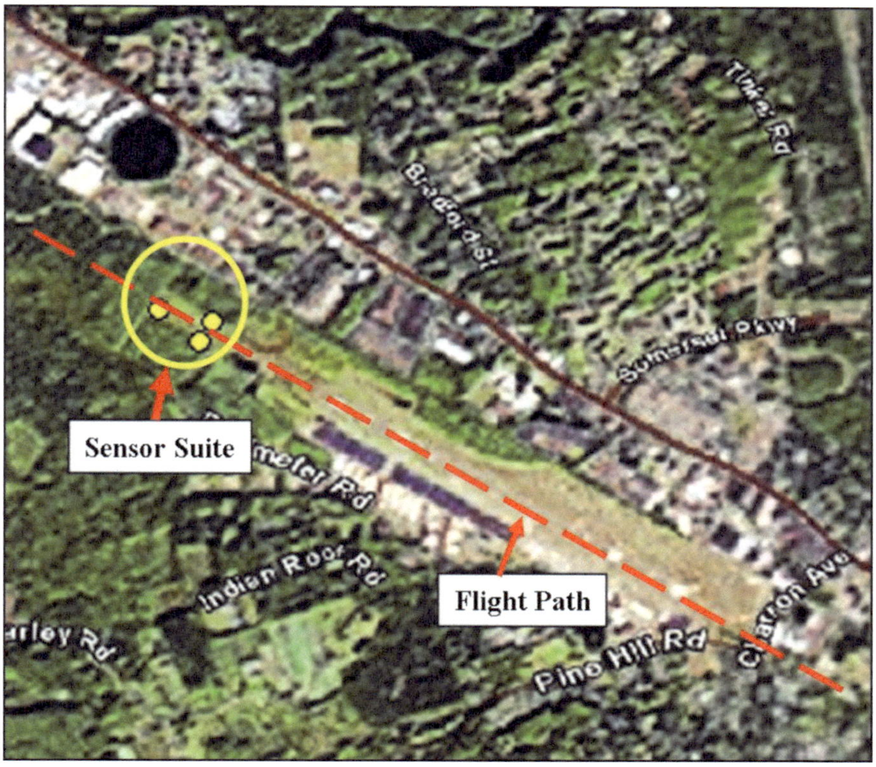

Fig. 15. Sensor configuration and layout in the field test.

A nearest neighbor classifier is then applied to train the extracted features and the final classification results can then be obtained. Euclidean distances between the moment vectors extracted from the observed images and those from the suspected 3D numerical models can be used to measure the confidence level of the classification results.. An image classification example is shown in Figure 17. In this case, a twin engine Beech BE-76 Duchess was correctly identified.

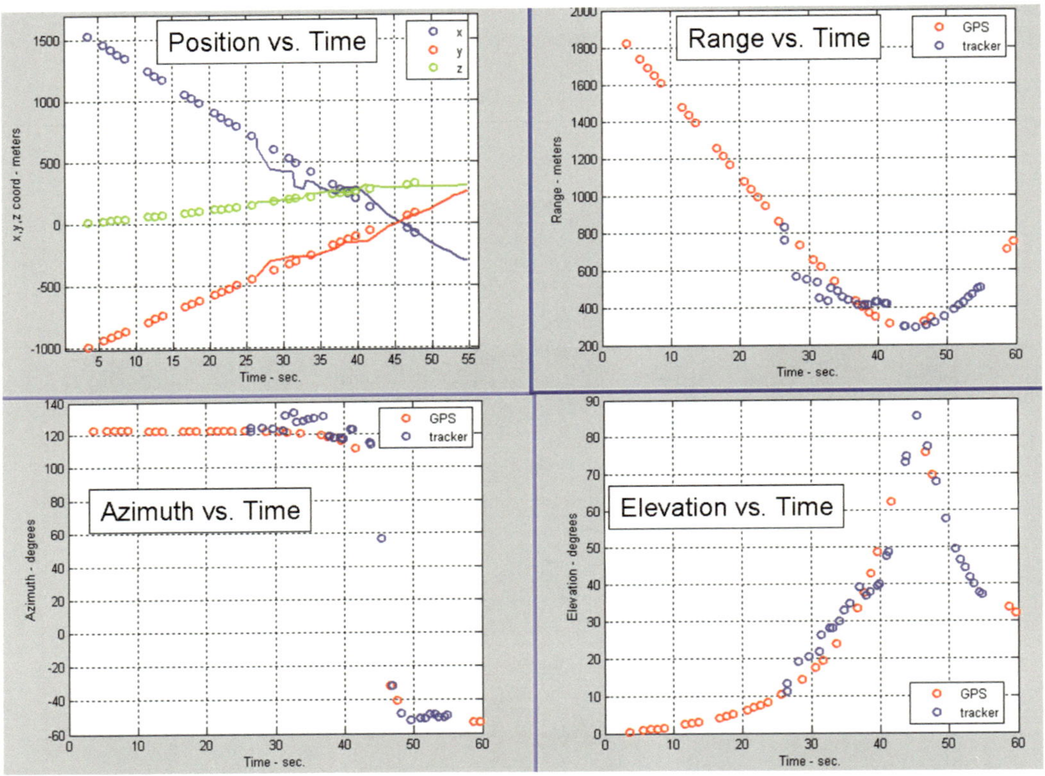

Fig. 16. Comparisons of tracker results with the ground truth GPS recording as a function of time including plots of: the target position (upper left, note: GPS recordings are denoted in circle dots, whereas the tracker results are denoted by lines); the range (upper right); the azimuth (lower left), and the elevation (lower right)

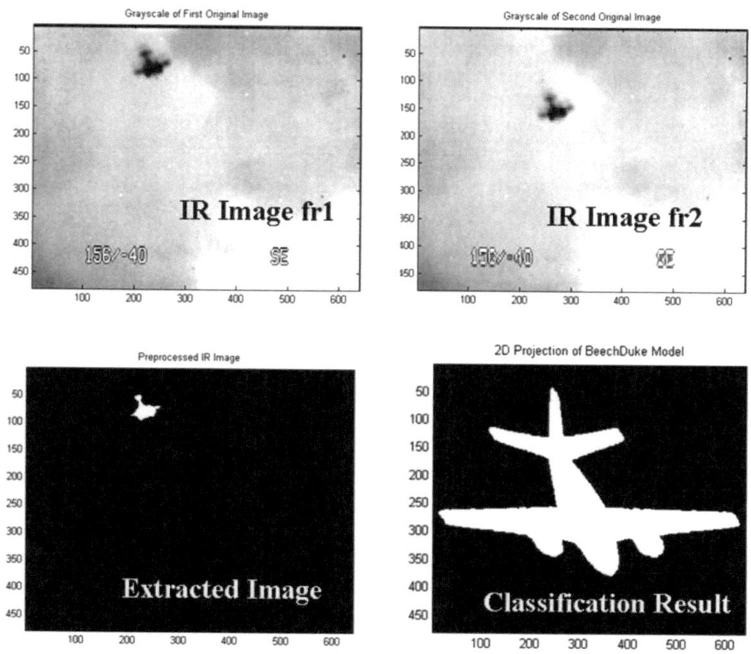

Fig. 17. An image classification example shows a pair of collected IR image frames (upper), the extracted target silhouette (lower left) and the classification result (lower right).

7. Conclusion

Small, low-flying airborne vehicles may pose an imminent threat to homeland security and border integrity. Using a forward-based fence that contains a mix of low cost, low power radar, acoustic and optical (Infrared and visible) sensors by appropriate sensor fusion methodologies it is feasible to detect, track and discriminate small, low flying airborne targets and provide 24/7 sentry functions to protect critical civilian and military infrastructure. We have demonstrated the technical feasibility of the netted sensor fence approach. A proof-of-concept initial experimental prototype has been built and tested using COTs components. The technology is highly modular by modality, and adaptable to potential customer needs and requirements.

8. Acknowledgment

The authors thank Greg Crawford, Ron Fante, Chris Bas, Jeff Atwood, Bryan George, Garry Jacyna, Mike Jeffris, Walter Kuklinski, Tim Nadeau, Michael Otero, Dennis Reeves, Lucien Teig, and Stephen Theophanis for their contributions to this project.

9. References

Skolnik, M., Radar Handbook (Second Edition) ,Chapter 2,McGraw-Hill(New York) 1990

Ferguson BG, Lo KW., "Turboprop and rotary-wing aircraft flight parameter estimation using both narrow-band and broadband passive acoustic signal-processing methods", J Acoust Soc Am. 2000 Oct;108(4):1763-71

T. Pham, N. Srour, "TTCP AG-6: Acoustic Detection and Tracking of UAVs", Proc. Of SPIE Vol. 5417, 24-30

MODTRAN code, Air Force Research Laboratory, URL site : http://www.vs.afrl.af.mil/Division/VSBYB/modtran4.html

R.H. Kingston "Detection of Optical and Infrared Radiation" by, Springer-Verlag, Pages 15-17, 1978

James D. Howe, "Electro-Optical Imaging System Performance Prediction", Volume 4, Infrared and Electro-Optical Systems Handbook, Page 85 SPIE Engineering Press, Second Printing (1996)

Anderson, Brian D.O. and Moore, John B., Optimal Filtering, Prentice-Hall, Inc., Englewood Cliffs, NJ, 1979.

Grewal, Mohinder S., and Andrews, Angus P., Kalman Filtering: Theory and Practice Using MATLAB, John Wiley & Sons, Inc., 2001.

Laura Keyes, Adam Winstanley, "APPLYING COMPUTER VISION TECHNIQUES TO TOPOGRAPHIC OBJECTS", IAPRS, Vol. XXXIII, Amsterdam, 2000

S. A. Dudani, K. J. Breeding, and R. B Mcghee, "Aircraft Identification by Moments Invariants", IEEE Transactions on Computers, Vol. C-26, No. 1 1977

M. K. Hu, "Visual Pattern Recognition by Moment Invariants", IRE Trans. Inform. Theory, Vol. IT-8, 179-187, Feb. 1962

9

Design, Implementation and Evaluation of a Multimodal Sensor System Integrated Into an Airplane Seat

Bert Arnrich, Cornelia Kappeler-Setz,
Johannes Schumm and Gerhard Tröster
ETH Zurich, Wearable Computing Lab
Switzerland

1. Introduction

Air travel has become the preferred mode of long-distance transportation for most of the world's travelers. People of every age group and health status are travelling by airplane and thus the airplane has become part of our environment, in which passengers could benefit from assistive support. In this regard, the European research project SEAT has investigated sensor technologies to provide assistive support related to the health and well-being of airplane passengers. Since the main interaction point between a passenger and the airplane is the seat, a seat-integrated sensor system was developed to measure health and affect-related signals of a passenger. The measured signals include the electrocardiogram (ECG), electrodermal activity (EDA), skin temperature, respiration as well as movement of the passenger. In this chapter we describe the design, implementation and evaluation of the seat-integrated sensor system. In particular we highlight two approaches of sensor fusion in order to appraise the signal quality in an airplane scenario and to identify the passenger's affective state.

In the first part, we show how the design of the seat-integrated sensor system is influenced by the trade-off between sensor comfort and signal quality: To achieve the acceptance and hence the use of the system, the sensors need to be attached in a comfortable and non-obtrusive way or even be totally integrated into the seat. On the other hand, a comfort-optimized sensor placement usually limits the signal quality. We argue that not only the development of comfortable and reliable sensor technology but also the quality appraisal of the data generated by the sensors needs to be addressed. Artifact detection through sensor fusion is presented and the working principle is shown in a feasibility study, in which normal passenger activities were performed. Based on the presented method, we are able to identify signal regions in which the accuracies for detecting the heart-rate is 88% compared to 40% without any artifact removal [Schumm et al., 2010].

In the second part, we will explain another sensor fusion approach in the context of emotion recognition. Previous work on emotion recognition from physiology has rarely addressed the problem of missing data. However, data loss due to artifacts is a frequent phenomenon in practical applications. Discarding the whole data instance if only a part is corrupted results in a substantial loss of data. To address this problem, we investigated two methods for handling missing data: imputation and reduced-feature models using ensemble classifier systems. The five emotions amusement, anger, contentment, neutral and sadness were elicited in 20 subjects by film clips. The following six physiological signals were recorded by means of the seat-integrated sensor system and an external recording device: ECG, electromyogram (EMG), electrooculogram (EOG), EDA, respiration and finger temperature. Results show that classifier fusion increases the recognition accuracy in comparison to single classifiers using imputation by up to 14%. We were able to analyze 100% of the data even though only 47% of the data was artifact free. Since more artifacts are expected in the "field" than in the laboratory, the proposed methods are especially beneficial for practical applications, e.g. in the airplane [Setz et al., 2009].

2. Design and implementation of a seat-integrated multimodal sensor system

Since air travel has become the preferred mode of long-distance transportation, the airplane will soon be part of our environment in which people may need assistive support. The European research project SEAT aims at extending existing airplane seats with new sensor technologies in order to assess and improve health and well-being of the passengers. A first step in this direction is the reliable and unobtrusive recording of relevant physiological signals. In order to achieve the acceptance and hence the use of the system, the sensors need to be attached in a non-obtrusive way or even be totally integrated into the environment. Thus, conventional sensors and measurement locations, such as wet electrodes at the chest for cardiovascular monitoring, are not feasible due to the lack of acceptance by the passengers.

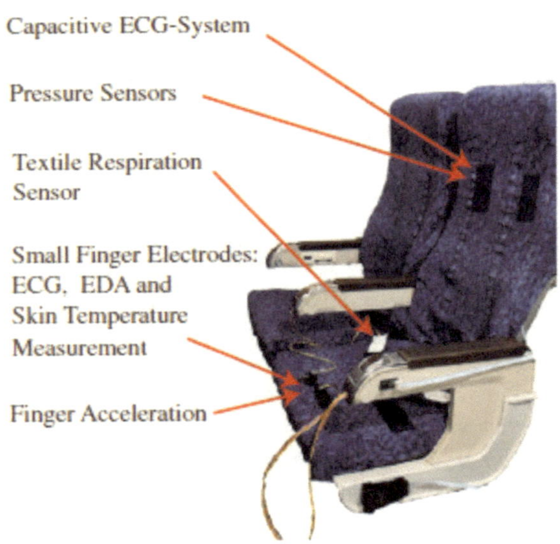

Fig. 1. Airplane seat with integrated unobtrusive sensors [Schumm et al., 2010]

In order to record health and affect-related signals of a passenger in an unobtrusive way, the following signal modalities were measured by sensors integrated into an airplane seat: electrocardiogram (ECG), electrodermal activity (EDA), skin temperature and respiration (see Figure 1). The comfort aspect was targeted by (i) incorporating a contact-less ECG measurement system; (ii) using dry electrodes at the fingers instead of wet electrodes at the chest; (iii) combining ECG, EDA and temperature sensors into one setup in order to reduce the number of electrodes; and (iv) incorporating the respiration measurement into the safety belt of the airplane seat. A main concern of the system design was to make sure that signal disturbances can be detected. Since passenger's movements induce signal disturbances, additional sensors for measuring the movements of the fingers and the contact pressure at the back rest were integrated into the seat. These additional sensors, referred to as artifact sensors, are able to measure movement patterns that provoke artifacts, which influence the signal quality of the physiological signals.

2.1 Sensor system

For the ECG measurement, two measurement systems were incorporated into the seat. The first system consists of a contactless capacitive ECG system developed by RWTH Aachen University [Steffen et al., 2007]. This system, referred to as "Contactless-ECG", allows measuring the ECG as unobtrusively as possible. It was incorporated in the backrest of the airplane seat and measures the ECG capacitively without direct skin contact. Since the Contactless-ECG system is sensitive to body movements, additional pressure sensors were incorporated into the backrest of the seat in order to measure the contact pressure between the body and the backrest. In order to find adequate locations for the pressure sensors, a feasibility study was conducted. The contact pressure at the back of the seat was recorded with a pressure mat, simultaneously with the contactless ECG signal. Based on a visual inspection of both signals, it was decided to place one pressure sensor at the top and another one at the bottom of each ECG electrodes. The second ECG system, referred to as "Finger-ECG", measures the ECG at the index finger of both hands. The used dry electrodes that are fixed to the fingers provide a higher user comfort compared to wet electrodes attached to the chest. Due to the direct skin contact, this system is clearly more obtrusive than the "Contactless-ECG" but also more reliable. However, finger movements can also evoke artifacts in the ECG signal. In order to spot these finger movements, the finger stripes for fixating the electrodes are equipped with 3-axis accelerometer sensors.

Since the EDA measurement requires direct skin contact, the EDA is recorded at the index and middle finger as proposed in literature [Boucsein, 1992]. The implemented measurement principle is referred to as an exosomatic quasi constant voltage method. Hereby, a constant voltage is applied to the electrode at the index finger, leading to a current flowing through the skin to the other electrode. This current is measured and thereby the

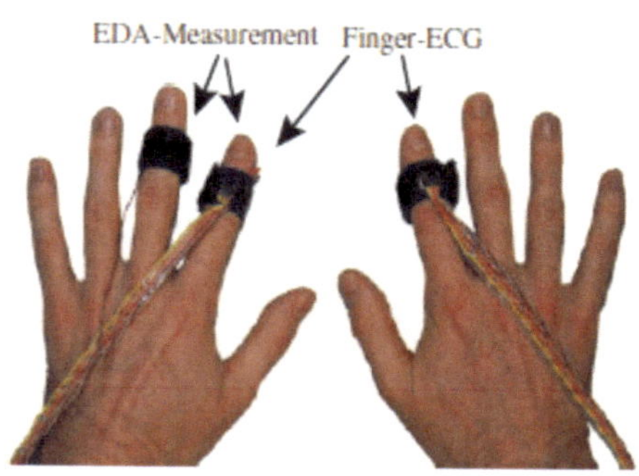

Fig. 2. Combining the measurement of EDA and ECG at the fingers [Schumm et al., 2010]

skin conductance can be assessed. In order to reduce the overall number of electrodes attached to the fingers, a novel concept for combining the measurement of EDA and ECG at the fingers was developed: the electrode at the index finger of the left hand was used for both measurements, ECG and EDA. As a result, the number of electrodes could be decreased from four to three (see Figure 2). For the ECG measurement, the high offset voltage at the left index finger, caused by the parallel EDA measurement, had to be considered.
For the measurement of the respiration, common approaches require a belt strapped around the chest. In our setting, a textile resistive sensor was incorporated in the seatbelt. During the breathing cycle, the sensor is periodically stretched and the resulting resistance change is measured.
For the measurement of the skin temperature, a commercially available sensor was chosen. Due to the small size of the sensor (5mm x 9.5mm x 9.1mm), it was incorporated beside the left electrode of the EDA measurement and does therefore not further decrease the perceived comfort.

2.2 Artifact detection through sensor fusion

Artifacts lead to signal disturbances and thereby might lead to a wrong interpretation of extracted features from the physiological signals. The proposed multi-modal sensor system allows artifact detection. The artifact sensors are used to spot regions of artifacts and consequently the corrupted sequences are not considered for further processing. In Fig. 3 the working principle is shown.

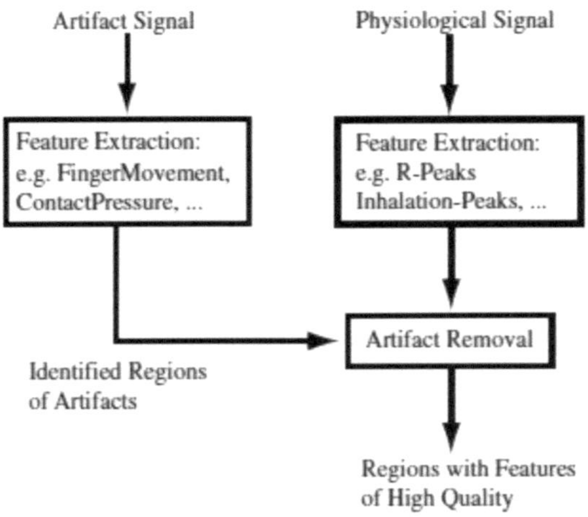

Fig. 3. Working principle of artifact removal through sensor fusion [Schumm et al., 2010]

For the Contactless-ECG and the respiration signals, movements of the upper body lead to artifacts. The Contactless-ECG system delivers good signal quality as long as a constant contact pressure exists between the upper body part and the back of the seat. Movements of a person like leaning forward change the elongation of the respiration sensor without being related to breathing. This information can be measured using pressure sensors incorporated into the back of the seat. By summing up the four pressure signals, a single artifact signal referred to as "contact pressure" is obtained. If no contact or fast changes due to movements of the upper body signal are detected, the signal is marked as corrupted. In

Figure 4, a signal example of the contactless ECG and the contact pressure is shown. It can be observed that changes in the contact pressure coincide with disturbances in the contactless ECG signal. In particular, it can be seen that the R-peaks, which are the most characteristic features of the ECG signal, are no longer visible during periods of varying contact pressure. As a result, it would not be possible to calculate the heart rate in those periods.

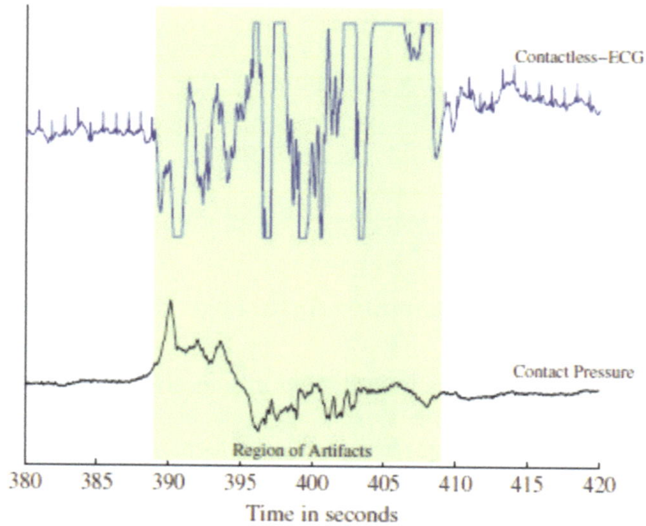

Fig. 4. Changes in the contact pressure due to upper body movements lead to a temporary disturbance of the contactless ECG signal [Schumm et al., 2010]

For the EDA and ECG measurements at the fingers, movements of the fingers or the hand can evoke artifacts. In order to spot these finger movements, the finger stripes of both index fingers are equipped with 3-axis accelerometers. The three dimensions of both accelerometers are added up to form a single artifact signal referred to as "finger movement". In Figure 5, a signal example of the finger ECG and the corresponding artifact signal is shown. It can be observed that the ECG signal is corrupted in regions with strong finger movements. Similar to the contactless ECG, it can be seen that the R-peaks are no longer visible during periods of movements.

In order to evaluate the performance of the proposed artifact detection through sensor fusion, we first define a quality measure to appraise the signal quality before and after artifact removal. The Quality Index (QI) describes the correctness of a characteristic feature extracted from the ECG. As already mentioned, the most characteristic feature of the ECG is the R-peak. We therefore compare the R-peaks extracted from the contactless ECG with the R-peaks extracted from a simultaneously recorded ground truth ECG signal. The ground truth signal was recorded from the chest using wet electrodes. For each detected R-peak in the ground truth ECG, we searched for a single corresponding R-peak in the contactless ECG, within a time interval of 150ms. The overall quality measure is defined as the ratio of correctly identified R-peaks in the contactless ECG divided by the total number of R-peaks detected in the ground truth ECG signal.

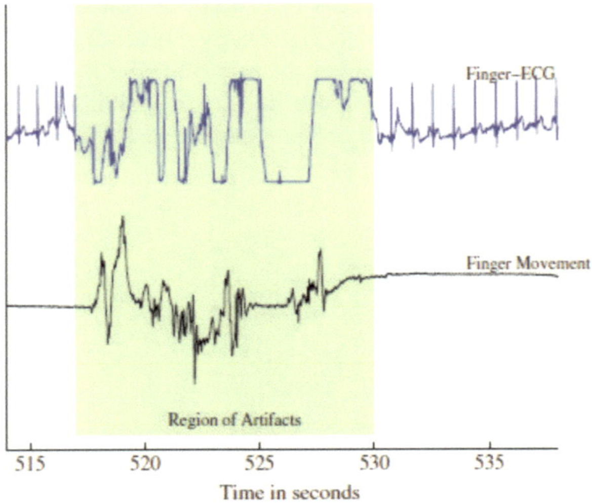

Fig. 5. Movements of the fingers evoke artifacts in the finger ECG measurement [Schumm et al., 2010]

The benefits of artifact removal were investigated in a feasibility study. A test subject was sitting in the seat while performing typical airplane activities such as entertainment, working, reading, sleeping and eating. Each activity was performed for 10 minutes. During these activities, the contactless ECG and the ground truth ECG were measured synchronously. After data recording, the quality measure was calculated before and after artifact removal. The resulting quality measures are presented in Table 1. First, it can be observed that the quality measures are substantially higher for the calm activities (entertainment and sleeping) in comparison to the more active activities (working, reading and eating). This reflects an inherent constraint of the contactless ECG system: it only delivers good signal quality if a good and constant contact pressure exists between the upper part of the body and the back of the seat. Second, it can be observed that regions with high activity are almost totally discarded by the artifact removal method: for working and reading all the data is discarded whereas for reading the remaining signal length is only 7%.

	Entertainment (%)	Working (%)	Reading (%)	Sleeping (%)	Eating (%)
Quality measure before artifact removal	71	16	32	63	17
Quality measure after artifact removal	88	-	47	97	-
Remaining signal length	38	0	7	35	0

Table 1. Quality measure before and after artifact removal for typical airplane activities

Taking all activities together, almost 84% of the data is discarded while the mean quality measure (weighted by the remaining signal length) rises from 40% to 88%. If the remaining signal length found in our study is scaled up to a 12h flight, 2.8h of the Contactless-ECG data is expected to be almost artifact free and would be available for computation of health indicators.

3. Sensor fusion in emotion recognition

Applications for emotion recognition are predominantly found in the field of Human-Computer Interaction (HCI). By including emotions, HCI shall become more natural, i.e. more similar to human-human interactions where information is not only transmitted by the semantic content of words but also by emotional signaling in prosody, facial expression and gesture. In recent years several research groups have employed pattern recognition methods in order to automatically detect different affective states of a subject. Modalities which have been used to detect affective states include facial expression [Busso et al., 2004], speech [Neiberg et al., 2006] and physiological signals [Kim & André, 2008].
Previous work on emotion recognition from physiology has rarely addressed the problem of missing sensor data. In general, a multi-modal data set is recorded, e.g. simultaneous recordings of ECG, EDA and respiration. If a single modality fails, the entire data instance, i.e. all the remaining signal modalities, are usually discarded. This results in a substantial amount of unusable data for classifier training. Moreover, predicting emotions during run-time becomes impossible, if any of the signal modalities that were used to train the classifier fails. Since data loss due to artifacts is frequently encountered not only in our airplane scenario but also in many practical applications, missing values represent a serious problem that needs to be addressed but has so far gained little attention in emotion recognition from physiology. We therefore investigated two methods for handling missing data: imputation and reduced-feature models using ensemble classifier systems.

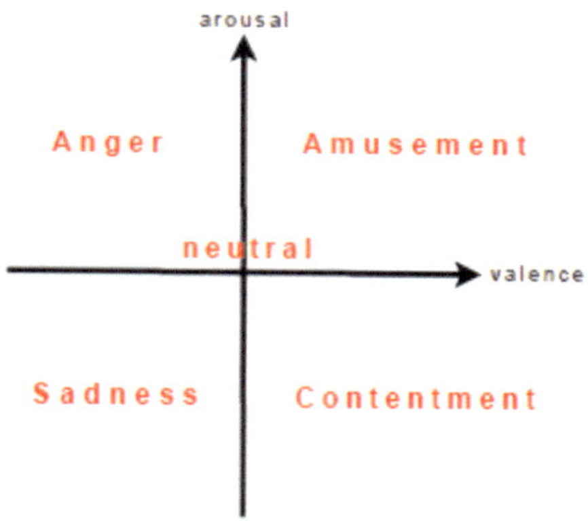

Fig. 6. Emotions to be recognized in the 2-dimensional emotion model of arousal and valence [Setz et al., 2009]

3.1 Emotion elicitation

The emotions to be recognized were chosen according to the well-known 2-dimensional emotion model of arousal and valence often used in emotion recognition studies [Cowie et al., 2001]. One emotion in each quadrant plus neutral were selected as shown in Figure 6: amusement (high arousal, positive valence), anger (high arousal, negative valence), contentment (low arousal, positive valence), sadness (low arousal, negative valence) and neutral (medium arousal, zero valence).
For each emotion and the neutral state one film clip has been chosen since films are capable of eliciting strong emotional responses under highly standardized conditions which enables replication studies [Rottenberg et al., 2007]. Twenty (12 male, 8 female) participants with a

mean age of 28.6 years were recruited for the experiment. After sensor attachment the subjects had 10 minutes to rest. Afterwards, the five film clips were presented to each subject in a constant order. Between two film clips a recovery time of 3 minutes was introduced. During the experiment the following physiological signals were recorded with the seat-integrated sensor system explained above: ECG, EDA, respiration and finger temperature (see Figure 7). In addition, two sensor modalities were simultaneous recorded with an external device: vertical component of the Electrooculogram (EOG) and the Electromyogram (EMG) of the muscle between mouth and eye (see Figure 8).

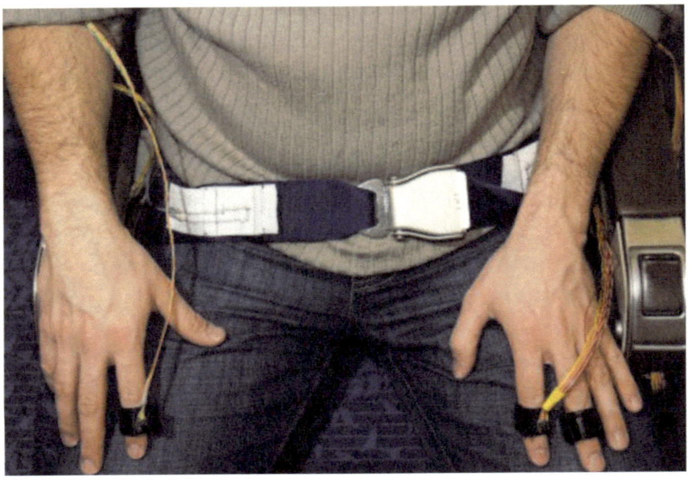

Fig. 7. ECG and EDA measurements at the hand, respiration measurement in the seat belt [Setz et al., 2009]

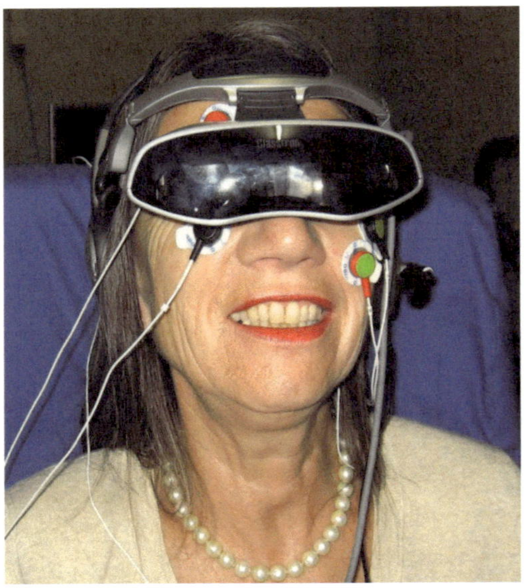

Fig. 8. Subject wearing video glasses to watch emotion eliciting film clips. Vertical EOG and EMG measurement [Schumm et al., 2010].

3.2 Emotion recognition

From all six physiological signals, a total of 53 features were extracted. In a first step, the following erroneous signal modalities were detected: (i) EDA signal reached saturation, (ii) erroneous R-peak detection from the ECG due to motion artifacts, and (iii) eye blinking is not visible in the EOG due to dry skin. This resulted in 96% correct data for EOG, 78% for EDA, 64% for ECG and 100% for the remaining modalities. The percentage of data containing only valid features of all modalities amounted to 47%. In order to tackle the low amount of data containing only valid features, we investigated imputation of missing data in combination with two classifier fusion approaches as outlined in the following.

Single classifier with imputed features: Missing feature values in the training data are imputed by the mean value of the available training instances belonging to the same emotion class. All signal modalities are used to train a classifier in order to discriminate the emotion classes. Missing features in the test data are imputed by the mean value of the entire training feature vector independent of the class membership of the test sample.

Fusion of classifiers with imputed features: Missing feature values are imputed in the same way as for the single classifier. For each signal modality, a separate classifier is trained. The decisions of those classifiers are fused according to majority or confidence voting. In majority voting, the class that receives the highest number of votes is chosen. In confidence voting, the class of the classifier with the highest confidence is chosen.

Fusion of classifiers with reduced-feature models: Missing feature values are not imputed. Like above, a separate classifier is trained for each signal modality. In case a missing feature is encountered in a certain signal modality, this modality is not used for training the classifier. In the test set decisions of all the available classifiers are fused according to majority or confidence voting.

Fusion Method	Classifier	Imputation	Accuracy 5 classes	Accuracy Arousal	Accuracy Valence
None	LDA	Yes	45.0	62.5	71.3
Majority	LDA	Yes	41.0	62.5	68.8
Confidence	LDA	Yes	50.0	63.8	72.5
Majority	LDA	No	41.0	67.5	70.0
Confidence	LDA	No	47.0	61.3	73.8
None	QDA	Yes	35.0	57.5	60.0
Majority	QDA	Yes	49.0	58.8	63.8
Confidence	QDA	Yes	45.0	60.0	63.8
Majority	QDA	No	49.0	58.8	68.8
Confidence	QDA	No	48.0	60.0	68.8

Table 2. Comparison of classifier fusion methods and single classifiers with and without imputation for LDA and QDA.

As classifiers, Linear and Quadratic Discriminant Analysis (LDA and QDA) were investigated. The performance of the classifier fusion methods in comparison to single classifiers with and without missing value imputation are shown in Table 2.
It can be observed that classifier fusion always yields a considerable benefit in comparison to single classifiers using all features (with imputation) for discriminating the 5 emotion classes. A maximum increase in accuracy of 14% was observed when comparing an QDA ensemble classifier system to a single classifier using imputation. When comparing the ensemble classifiers using imputation with the corresponding ones using reduced-feature models, the reduced-feature model ensembles perform once better, twice equally and once worse. Based on these result, it is difficult to decide whether to use imputation or not. The two strategies seem to be competitive. However, using no imputation is computationally less expensive and might be preferred in practical applications. Another interesting observation is that confidence voting always performs better then majority voting for LDA, while the tendency is reversed for QDA. The results for discriminating arousal and valence again indicate that classifier fusion outperforms single classifiers which use all the features (with imputation). When comparing the ensemble classifiers using imputation with the corresponding ones employing reduced-feature models, the reduced-feature model ensembles perform once better, once worse and twice equal for arousal. Considering valence, reduced-feature models yield better results in all four cases.

4. Conclusion

Due to the evolution of long-distance transportation, the European Commission has identified the airplane as part of our environment in which passengers could benefit from assistive support. The European research project SEAT aims at extending existing airplane seats with new sensor technologies in order to assess and improve health and well-being of the passengers. A first step in this direction is the reliable and unobtrusive recording of relevant physiological signals. In the first part of this chapter we have presented our approach to incorporate sensor technology into an airplane seat aiming at unobtrusively measuring physiological signals. The trade-off between passenger's comfort and signal quality was identified to be an important issue to achieve the acceptance and hence the use of the system. The proposed multi-modal sensor system allows automatic artifact detection of physiological signals through sensor fusion. For validation purposes, we proposed a quality measure that appraises the signal of interest based on a ground truth signal. In a feasibility study the benefits of artifact removal were investigated. During typical airplane activities, the contactless ECG and the ground truth ECG were measured synchronously and the quality measure was calculated before and after artifact removal. It could be shown that the quality measures are substantially higher for the calm activities in comparison to the more active activities. Taking all activities together the mean quality measure rises from 40% to 88%. If the remaining signal length after artifact removal is scaled up to a 12h flight, 2.8h of the contactless ECG data is expected to be almost artifact free and would be available for computation of health indicators.
Since data loss due to artifacts is frequently encountered not only in our airplane scenario but also in many practical applications, in the second part of this chapter we have investigated two methods for handling missing data: imputation and reduced-feature

models using ensemble classifier systems. In our emotion experiment, more than half of the data would have been lost if no strategy to handle missing values had been employed. With the proposed methods we were able to analyze 100% of the data. Classifier fusion has shown to substantially increase the recognition accuracies. A maximum increase in accuracy of 14% was observed when comparing an ensemble classifier system to a single classifier using imputation. Whether majority or confidence voting performs better depends on the underlying classifier.

5. Acknowledgment

This project is partly funded by the EU research project SEAT (http://www.seat-project.org), contract number: 030958. All views here reflect the author's opinion and not that of the European Commission.

6. References

[Boucsein, 1992] Boucsein, W. 1992. *Electrodermal Activity*. Plenum Press, New York.

[Busso *et al.*, 2004] Busso; Carlos; Deng, Zhigang; Yildirim; Serdar, Bulut; Murtaza; Lee; Chul Min; Kazemzadeh; Abe; Lee; Sungbok; Neumann; Ulrich; & Narayanan; Shrikanth. 2004. Analysis of emotion recognition using facial expressions, speech and multimodal information. *Pages 205–211 of: ICMI '04: Proceedings of the 6th international conference on Multimodal interfaces*. New York, NY, USA: ACM.

[Cowie *et al.*, 2001] Cowie; R., Douglas-Cowie; E., Tsapatsoulis; N., Votsis; G., Kollias; S., Fellenz; W., & Taylor, J.G. 2001. Emotion recognition in human-computer interaction. *Signal Processing Magazine, IEEE*, 18(1), 32 -80.

[Kim & André, 2008] Kim, Jonghwa; & André, Elisabeth. 2008. Emotion Recognition Based on Physiological Changes in Music Listening. *IEEE Trans. Pattern Anal. Mach. Intell.*, 30(12), 2067-2083.

[Neiberg *et al.*, 2006] Neiberg, Daniel, Elenius, Kjell, & Laskowski, Kornel. 2006. Emotion Recognition in Spontaneous Speech Using GMMs. *In: Proceedings of the 9th ISCA International Conference on Spoken Language Processing (Interspeech)*.

[Rottenberg *et al.*, 2007] Rottenberg, J.; Ray, R. R.; & Gross, J. J. 2007. *The handbook of emotion elicitation and assessment*. New York: Oxford University Press. Chap. Emotion elicitation using films, page 9-28.

[Schumm *et al.*, 2010] Schumm, Johannes; Setz, Cornelia; Bächlin, Marc; Bächler, Marcel; Arnrich, Bert, & Tröster, Gerhard. 2010. Unobtrusive Physiological Monitoring in an Airplane Seat. *Personal and Ubiquitous Computing*, 14(6), 541-550.

[Steffen *et al.*, 2007] Steffen, M.; Aleksandrowicz, A., & Leonhardt, S. 2007. Mobile Noncontact Monitoring of Heart and Lung Activity. *Biomedical Circuits and Systems, IEEE Transactions on*, 1(4), 250 -257.

[Setz et al., 2009] Setz, C.; Schumm, J.; Lorenz, C.; Arnrich, B. & Tröster, G. Using Ensemble Classifier Systems for Handling Missing Data in Emotion Recognition from Physiology: One Step Towards a Practical System. Proceedings 2009 International Conference on Affective Computing & Intelligent Interaction ACII 2009, 2009, 187-194

10

Sensor Fusion-Based Activity Recognition for Parkinson Patients

Majid Bahrepour, Nirvana Meratnia,
Zahra Taghikhaki and Paul J. M. Havinga
Pervasive Systems Group, University of Twente,
Netherlands

1. Introduction

Parkinson disease (PD) is a slow destructive disorder of the central nervous system in which dopamine, i.e., catecholamine neurotransmitter in the central nervous system is lost. PD hurts patients' movement and speech ability. Sometimes, it can also affect patients' mood, behavior, and thinking ability. Falling down is a common problem in PD patients and on time fall detection is important to assist PD patients and prevent them from being injured. To this end, being able to correctly distinguish various activities, e.g. walking, sitting, standing still, is a must. To monitor activities and moving patterns of PD patients, a wireless body sensor network (BSN) may prove to be useful. By attaching various wireless sensor nodes on the body of PD patients or integrating them into their shoes or cloths, their activities and physiological conditions can be checked regularly and an alarm can be generated in case of emergency or need for additional assistances.

A wireless body sensor network consists of a number of wireless sensor nodes that cooperatively monitor physical (e.g. motion) and physiological (e.g. heart rate) conditions of a person. In addition to sensors, each sensor node is typically equipped with a radio transceiver or other wireless communication devices, a small microcontroller as processing unit, and an energy source in a form of a battery. Sensor nodes may vary in size and type of sensors they are equipped with. Size and cost constraints on sensor nodes cause limitations on their resources in terms of energy, memory, and computational processing. Figure 1 shows an example of a body sensor network.

Previous studies for activity recognition of PD patients mostly use accelerometer and occasionally gyroscope sensors attached to various parts of patients' body (JJ., HA. et al. 1991; Aminian, Robert et al. 1999; JI., AA. et al. 2001; JI., V. et al. 2004; N., T. et al. 2004; White, Wagenaar et al. 2006; Moorea, MacDougalla et al. 2007; Salarian, Russmann et al. 2007). One of the main criticisms on the previous studies is that they use centralized techniques which not only require expensive equipments to monitor physiological conditions and activities of patients [e.g. Vitaport 3 (White, Wagenaar et al. 2006)] but also introduce delays in the detection process. Also due to having a single point of failure they are more prone to failures and crashes. In contrary, we propose a fusion-based distributed algorithm which can be easily implemented on resource constrained wireless sensor nodes and detect and distinguish activities in (near) real-time. Our approach offers three main advantages: (i) distributed processing and reasoning which decreases the data processing time and provides fast activity detection and classification, (ii) robustness against sensors failure, and (iii) accurate detection through use of sensor fusion.

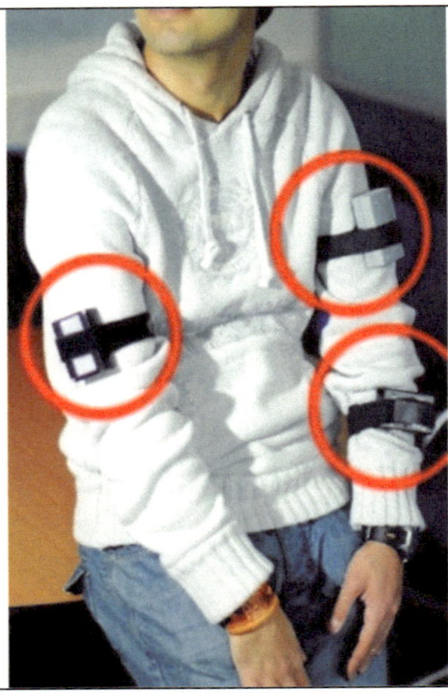

Fig. 1. An example of a body sensor network.

Our approach is based on implementing classification techniques on wireless sensor nodes attached to patients' body, online evaluation of the classification results on individual nodes, and fusing results of various nodes to resolve possible conflicts between sensor nodes and reach a consensus. Previously in (Bahrepour, Meratnia et al. 2009; Bahrepour, Meratnia et al. 2009; Bahrepour, Zhang et al. 2009; Bahrepour, Meratnia et al. 2010; Bahrepour, Meratnia et al. 2010; Bahrepour, van der Zwaag et al. 2010), we have shown capability of machine learning based classification techniques in distributed detection of environmental events such as fire. There is no reason to believe that classification techniques used in other domains are not applicable for medical domain. The challenge here, however, is twofold: (i) investigating capabilities of these classification techniques for a more complex data such as activity data, and (ii) being able to determine the most relevant sensor data among a large number of sensor types for a specific purpose, e.g., fall detection of PD patients. In what follows we address both challenges.

The organization of this paper is as follows. Section 2 provides an overview of the related work. Section 3 introduces the machine learning based classification and sensor fusion techniques that will be used in this study for the purpose of activity recognition. In section 4, our processing models are introduced, which will be followed by our proposed approach explained in Section 5. Section 6 describes the dataset and presents experimental results. Conclusions are drawn in Section 7 and the obtained results are discussed.

2. Related work

Related work on use of wireless body sensor network in medical domain can be generally classified into three groups:

i. those related to hardware platform design [e.g. (Kern, Schiele et al. 2003; Park, Liu et al. 2005; Lorincz, Kuris et al. 2007; Ying, Schlösser et al. 2008; Nabar, Banerjee et al. 2010)];
ii. those related to activity recognition methods and algorithms (Aminian, Robert et al. 1999; Veld, M.H.A. et al. 2005; Moorea, MacDougalla et al. 2007; Osmani, Balasubramaniam et al. 2007; Salarian, Russmann et al. 2007; Osmani, Balasubramaniam et al. 2008; Lee, Kim et al. 2009; Khattak, Vinh et al. 2010);
iii. those related to making the information flow between patients and medical team more efficient [e.g. (Centeno, Giachetti et al. 2003; Martinez-Garcia and Menndez-Olague 2003; Wijewickrama and Takakuwa 2006; Sanchez, Tentori et al. 2008)].

While the first group focuses on design and implementation of small, easy to wear, and cheap sensor node platforms equipped with a set of sensors, the second group focuses on design of computationally light signal processing, feature extraction, classification, and pattern recognition techniques. The focus of the third group is on providing timely information to the doctors in charge, reducing patients waiting time, and improving interaction between doctors and patients, etc.

The main objectives of performing activity recognition in medical domain are:

- Prevention (e.g., (Steele, Belza et al. 2003; Wu, Bui et al. 2008; Le and Pan 2009; Benocci, Tacconi et al. 2010)
- Rehabilitation (e.g., (Jarochowski, Shin et al. 2007; Soini, Nummela et al. 2008; Zhang and Sawchuk 2009)
- Assistance and care giving (e.g., (Hou, Wang et al. 2007; S. Bosch 2009)

To perform activity recognition usually the following steps are taken (Krishnan, Juillard et al. 2009; Avci, Bosch et al. 2010; Horst and Meratnia 2011):

- Sampling, which refers to the process of taking measurements from the body sensors. Success of this process heavily depends on robustness of the sensor node platform, radio communication, as well as availability of accurate timing.
- Preprocessing, which involves refining the sensor data. More specifically this step deals with noise reduction through for example filtering (Yun, Lizarraga et al. 2003) as well as signal transformation through use of for example Fourier transformation (Wu, Pan et al. 2009).
- Segmentation, which is the process of identification of beginning and end of an important feature in the sensor data stream. Segmentation can be performed manually, e.g., (Jafari, Li et al. 2007) or automatically, e.g., (Guenterberg, Bajcsy et al. 2007; Guenterberg, Ostadabbas et al. 2009).
- Feature extraction, classification, and pattern recognition, which refer to the process of giving semantic to the identified feature and identification of repeated patterns and trends. During feature extraction, time-domain, frequency-domain, or time-frequency domain features can be identified. Techniques such as neural network, Bayes network, hidden markov model, support vector machine, and decision trees are often used for the classification purpose.
- Pattern matching, which deals with comparing the identified patterns and features either with those identified on other nodes or with pre-defined templates.
- Feedback, which is the process of taking appropriate action based on output of classification.

For a more elaborate overview of state of the art on use of wireless body sensor networks in medical domain, reader is referred to (Abbate, Avvenuti et al. 2010; Avci, Bosch et al. 2010; Horst and Meratnia 2011).

3. Classification and fusion techniques

For the purpose of activity recognition and classification, we use Feed Forward Neural Network (FFNN), Naïve Bayes (NB), and Decision Tree (DT). For the purpose of sensor fusion, we use reputation-based voting and majority voting. In this section, we first provide a short explanation of the concepts used by these techniques.

3.1 Feed forward neural network (FFNN)

Feed forward neural network (FFNN) is a type of the neural network, in which each layer is fed by its back layer (Mehrotra, Mohan et al. 1996). FFNN consists of one input layer, one or more hidden layers and one output layer. The challenge faced in using FFNN as a classifier is finding correct weights. The process of finding these weights, which is called 'learning', can be carried out using algorithms such as gradient descent (GD) algorithm (Wikipedia ; ALPAYDIN 2004).

3.2 Naïve bayes classifier (NB)

A Naïve Bayes classifier uses Bayesian statistics and Bayes theorem to find the probability of each instance belonging to a specific class. It is called Naïve because of its emphasis on independency of the assumptions. Naïve Bayes classifier finds the probability of belongingness of each instant to a specific class.

3.3 Decision tress

A decision tree is a learning algorithm that uses tree-like graphs to model and evaluate discrete functions (Russell and Norvig 2003; ALPAYDIN 2004). Construction of a decision tree for classification purpose requires a training phase (ALPAYDIN 2004), which employs a set of data and a learning algorithm to find a minimum depth decision tree. The tree should contain the minimum required nodes (or minimum depth) to reduce time and memory complexities. Therefore, the training algorithm is usually a local search greedy algorithm to find an optimum decision tree (ALPAYDIN 2004). Once the decision tree is created, the tree can be used by evaluating nodes from the root down to the leaves. The final leaf contains a value that shows the result of the classification. After the decision tree is constructed in the learning phase, it can be pruned to save memory. Pruning is only required for large trees to alleviate computational complexity.

3.4 Reputation-based voting

Reputation based voting is a fusion technique that happens after classification. Once each classifying entity makes its individual decision about belogness of an instance to a class, a consensus needs to be reached among the classifying entities. Reputation-based voting approaches are based on finding reputation of individual classifying entity and choosing the decision made by the classifying entities having the highest reputation. Assuming that classifying entities have correctly classified instances, they should judge how well the other entities have performed the classification. To do the judgment, each entity first sends its classification result, called Detection Value (DV), to all other entities in its neighborhood. The DVs received from the neighbors will be stored in a table called Neighbors Detection Value Table (NDVT). In the next step, each entity should judge about its neighboring entity by considering itself as the reference. The judgment is accomplished by comparing the difference between value of entity itself and value of the other neighboring classifying entities. If the difference is less than a threshold value θ (which is chosen based on the context), the "judging" entity gives a positive vote ($V^{new} = V^{old} + 1$) to the other entities. Otherwise, the "being judged" entity receives a negative vote ($V^{new} = V^{old} - 1$). Finally, NDVT tables are sent to a voter to reach a consensus among different opinions. The challenging part of reputation-based voting is how to assign a global reputation value to each entity in order to choose high reputed entity and its classification result. There are various ways to do so, two of which are explained here:

3.4.1 Reputation technique 1

This reputation technique checks local reputation of every individual entity from the other entities' perspective. The local reputation value is obtained based on average value of V_i (positive or negative votes which were given by the other entities) for each classifying entity. Then, the average local reputation is multiplied by the weight of sensor nodes calculated using Equation 1 to assign global reputation values. The class with the highest reputation weight (W) is the result of the voting procedure. Equation 1 shows how the weights are calculated.

$$W_i = R_i \times Acc_i \tag{1}$$

where W_i is the reputation value corresponding to classifying entity i, R_i is the local reputation value of classifying entity i from other entities' perspective, and Acc_i is weight of classifying entity *i* (Bahrepour, Meratnia et al. 2010).

3.4.2 Reputation technique 2

In this reputation technique, two threshold values, i.e., θ_1, θ_2 are used. Comparing the local reputation value (R_i) with θ_1 and θ_2 gives an insight about how well classification is performed. If ($R_i \geq \theta_1$), then classifying entities have made perfect decisions, if ($\theta_1 > R_i \geq \theta_2$) then classifying entities have made OK decisions, and if ($\theta_2 \geq R_i$) then classifying entities have made poor decisions. We assign 0.5 to poor classification performance, 1 to normal classification performance, and 2 to perfect classification performance. Based on these values, this reputation technique uses Equation 2 to assign reputation to each classifying entity.

$$W_i = S_i \times Acc_i \tag{2}$$

where W_i is the reputation value corresponding to classifying entity i, S_i is obtained from Eq. 3, and Acc_i is weight of classifying entities (Bahrepour, Meratnia et al. 2010).

$$S_i = \begin{cases} 2 & if\ (R_i \geq \theta_1) \\ 1 & if\ (\theta_1 > R_i \geq \theta_2) \\ 0.5 & if\ (\theta_2 \geq R_i) \end{cases} \tag{3}$$

θ1 and θ2 are application dependant.

3.5 Majority voting

Majority voting is a simple voting technique, which selects the classification result obtained from majority of sensor nodes.

4. Classification models

In this section, we present our general classification models regardless of which classifier is being used for the classification purpose. For more information about properties of these models and their performance, reader is referred to (Bahrepour, Meratnia et al. 2009; Bahrepour, Meratnia et al. 2009).

4.1 Local classification model

Local model assumes that each sensor node performs classification individually without communicating and cooperating with others. Figure 2 illustrates processing model of local classification, which consists of (i) a number of sensors providing input to the classifier, (ii) the classifier, which is responsible for activity recognition and determining the belogness of each instance to an activity class, and (iii) classification output, which is called activity. One should note that not all sensor nodes need to have the same classifiers.

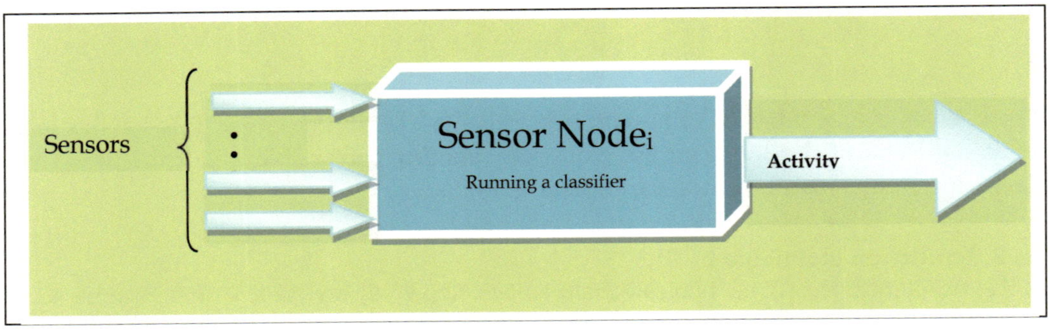

Fig. 2. Local classification model

4.2 Fusion-based classification model

The local approach is simple and works fine in situations, in which the sensor nodes are highly accurate and not prone to noises. However, generally speaking sensors, sensor nodes, and communication links are not always reliable and their failure is a common practice. Fusion-based classification model tolerates individual sensor and sensor node failures and involves more than one sensor node in the classification process. By doing so, it ensures that there are always some sensor nodes contributing to the classification process and compensating for the errors. The fusion-based approach uses the basic notions of the local approach and lets individual sensor nodes first classify and detect activities on their own. Then, the classification results are all sent to a fuser/ voter node (e.g., a cluster head) to reach a consensus. Figure 3 illustrates processing model of fusion-based classification. Similar to the local model, not all sensor nodes (including the fuser node) need to have the same classifiers.

In this study we use fusion-based classification model for activity recognition and investigate its performance by making use of various classifiers. Choice of classifier not only has direct effect on classification accuracy but also on complexity, as presented in the following sections.

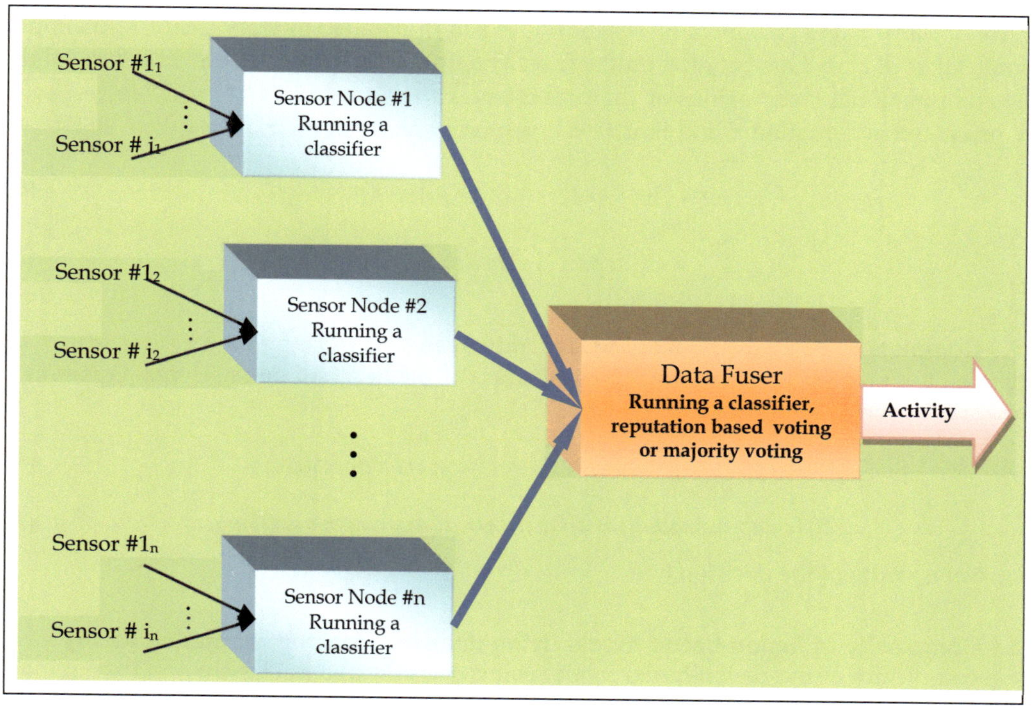

Fig. 3. Fusion-based classification model

4.3 Computational complexity consideration

4.3.1 Complexity of local model using FFNN and naïve bayes

Complexity of FFNN is function of m(number of features or number of nodes in the input layer),z (number of neurons in the hidden layer), and p(number of neurons in output layer). Equation 4 presents time complexity of FFNN in the local model (Bahrepour, Meratnia et al. 2009):

$$O_{FFNN} = O(m \times z \times p) \tag{4}$$

Complexity of Naïve Bayes is a function of m (number of features), c (number of classes), and j (number of partitions for distribution estimation). Equation 5 presents time complexity of Naïve Bayes classifier (Bahrepour, Meratnia et al. 2009) in the local model:

$$O_{NaiveBayes} = O(m \times c \times j) \tag{5}$$

Since training phase of FFNN and Naïve Bayes is conducted offline and only once, the computational complexity of training part can be disregarded.

4.3.2 Complexity of fusion-based model using FFNN and naïve bayes

When naïve Bayes or FFNN are used in the fusion-based model, the data-fuser has to wait till all classification results from the sensor nodes are available. In the worst case for n time duration (where, n is the number of sensor nodes involved in the classification process)

FFNN or naïve Bayes classifiers on sensor nodes run in parallel. In this case, computational complexities of classifiers running on the fuser and on the sensor nodes are added together. The computational complexities of the fusion-based model using FFNN and Naïve Bayes are presented by Equation 6 and Equation 7, respectively (Bahrepour, Zhang et al. 2009).

$$O_{FFNN} = O[(m \times z \times p) + (m \times z \times p)] = O(m \times z \times p) \tag{6}$$

$$O_{NaiveBayes} = O[(m \times c \times j) + (m \times c \times j)] = O(m \times c \times j) \tag{7}$$

4.3.3 Complexity of local model using Decision Tree

Complexity of decision tree appraisal is a function of the depth of decision tree. Equation 9 presents this time complexity:

$$O(local\ approach) = O(Decision\ tree\ appraisal) \tag{8}$$

$$O(Local\ approach) = O(m)O(Local\ approach) = O(m) \tag{9}$$

where m is depth of the decision tree

4.3.4 Complexity of fusion-based model using decision tree and reputation theory

Time complexity of the fusion-based model using decision tree as classifier and reputation theory as fuser is a function of three parameters: (i) complexity of making the decision tree on each node, (ii) complexity of performing local classification on each node, and (iii) complexity of performing fusion and reaching consensus between classification results. Equation 12 presents the final complexity.

$$O(Fusion\ based\ decision\ tree\ with\ reputation) \\ = \max[O(Decision\ tree\ appraisal) + O(process\ on\ the\ node) + O(reputation\ voting)] \tag{10}$$

$$O(Fusion\ based\ decision\ tree\ with\ reputation) = \\ Max[O(m) + O(n(n-1)) + O((n(n-1)) + n + c)] \tag{11}$$

$$O(Fusion\ based\ decision\ tree\ with\ reputation) = O(n^2) \tag{12}$$

where n is the number of nodes

4.3.5 Complexity of fusion-based model using decision tree and majority voting

Complexity of classification process of fusion-based model using decision trees is $O(m_1 + m_2 + \cdots + m_n) = O(m)$; where n is the number of nodes involved in the classification and m is depth of the decision tree. Since the voting is independent from the classification, its time complexity is added to the classification time as shown in Equation 13. Equation 16 presents the final complexity of fusion-based model using decision trees and majority voting.

$$OO(Distributed\ approach\ using\ majority\ voting) = \\ O([m]) + O[Majority\ voting 1]) = Max[(m), Majority\ voting] \tag{13}$$

$$O(Distributed\,approach\,using\,majority\,voting) = O(Max\,finding) + \mathrm{O(m)} \tag{14}$$

$$O(\max finding) = O(\mathrm{c}) \tag{15}$$

$$O(O(Distributed\ approach\ using\ majority\ voting) = O(c) + \mathrm{O(m)}) = O(m) \tag{16}$$

where m is depth of the decision tree, and n is number of nodes in the network.

4.4 Computational comparison

The time complexities of local and fusion-based models using different classifiers and fusers are summarized in Table 1.

Model	Time complexity
Fusion based model using FFNN	$O\ = O(m \times z \times p)$
Fusion based model using Naïve Bayes	$O = O(m \times c \times j)$
Fusion based model using Decision Tree and Reputation-based voting	$O(n^2)$
Fusion based model using Decision Tree with majority voting	$O(m)$

Table 1. Complexity comparison of local and fusion-based classification models with different classifiers

where n is the number of sensor nodes in the network, c is number of classes, m is number of features, and j is number of partitions for distribution estimation.

5. Data description and empirical results

5.1 Activity dataset and feature reduction process

To investigate applicability of the aforementioned classifiers for the activity recognition task in the medical domain, we used an activity dataset provided by Enschede Hospital (Medisch Spectrum Twente). The dataset consists of data from sensors attached to a 25 year old person, while he had been walking, sitting and standing still. The dataset contains 30 features from 5 tri-axial accelerometers and 5 tri-axial gyroscope sensors located on feet, shank, thigh and trunk. To find the most important features of the dataset, we first developed a feature reduction technique using genetic algorithm and decision tree. Feature reduction is the technique of selecting a subset of relevant features for making robust learning models. By removing redundant features from the dataset, feature selection helps improve the performance of learning models by (i) alleviating the effect of the curse of dimensionality, (ii) enhancing generalization capability, and (iii) speeding up learning process (Isabelle Guyon 2006). Genetic algorithms (GAs) belong to the larger classes of evolutionary algorithms (EAs) that generate solutions to optimization problems using techniques inspired by natural evolution, e.g., inheritance, mutation, selection, and crossover (Goldberg 1989). GA optimizes the features in order to find those most contributing to the classification process. Using GA for feature reduction, we first selected four most contributing features from the dataset. A decision tree was then created using these four features and its classification accuracy was considered as the fitness value for the four features. After the feature reduction process, our activity dataset contained four features, 8330 instances of data and three classes namely walking, sitting, and standing still. Figure 4 shows data distribution for these three classes and Table 2 presents statistical information of dataset, while the four features of the dataset are: 'Z' vector of gyroscope installed on the right foot, 'Y' vector of accelerometer installed on trunk, 'Z' vector of accelerometer installed on trunk, and 'X' vector of accelerometer installed on left foot. As it can be seen from Figure 4, data representing three classes have a high degree of overlap, which makes the dataset very complex for the classification process.

Feature	Min	Max	Mean	STD	Info
1	9.233	10.7756	-0.040	1.4355	right foot gyroscope Z vector
2	78.7644	50.0057	-3.9921	6.942	trunk accelerometer Y vector
3	-24.5415	20.3732	-2.8579	2.2486	trunk accelerometer Z vector
4	29.8074	1.8017	-4.4615	2.753	left foot gyroscope X vector

Table 2. Activity data set information (8360 instances, 3 classes: stand still, walking, sitting)

To simulate the classification process, we use Matlab® to train and test every individual node separately. In the training phase, first the classifiers (local model) running on the sensor nodes are trained. Then, according to the trained classifiers on sensor nodes, the fuser (fusion-based model) is trained. In the test phase, an instance of data is given to all sensor nodes and then their outputs are fused using either another classifier or a voting technique. Training phase is conducted with 2/3 of data and testing is performed with 1/3 of data.

5.2 Empirical results

We perform each activity classification experiment ten times and accuracy ratios are reported in Table 3.

According to Table 3, fusion based models using decision trees and reputation-based voting (technique 1 and 2) and using majority voting provides more accurate activity recognition. Additionally, in this specific dataset (activity dataset) reputation based voting technique 1 and 2 provides the same accuracy ratio.

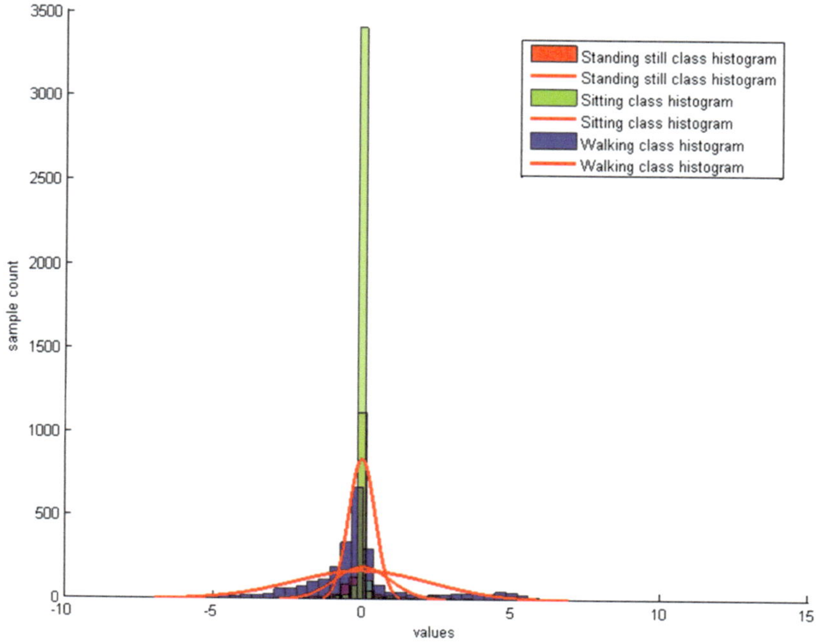
Standing still class histogram
Standing still class histogram
Sitting class histogram
Sitting class histogram
Walking class histogram
Walking class histogram
sample count
values
3500
3000
2500
2000
1500
1000
500
0
-10
-5
0
5
10
15

(a)

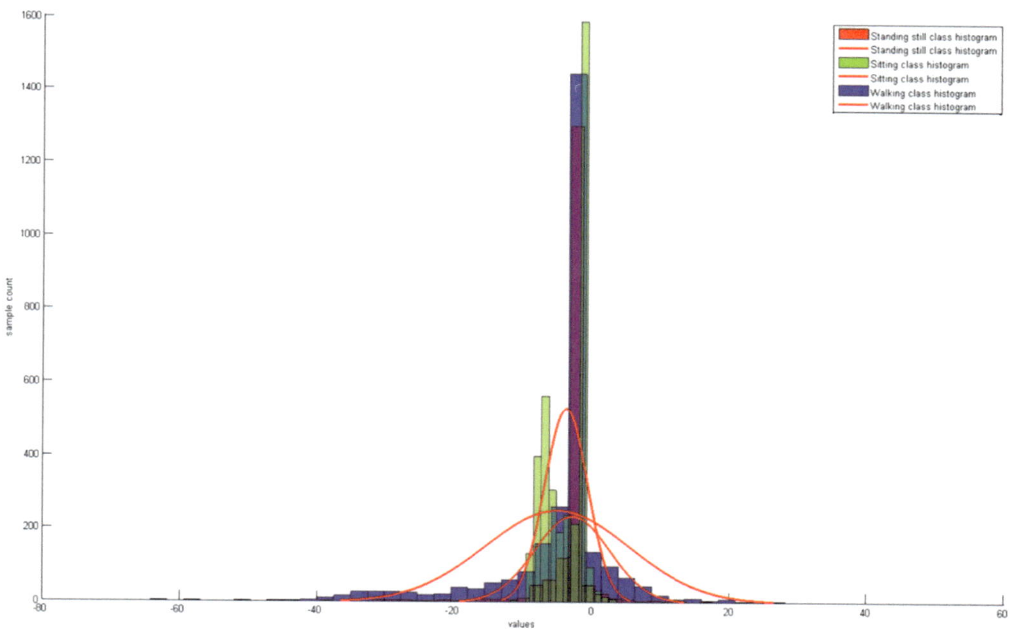
Standing still class histogram
Standing still class histogram
Sitting class histogram
Sitting class histogram
Walking class histogram
Walking class histogram
sample count
values
1600
1400
1200
1000
800
600
400
200
0
-80
-60
-40
-20
0
20
40
60

(b)

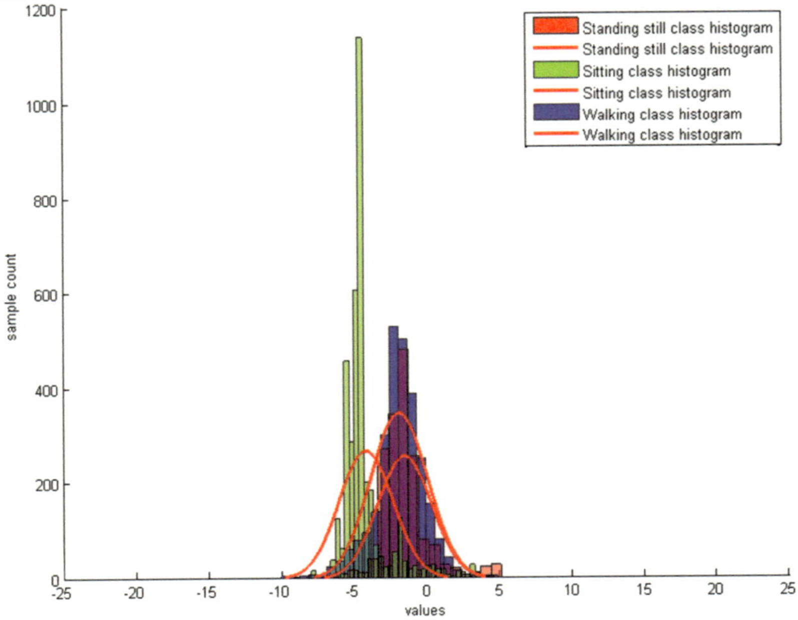

(c)

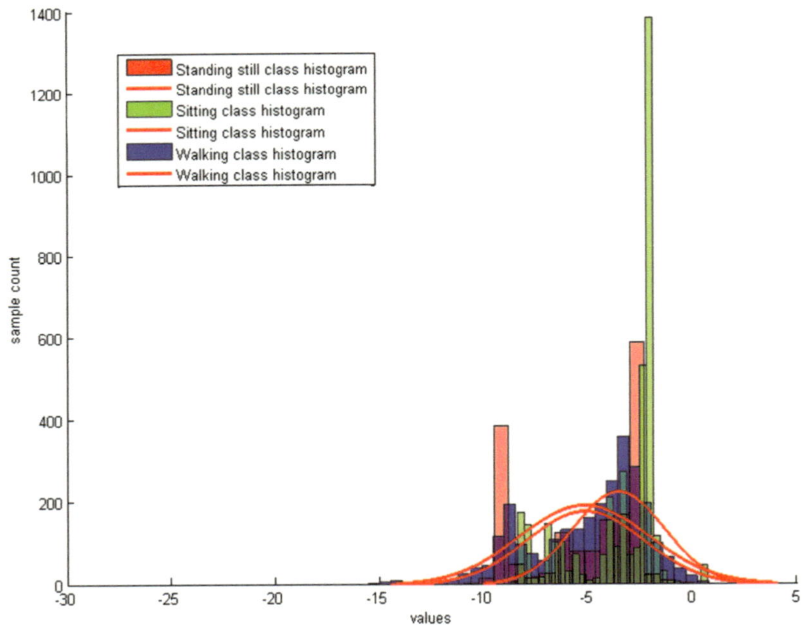

(d)

Fig. 4. Data distribution of three classes per each feature: (a) 'Z' vector of gyroscope installed on the right foot, (b) 'Y' vector of accelerometer installed on trunk, (c) 'Z' vector of accelerometer installed on trunk, (d) 'X' vector of accelerometer installed on left foot

Model	Classification Accuracy (Mean)	Standard Deviation
Fusion based model using FFNN	73.6022	1.4858
Fusion based model using Naïve Bayes	73.4624	1.2771
Fusion based model using Decision Tree and Reputation-based voting (technique 1)	78.9881	0.7820
Fusion based model using Decision Tree and Reputation-based voting (technique 2)	78.9881	0.7820
Fusion based model using Decision Tree with majority voting	78.6712	0.9836

Table 3. Accuracy and standard deviation of activity classification and recognition

5.3 Parameter study

There are some parameters in our classification process, which can affect classification accuracy. These parameters include (i) type of the classifiers, (ii) internal parameters of the classifiers (e.g. number neurons in hidden layer for FFNN), (iii) number of sensors, and (iv) type of sensors. In (Bahrepour, Meratnia et al. 2010), we have shown that the most effective parameter in fusion-based classification is the sensor types. This means that it is important to identify the most contributing sensors and ensure their presence in the classification process. As mentioned in Subsection 5.1, using a genetic algorithm and decision tree in the feature reduction process, we have identified the most contributing features of the dataset. In this section we study internals parameters of classifiers and their effects on classification accuracy.

5.3.1 Effects of internal parameters of feed forward neural network (FFNN) on classification

Internal parameters for FFNN are number of hidden layers and number of neurons in each hidden layer. The hidden layer in FFNN can grow vertically, horizontally and bi-directional (in both horizontal and vertical). The vertical growth means keeping number of hidden layers as low as possible and increasing the number of neurons in hidden layers. Number of neurons can also be kept as low as possible while number of hidden layers is increased. This process is called horizontal growth. A combination of both vertical and horizontal growth is also possible (bi-directional growth). To see the effects of growth in different directions, we performed several experiences, whose results are reported in Table 4.

Generally speaking, vertical growth of neural network leads to more accurate results comparing to horizontal growth. Moreover, by increasing number of hidden layers and/or neurons in the hidden layers, the accuracy enhances; however, at a certain point increasing hidden layer elements decreases the accuracy. Therefore, number of hidden layers as well as their neurons in each layer should be increased to the point which generates the highest accuracy. This is obtained experimentally.

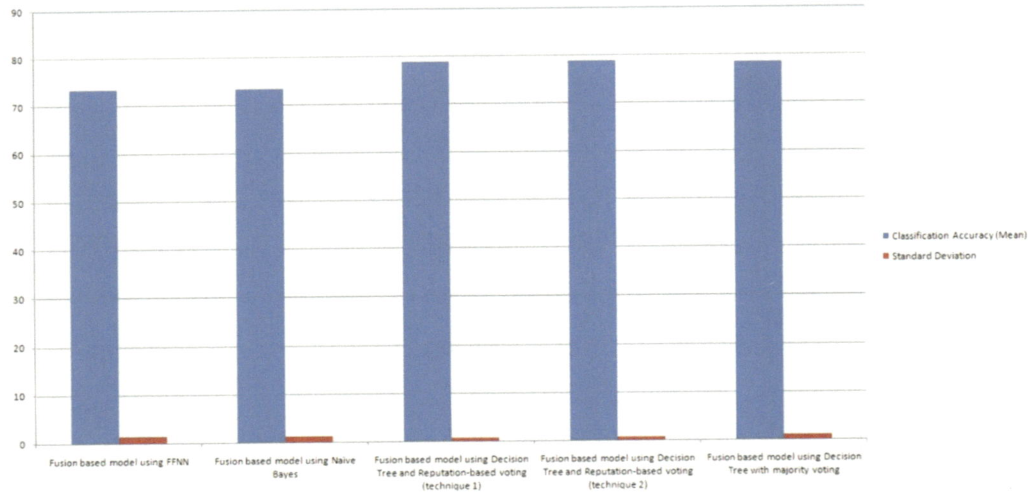

Fig. 5. Accuracy and standard deviation of activity classification and recognition.

	Hidden Layers	Neurons	Accuracy	STD
Horizontal growth	1	1	45.52	0.92
	10	1	46.33	0.61
	100	1	46.59	1.01
	1000	1	64.27	13.07
Vertical growth	1	10	55.77	1.37
	1	100	58.53	1.22
	1	1000	44.57	18.49
Bi-directional growth	10	10	51.44	0.92

Table 4. Effect of neural network's internal parameters on classification accuracy

5.3.2 Effects of internal parameters of naïve bayes (NB) classifier on classification

The only parameter for Naïve Bayes classifier is number of partitions for making a histogram. Table 5 shows how this parameter affects classification accuracy.

Partitions	Accuracy	STD
10	55.81	1.61
100	66.34	2.00
1,000	67.81	1.01
10,000	66.51	0.76
100,000	57.99	1.2

Table 5. Effect of Naïve Bayes' internal parameters on classification accuracy

As it can be seen in Table 5, generally speaking increasing number of partitions leads to increase of the classification accuracy. However, at a certain point either the accuracy stays roughly the same or drops.

5.3.3 Effects of internal parameters of decision tree (DT) on classification

Decision trees have almost no internal parameter because the whole tree is made during the training phase. However, after the training phase, the obtained tree can be pruned and some less-contributing branches can be removed. We performed a number of experiments, in which we prune the decision tree created for the activity dataset to a degree that tree has only one node. Figure 6 illustrates the effect of this pruning on classification accuracy. As it can be seen, pruning causes degradation of classification accuracy, in general. One also notices that for our activity dataset, pruning first increases the accuracy to some extent and then decreases it. The reason for this behavior is that the tree is over branched (i.e., almost 240 level tree is sophisticated enough to create some ambiguity and inaccuracy in the classification). Reducing some branches helps tree be simpler and perform classification slightly more accurate. However, at a certain point the accuracy drops. A general conclusion is, pruning causes reduction in accuracy ratio because the decision tree has to perform classification with fewer nodes.

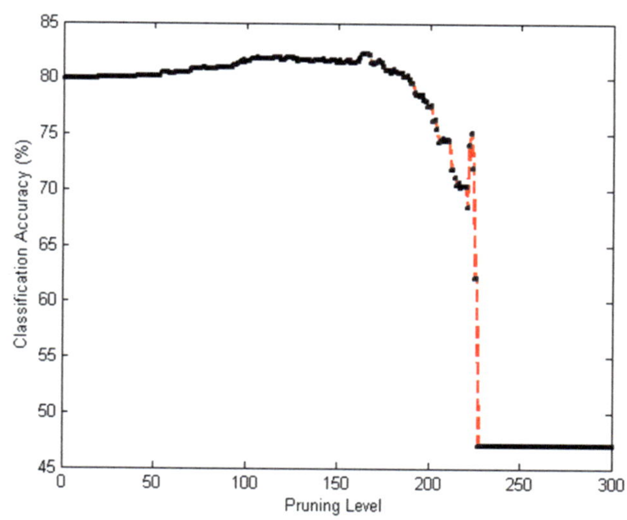

Fig. 6. Pruning and its effect on classification accuracy

6. Conclusion

Fall detection is a common problem in PD patients. Accurate activity classification and recognition is one of the first steps towards on time fall detection. Body senor networks are an emerging technology enabling online and distributed monitoring of both physical and physiological conditions of patients. Integrating this monitoring capability with (near) real-time activity recognition and classification can speed up the fall detection process and enable provision of timely feedback and assistance. To this end, having a thorough understanding of capabilities and performance of classification techniques is essential.

In this paper we investigated applicability of three classification techniques, i.e., Naïve Bayes, Feed Forward Neural Network, and Decision Tree for activity recognition in medical domain and showed how fusion-based classification can improve classification accuracy. In addition to considering classification accuracy, we studied effects of internal parameters of the classifiers on the classification performance and compared various techniques in terms of their complexity.

Activity recognition in medical domains requires inexpensive and easy wearing hardware components and computationally-light algorithms. To this end, our fusion-based classification technique offers three main advantages: (i) distributed processing and reasoning which decreases the data processing time and provides fast activity detection, (ii) robustness against sensors failure, and (iii) accurate detection through use of sensor fusion.

We performed a number of experiments using an activity dataset collected from a 25 year old person walking, sitting, and standing still. Performing an offline analysis shows high complexity of the dataset, as data representing three classes have a high degree of overlap. The evaluation results show that, considering this high complexity of the dataset, our fusion-based technique can reach reasonable classification accuracy.

Our future plans include design of temporal learning techniques to increase detection accuracy over time.

7. Acknowledgment

The authors would like to acknowledge Dr. J. P. P. van Vugt from Medisch Spectrum Twente and Sarvi Sharifi from the University of Twente for sharing their activity dataset. This work is supported by the EU Seventh Framework Programme and the SENSEI project.

8. References

Abbate, S., M. Avvenuti, et al. (2010). Monitoring of Human Movements for Fall Detection and Activities Recognition in Elderly Care Using Wireless Sensor Network: a Survey. *Wireless Sensor Networks: Application-Centric Design Book.*

ALPAYDIN, E. (2004). *Introduction to Machine Learning,* The MIT Press.

Aminian, K., P. Robert, et al. (1999). "Physical activity monitoring based on accelerometry: validation and comparison with video observation " *Medical & biological engineering & computing* 37(3).

Avci, A., S. Bosch, et al. (2010). Activity Recognition Using Inertial Sensing for Healthcare, Wellbeing and Sports Applications: A Survey. *23rd International Conference on Architecture of Computing Systems, ARCS 2010.* Hannover, Germany.

Bahrepour, M., N. Meratnia, et al. (2009). Sensor Fusion-based Event Detection in Wireless Sensor Networks. *SensorFusion'09.* IEEE. Toronto, Canada.

Bahrepour, M., N. Meratnia, et al. (2009). Use of AI Techniques for Residential Fire Detection in Wireless Sensor Networks. *Artificial Intelligence Applications in Environmental Protection Workshop, AIAEP 2009*, Thessaloniki, Greece.

Bahrepour, M., N. Meratnia, et al. (2010). "Fast and Accurate Residential Fire Detection Using Wireless Sensor Networks." *Environmental Engineering and Management Journal* 9(2): 215-221.

Bahrepour, M., N. Meratnia, et al. (2010). Distributed Event Detection in Wireless Sensor Networks for Disaster Management. *International Conference on Intelligent Networking and Collaborative Systems, INCoS 2010.* Thessaloniki, Greece, IEEE Computer Society.

Bahrepour, M., B. J. van der Zwaag, et al. (2010). Fire data analysis and feature reduction using computational intelligence methods. *Advances in Intelligent Decision Technologies - Proceedings of the Second KES International Symposium IDT 2010.* Baltimore, Maryland, USA: 289-298.

Bahrepour, M., Y. Zhang, et al. (2009). Use of Event Detection Approaches for Outlier Detection in Wireless Sensor Networks. *Proceedings of Symposium on Theoretical and Practical Aspects of Large-scale Wireless Sensor Networks, The 5th International Conference on Intelligent Sensors, Sensor Networks and Information Processing 2009 (ISSNIP 2009)*, Melbourne, Australia.

Benocci, M., C. Tacconi, et al. (2010). "Accelerometer-based fall detection using optimized ZigBee data streaming." *Microelectron Journal.*

Centeno, M., R. Giachetti, et al. (2003). A simulation-ILP based tool for scheduling ER staff. *2003 winter simulation conference.*

Goldberg, D. E. (1989). *Genetic Algorithms in Search, Optimization, and Machine Learning*, Addison-Wesley Professional.

Guenterberg, E., R. Bajcsy, et al. (2007). A segmentation technique based on standard deviation in body sensor networks. *EEE Dallas Engineering in Medicine and Biology Workshop.*

Guenterberg, E., S. Ostadabbas, et al. (2009). "An automatic segmentation technique in body sensor networks based on signal energy." *BodyNets '09 - Fourth International Conference on Body Area Networks.*

Horst, A. and N. Meratnia (2011). Online Activity Matching using Wireless Sensor Network. *1st International Conference on Pervasive and Embedded Computing and Communication Systems (PECCS 2011), In conjunction with VISIGRAPP 2011,.*

Hou, J., Q. Wang, et al. (2007). Pas: A wireless-enabled, sensor-integrated personal assistance system for independent and assisted living. *HCMDSS-MDPnP.*

Isabelle Guyon , S. G., Masoud Nikravesh , Lofti A. Zadeh (2006). *Feature Extraction: Foundations and Applications (Studies in Fuzziness and Soft Computing)*, Springer.

Jafari, R., W. Li, et al. (2007). Physical activity monitoring for assisted living at home. *4th International Workshop on Wearable and Implantable Body Sensor Networks*. Berlin.

Jarochowski, B. P., S. Shin, et al. (2007). Ubiquitous Rehabilitation Center: An Implementation of a Wireless Sensor Network Based Rehabilitation Management System. *International Conference on Convergence Information Technology*, IEEE Computer Society.

JI., H., v. d. P. AA., et al. (2001). "Accelerometric assessment of levodopa-induced dyskinesias in Parkinson's disease." *Mov Disord* 16: 58-61.

JI., H., v. d. M. V., et al. (2004). "Accuracy of objective ambulatory accelerometry in detecting motor complications in patients with Parkinson's disease." *Clin Neuropharmacol* 27: 53-57.

JJ., v. H., M. HA., et al. (1991). "A new approach in the assessment of motor activity in Parkinson's disease." *Neurol Neurosurg Psychiatry* 54: 976–979.

Kern, N., B. Schiele, et al. (2003). *Multi-sensor activity context detection for wearable computing*. Proceedings of European symposium on ambient intelligence (EUSAI03).

Khattak, A. M., L. T. Vinh, et al. (2010). Context-aware Human Activity Recognition and Decision Making. *12th International Conference on e-Health Networking, Application Services (IEEE HealthCom 2010)*. Lyon, France.

Krishnan, N. C., C. Juillard, et al. (2009). "Recognition of hand movements using wearable accelerometers." *JAISE*.

Le, T. M. and R. Pan (2009). Accelerometer-based sensor network for fall detection. *Biomedical Circuits and Systems Conference*.

Lee, M.-H., J. Kim, et al. (2009). Physical Activity Recognition Using a Single Tri-Axis Accelerometer. *WCECS 2009*. San Francisco, USA.

Lorincz, K., B. Kuris, et al. (2007). Wearable wireless sensor network to assess clinical status in patients with neurological disorders. *6th international conference on Information processing in sensor networks*, ACM.

Martinez-Garcia, A. and R. Menndez-Olague (2003). "Process improvement with simulation in the health sector." *e-Health:application of computing science in medicine and health care*.

Mehrotra, K., C. K. Mohan, et al. (1996). Elements of Artificial Neural Networks, MIT Press.

Moorea, S. T., H. G. MacDougalla, et al. (2007). "Long-term monitoring of gait in Parkinson's disease." *Gait & Posture* 26(2): 200-207.

N., S., Y. T., et al. (2004). "Lifecorder: A new device for the long-term monitoring of motor activities for Parkinson's disease." *Intern Med* 43: 685-392.

Nabar, S., A. Banerjee, et al. (2010). Evaluation of body sensor network platforms: a design space and benchmarking analysis. *In Wireless Health 2010 (WH '10)*. New York, NY, USA: 118-127.

Osmani, V., S. Balasubramaniam, et al. (2007). Self-organising object networks using context zones for distributed activity recognition. *ICST2007*. Florence, Italy.

Osmani, V., S. Balasubramaniam, et al. (2008). "Human activity recognition in pervasive health-care: Supporting efficient remote collaboration." *Journal of Network and Computer Applications:* 628-655.

Park, C., J. Liu, et al. (2005). Eco: an ultra-compact low-power wireless sensor node for real-time motion monitoring. *IPSN 2005*: 398-403.

Russell, S. J. and P. Norvig (2003). *Artificial Intelligence: A Modern Approach, 2 edition*, Pearson Education

S. Bosch, M. M.-P., R. S. Marin-Perianu, P. J. M. Havinga, and H. J. Hermens, Keep on moving! activity monitoring and stimulation using wireless sensor networks, in the Proceedings of the 4th EuroSSC conference, vol. 5741, September 2009, pp. 11–23 (2009). Keep on moving! activity monitoring and stimulation using wireless sensor networks.

Salarian, A., H. Russmann, et al. (2007). "Ambulatory Monitoring of Physical Activities in Patients With Parkinson's Disease." *IEEE Transactions on Biomedical Engineering* 54(12): 2296 - 2299.

Sanchez, D., M. Tentori, et al. (2008). Activity Recognition for the Smart Hospital. *IEEE Intelligent Systems*: 50-57.

Soini, M., J. Nummela, et al. (2008). "Wireless Body Area Network for Hip Rehabilitation System." *Ubiquitous Computing and Communication Journal* 3(5).

Steele, B. G., B. Belza, et al. (2003). "Bodies in motion: Monitoring daily activity and exercise with motion sensors in people with chronic pulmonary disease." *Rehabilitation Research and Development* 40(5).

Veld, H. i. t., O. M.H.A., et al. (2005). "Context aware algorithm for epileptic seizure detection." *Awareness deliverables.*

White, D. K., R. Wagenaar, et al. (2006). "Monitoring Activity in Individuals with Parkinson Disease: A Validity Study." *Journal of Neurologic Physical Therapy* 30(1): 12-21.

Wijewickrama, A. and S. Takakuwa (2006). Simulation analysis of an outpatient department of internal medicine in a university hospital. *2006 winter simulation conference WSC 06.*

Wikipedia Decision tree, http://en.wikipedia.org/wiki/Decision_tree.

Wu, J., G. Pan, et al. (2009). Gesture recognition with a 3-d accelerometer. *UIC 2009*: 25-38.

Wu, W. H., A. A. Bui, et al. (2008). "MEDIC: medical embedded device for individualized care." *Artificial Intelligence in Medicine* 42(2): 137-152.

Ying, H., M. Schlösser, et al. (2008). Distributed Intelligent Sensor Network for Neurological Rehabilitation Research. *4th European Conference of the International Federation for Medical and Biological Engineering (EMBEC 2008 - ESEM 2008)*. Antwerp, Belgium.

Yun, X., M. Lizarraga, et al. (2003). An improved quaternion-based kalman filter for real-time tracking of rigid body orientation. *International Conference on Intelligent Robots and Systems.*

Zhang, M. and A. Sawchuk (2009). A Customizable Framework of Body Area Sensor Network for Rehabilitation. *2nd International Symposium on Applied Sciences in Biomedical and Communication Technologies (Isabel).*

11

A Possibilistic Framework for Sensor Fusion with Monitoring of Sensor Reliability

Volker Lohweg[1], Karl Voth[2] and Stefan Glock[3]
[1,2]*Ostwestfalen-Lippe University of Applied Sciences, inIT - Institute Industrial IT, Lemgo*
[3]*Friedrich-Alexander University of Erlangen-Nuremberg, Institute for Electronics Engineering, Erlangen*
Germany

1. Introduction

Sensor and Information fusion is recently a major topic, not only in traffic management, military, avionics, robotics, image processing, and e. g. medical applications, but becomes more and more important in machine diagnosis and conditioning for complex production machines and process engineering. Several approaches for multi-sensor systems exist in the literature, cf. (Hall & Llinas, 2001); (Bossé et al., 2007).
The diagnosis and flaw detection in machines and process plants is a complex task, which is dependent on many effects. In the past decades, autonomous and automatic systems have become ubiquitous in our daily life. As famous examples, one may cite advanced driver-assistance systems in vehicles, quality monitoring in production processes, or semi-autonomous unmanned aircraft systems in military forces. All these systems have the common feature that they (partially) capture the state of the environment and generate fused information based on the data gained. More technically, the analogous environment is captured applying several sources (sensors, experts, etc.). After obtaining these analogous signals and experts' information, they are preprocessed and transformed into the digital domain, so that they can be processed on computers. In many cases the information captured from the environment may be imprecise, incomplete or inconsistent. Furthermore, signal sources may be not reliable. Therefore, it is necessary to extend known fusion concepts insofar that they are able to handle and to measure imprecision and reliability. In this chapter we will highlight, to our best knowledge, a new human-centric based fusion framework. Additionally, an application of the fusion method is shown for printing machines in the area of quality inspection and machine conditioning. The goal is the early recognition of errors in machines in order to avoid flaws by combining measuring data from several sensors with expert knowledge for the improvement of quality by

- using several information sources,
- integrating expert knowledge and perception,
- extracting reasonable features, and
- generating intuitive results.

This chapter investigates *context-based anticipatoric multi-sensor fusion* (CBAMSF) using evidence-based theories. A new algorithmic concept which considers a sensor's imprecision and process vagueness is presented and compared with other Fuzzy and Evidence based algorithms. In particular, the algorithm is inspired by the human behaviour in group-decision situations. This behaviour enables the algorithm to discount defective sensors. The performance of the novel fusion algorithm is evaluated with reference to real-life applications. In this context, the conditions of an intaglio printing machine (banknote printing machine) are monitored whilst the machine is producing security prints.
The sections are organised as follows: After a short introduction aleatoric (statistical) uncertainty versus epistemic (subjective, human-oriented) uncertainty is defined and explained. Furthermore, uncertainty in diagnosis of technical systems is highlighted. Thereafter, the concept of evidence theories is pointed-out in a short review—namely: *Probability Theory, Belief (Dempster-Shafer) Theory, Fuzzy Set Theory and Possibility Theory*. In the next sub-section we describe fusion algorithms which are based on human decision processes (Two-Layer Conflict Solving). From these concepts we derive a possibilistic framework for sensor fusion with reliability strategies. The following sub-section describes and compares a certain application. The succeeding section concludes the chapter and gives an outlook on further research topics.

2. Taxonomy of uncertainty

In today's production world we are able to generate a huge amount of data from analogue or digital sensors, PLCs, middleware components, control PCs, and, if necessary, from ERP systems. However, creating reliable knowledge about a machine process is a challenge because it is a known fact that Data $\neq$ Information $\neq$ Knowledge.
Insofar, a fusion process must create a low amount of data which creates reliable knowledge. Usually the main problems in sensor fusion can be described as follows: Too much data, poor models, bad or too many features, and improperly analysed applications. One major mis-belief is that machine diagnosis can only be handled based on the generated data—knowledge about the technical, physical, chemical or other processes are indispensable for modelling a multi-sensor system.

2.1 Knowledge and ignorance

Specialists in Engineering and Computer Science world accentuate the value of data, assuming that all necessary information about behaviour of a system, etc. is sufficiently contained in the data. Insofar, it is believed that information is an equivalent for knowledge. Therefore, human beings tend intentionally or unintentionally to sweep aside uncertainty in information or ignorance about a state of affairs of a (technical) system. In this sense, we overestimate the amount of information in data which is usable. Following these explanations, it is straightforward to cite Stephen Hawking: *"The greatest enemy of knowledge is not ignorance, it is the illusion of knowledge"*. The human knowledge and ignorance behaviour is described in Fig. 1. Referring to (Ayyub & Klir, 2006), the evolutionary certain knowledge (ECK) about a system or state of affairs can intrinsically not be interfered or camouflaged due to its evolutionary nature of this kind of knowledge. The current state of reliable knowledge (RK) has to be interpreted as trained knowledge under the general condition of provable know-how within well-grounded reliability levels, cf. Eq. (1).
The intersection of ECK and RK defines the human knowledge base of certain claims and propositions (CCK). It is assumed that this kind of knowledge withstands any

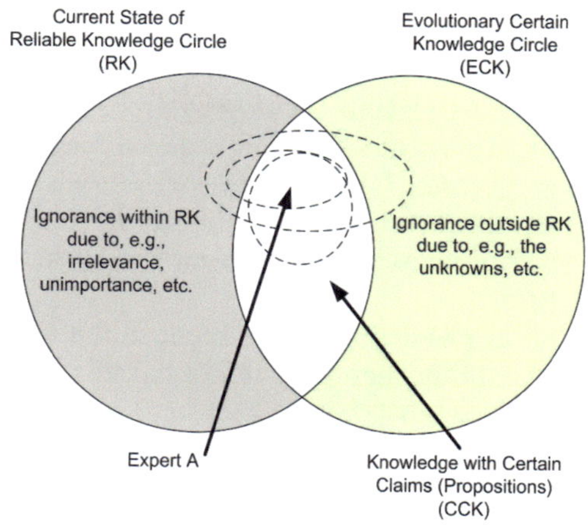

Fig. 1. Human evolutionary and trained knowledge. (Ayyub & Klir, 2006)

paradigm change or temporal opinion shifts. It can be formulated as:

$$\text{CCK} = \text{ECK} \cap \text{RK}. \tag{1}$$

Referring to Fig. 1, two types of ignorance can be distinguished: i) ignorance within the RK circle, based on irrelevance, etc.; and ii) ignorance outside RK, based on unknowns, missing know-how, etc. An expert (Expert A in Fig. 1), equipped with some know-how about a system or state of affairs, pictured with ellipses in Fig. 1, can be represented in three ways; 1) equipped with trained, created, or captured knowledge as a subset of ECK; 2) equipped with self-perceived knowledge by the expert; and 3) equipped with perception by others of the expert's knowledge. As (Ayyub & Klir, 2006) pointed out, the ECK of an expert may be smaller than the self-perceived knowledge, and therefore, the difference of the two types can be interpreted as a measure of overconfidence. Furthermore, an ellipse can also extend beyond the RK base as a result of creativity and imagination. The difference between the expert's knowledge and the ignorance space outside CCK can be interpreted as a human intuition process (creative process). It is unquestionable, that other experts will have other individual ellipses which may or may not overlap each other. If the ellipses of two or more experts do not overlap, based on a conjoint topic, a conflict situation may occur. Usually two types of ignorance can be identified: i) *blind ignorance* with its sub-categories *unknowable, irrelevance,* and *fallacy* (*camouflage*); ii) *conscious ignorance* with its sub-categories *inconsistency* (*confusion, conflict, inaccuracy*) and *incompleteness* (*absence, unknowns, uncertainty*).

In technical systems the latter ones are of main interest. Inconsistency occurs mainly in the case of inaccurate measurements and conflict-afflicted sensor information. Incompleteness of information, in the sense of uncertainty, causes in most instances problems in system modelling. Uncertainty is defined as *intrinsic absence of necessary knowledge.* In many cases, as, e. g. in complex systems, it is usually impossible to acquire all necessary information to create certain knowledge. Therefore, major attention should be given to this type of ignorance. As it was stated in (Klir & Wierman, 1998) uncertainty can rarely be avoided, when dealing with real-word problems. At sensory and pre-processing level, uncertainty is an inseparable companion of any measurement, resulting from measurement errors, stochastic effects, such as noise and resolution limits in terms of sampling and quantisation (Lohweg & Mönks, 2010a). At knowledge level (cognitive level), we are conditioned by vagueness and ambiguities in the description of processes or natural languages.

The concept of uncertainty was and is philosophically widely discussed in the community of Bayesian and Probability advocates.
Bayesians argue that it is invariably feasible to design a probability model for any on random variables based technical task. This argumentation is questionable in a sense that it does not fulfil a real-world scenario in all cases. Usually at least two kinds of uncertainty species have to be taken into account: i) the aleatoric uncertainty (random effects, signal noise, etc.); and ii) the epistemic uncertainty (lack of knowledge or incomplete knowledge regarding a complex system, etc.), cf. (Hacking, 1975).
The latter one is challenging in problem solving, because the unknown knowledge has to be modelled in different ways, other than probability based concepts. This argumentation leads to Shafer (Shafer, 1976), Zadeh (Zadeh, 2008) and others, who stated that the lack of knowledge is precisely reflected by the situation where the probability of events is ill-known or expert's knowledge is necessary. Furthermore, recently Salicone has presented a possibilistic approach to handle measurement uncertainty in electrical engineering applications (Salicone, 2007).

2.1.1 Aleatoric uncertainty

If data is complete and intrinsically non-deterministic in nature, it can be assumed as random (stochastic). The uncertainty is attributed to real-world phenomena and it can not be reduced or even eliminated by expanding an underlying knowledge base. Probabilistic approaches, such as classical Probability Theory (frequentist) and Bayesian Probability Theory are an effective way to model stochastic uncertainties, like measurement noise, etc., cf. (Ayyub & Klir, 2006). This type of uncertainty is referred to as *aleatoric uncertainty* (cf. Table 1).

2.1.2 Epistemic uncertainty

In many situations we lack information, that is, not all intrinsically necessary knowledge is available at state. In this case, the uncertainty range should be reduced by expanding the underlying knowledge base. When data is scarce the probabilistic approach may not be appropriate to reduce the system's uncertainty. Major types of this uncertainty are inconsistent and incomplete data as well as inconsistent information or knowledge. In many cases this uncertainty can be reduced by multi-sensory fusion and expert's knowledge, cf. (Ayyub & Klir, 2006). This type of uncertainty is referred to as *epistemic uncertainty* (cf. table 1).

	Aleatoric Uncertainty	*Epistemic Uncertainty*
Type	irreducible	reducible
Data	random, stochastic	scarce
Origin	intrinsic variations in data	inconsistent & incomplete data, lack of knowledge
Model	Probability Theory	Evidence and Fuzzy Theories

Table 1. Classification of aleatoric and epistemic uncertainty.

3. Basics on Evidence Theory

The Evidence Theory was introduced by Glenn Shafer in the 1970s, summarised and condensed in the book *A Mathematical Theory of Evidence* in 1976 (Shafer, 1976). He extended the seminal work of Arthur P. Dempster from the 1960s about upper and lower bounds of probability measures (Dempster, 1967). Today many quantitative approaches are available to model uncertainty and knowledge processing in the light of sensor and information fusion. It is an interesting fact that many approaches have a common theoretical base.

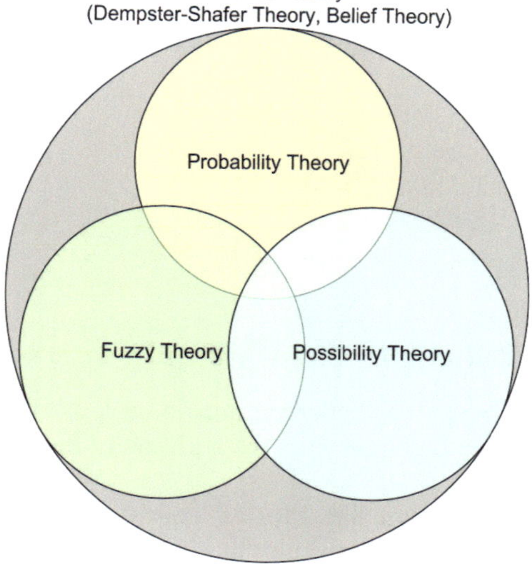

Fig. 2. Theories embedded in evidence theory.

Dempster-Shafer Theory (sometimes called *Belief Theory*) serves as an interconnection between many quantitative approaches, namely: Probability Theory, Possibility Theory, Fuzzy Theory, and others. The above mentioned theories are (partly) special branches of the Evidence Theory (Dempster-Shafer Theory). Fig. 2 illustrates symbolically the interconnection of different approaches insofar they are needed in this chapter.

3.1 Dempster-Shafer (Belief) Theory

Data fusion deals with data which is received from sensors, experts or human linguistic words, etc. Furthermore, much of such knowledge is cognitive and imprecise (incomplete) to some degree. To deal with uncertain knowledge, various strategies based on Dempster-Shafer Theory (DST) (Dempster, 1967), (Shafer, 1976), are researched and applied, because DST is capable of managing uncertainty due to its framework. DST acts as the pioneer in data fusion algorithms, which was proposed by Dempster and extended by Shafer subsequently. Serving as a seminal fusion approach, DST stirs up many discussions and studies in data fusion. DST is a mathematical theory of evidence, which combines independent sources of information (Dempster, 1967), (Shafer, 1976). With combination of evidence sources obtained from sensors (experts), more reliable and convincing fusion results are expected. First a finite frame of discernment, which forms a set Θ, is defined, $\Theta = \{\theta_1, \theta_2, ..., \theta_n\}$, with an arbitrary event variable (proposition, hypothesis) θ_i, $i \in \mathbb{N}$. A power set, $\mathcal{P}(\Theta) = 2^{\Theta}$, includes all the possible combinations of propositions θ_i. Propositions are regarded to be *mutually exclusive* and *exhaustive*. Thus, the power set contains 2^n elements. As an example, assume the following: $\Theta = \{A, B\} \rightarrow \mathcal{P}(\Theta) = \{\varnothing, A, B, \Theta\}$.

A function $m : 2^{\Theta} \rightarrow [0,1]$ is called a mass function, also known as Basic Probability (Belief) Assignment (BPA, BBA) of an event, e. g. A, with

$$m(\varnothing) = 0, \sum_{A \subseteq \mathcal{P}(\Theta)} m(A) = 1. \tag{2}$$

If there is no element in the BBA, then its mass is zero. On the other hand, as $\mathcal{P}(\Theta)$ is a power set composed of all the subsets, the sum of all the masses must be equal to one. Furthermore, the focal element (mass is larger than zero) is defined as:

$$\{(A, m(A)) | A \subseteq \Theta \, \text{and} \, m(A) > 0\}. \tag{3}$$

A *belief function*, Bel(·): $2^{\Theta} \rightarrow [0,1]$, satisfies the following three axioms (cf. Eq. (4) - Eq. (6)):

- Axiom 1:

$$\text{Bel}(\varnothing) = 0. \tag{4}$$

- Axiom 2:

$$\text{Bel}(\Theta) = 1. \tag{5}$$

- Axiom 3:

$$\text{Bel}(A \cup B) \geq \text{Bel}(A) + \text{Bel}(B) \, \text{if} \, A \cap B = \varnothing. \tag{6}$$

Contrary to the probability theory, the belief measure is *super-additive*. That is, the belief of the union of two propositions may be larger than the addition of the two belief measures of each proposition A and B.
Belief (Bel) and plausibility function (Pl), Pl(·): $2^{\Theta} \rightarrow [0,1]$, are essential concepts in DST, which are used in decision-making. Plausibility has to be interpreted as the dual function of the belief.

$$\text{Bel}(A) = \sum_{B \subseteq A} m(B) \, \text{and} \, \text{Pl}(A) = 1 - \text{Bel}(A^c), \tag{7}$$

where A^c is the complement of proposition A.
In DST, Bel(A) has to be interpreted after Shafer (Shafer, 1976) as *"a person's (sensor information) degree of belief that the truth lies in A"*, whereas Pl(A) can be considered as a measure of maximum support for a proposition A under the constraint of disbelief in A. Thus, Bel(A) is the total belief mapped to A; Pl(A) is the total mass assignment, which can move into A, and $m(A)$ is the basic belief assignment, which is exclusively dedicated to A. Furthermore, plausibility can be derived as follows:

$$\text{Pl}(A) = 1 - \text{Bel}(A^c) = 1 - \sum_{B \subseteq A^c} m(B) = \sum_{B \subseteq \mathcal{P}(\Theta)} m(B) - \sum_{B \subseteq A^c} m(B) = \sum_{B \cap A \neq \varnothing} m(B). \tag{8}$$

From Eq. (7) the following can be directly derived:

$$\text{Bel}(A^c) + \text{Pl}(A) = 1 \text{ and } \text{Bel}(A) + \text{Pl}(A^c) = 1. \tag{9}$$

Bel(·) is, in an early interpretation, called lower bound probability, while Pl(·) is the upper bound probability, for the reason that Bel(·) is the "must-be"-probability and on the other hand Pl(·) is the "might-be"-probability. Therefore, Pl(·) includes more mass than Bel(·), which is illustrated with

$$\text{Bel}(A) \leq \text{Pl}(A). \tag{10}$$

The *Ignorance* Igr(A) can be interpreted as

$$\text{Igr}(A) = \text{Pl}(A) - \text{Bel}(A). \tag{11}$$

After obtaining the above-mentioned concepts, we are able to use DST to fuse independent data sources by applying *Dempster's Rule of Combination*, cf. (Shafer, 1976). It is also called normalised conjunctive combination rule:

$$\bigoplus_{i=1}^{n} m_i(A) = K \cdot \sum_{A_1 \cap \ldots \cap A_n = A} \prod_{i=1}^{n} m_i(A). \tag{12}$$

The term $\sum_{A_1 \cap \ldots \cap A_n = A} \prod_{i=1}^{n} m_i(A)$ aggregates the consonant opinions (non-conflicting parts) from sensors which are subsequently multiplied with *conflicting factor* $K = (1 - k_c)^{-1}$, where

$$k_c = \sum_{A_1 \cap \ldots \cap A_n = \varnothing} \prod_{i=k}^{n} m_i(A_k). \tag{13}$$

The variable $m_i(A_k)$ denotes the mass of propositions from sensor i. According to the definition of k_c it defines the empty intersection of the propositions of all sensors. Therefore, it is also called *conflicting coefficient*.

The DST offers two main advantages: i) belief and possibility model two levels of probability, generating and assigning ignorance (range of uncertainty) or uncertainty to a proposition A, furthermore *total uncertainty* can be modelled by the vacuous belief, that is, $m(\Theta) = 1$; ii) as the measures are defined on subsets of $\mathcal{P}(\Theta)$, non-specificity and conflict can be modelled.

In order to demonstrate the concept of DST, let us assume a banknote was given to an expert for forgery analysis. Is it genuine or is it counterfeit? The proposition A corresponds to the possibility that the banknote is genuine, whereas the proposition B reflects the proposition that the banknote is counterfeit; then $\Theta = \{A, B\}$ is the frame of discernment and $\mathcal{P}(\Theta) = \{\varnothing, A, B, \Theta\}$ is the corresponding power set. The belief functions Bel(A) and Bel(B) represent the belief measures in genuineness and counterfeit of the banknote.

As, in this case, $B = A^c$, Eq. (6) changes to $\mathrm{Bel}(A \cup A^c) \geq \mathrm{Bel}(A) + \mathrm{Bel}(A^c)$ and $A \cap A^c = \varnothing, A \cup A^c = 1 \Rightarrow \mathrm{Bel}(A) + \mathrm{Bel}(A^c) \leq 1$. If the expert assumes the genuineness of the banknote, one would set Bel(A) near to 1 and—independently of Bel(A)—Bel(A^c) to a low value. In case one believes in the opposite, one sets Bel(A^c) to a high value and Bel(A) to a low value. If the expert is uncertain about the decision, but tends to genuineness, one might vote for Bel(A) = 0.5 and Bel(A^c)= 0.3, and in the desperate case of total uncertainty (total ignorance), one will set both belief measures to zero.

It has to be pointed out that the literature on DST is manifold. For deeper insights we refer to, e. g., (Dempster, 1967), (Shafer, 1976), (Yager, 1987), (Hall & Llinas, 2001), (Ayyub & Klir, 2006), (Campos, 2006), (Yager & Liu, 2008).

3.2 Probability Theory

Probability Theory (PT) is the oldest concept for quantifying uncertainty. It is one of the last major mathematical domains to be formed at the end of the seventeenth century. Based on a correspondence between Fermat and Laplace about gambling, Probability Theory is also today an active field of research. PT typically models *random phenomena* and *chance*.

Although, PT is an "old" theory, it was put into question in the twentieth century, whether the Probability Theory can only be viewed as a theory of chance. In every day situations it is used in two different ways: i) when randomness is assumed; ii) in cases of lacking knowledge. Therefore, two schools of interpretation can be found: objectivists (frequentists) and subjectivists (Bayesians). Hacking (Hacking, 1975) calls probability a "Janus-faced" theory. We refer to the philosopher Karl Popper (Popper, 1934) on the one hand and the

subjectivists John Maynard Keynes (Keynes, 1921) as well as to Rudolf Carnap (Carnap, 1947) or Frank P. Ramsey (Ramsey, 1960) on the other hand. The dispute in the community still exists.
Based on random experiments, like throwing the dice or tossing a coin, classical probability is defined by the relative frequency of events in the long run. If $\Theta = \{\theta_1, \theta_2, ..., \theta_n\}$ denotes the so called *sample space*, the discrete, finite, non-empty set of all possible outcomes, then $\theta_i, i \in \mathbb{N}$, are the elementary events (singletons). A non-negative number $P(A)$, $P : 2^{\Theta} \rightarrow [0,1]$, is assigned to each subset $A \in \Theta$—an event of any collection of outcomes: $A = \{\theta \in \Theta \,|\, \theta \in A\}$. This number is called *probability of the event* A. The axiomatic approach of probability was formulated by Kolmogorov in 1933 (Kolmogorov, 1933). The number $P(A)$ must satisfy the three following axioms; cf. Eq. (14) - Eq. (16):

- Axiom 1:
$$P(\varnothing) = 0. \tag{14}$$
- Axiom 2:
$$P(\Theta) = 1. \tag{15}$$
- Axiom 3:
$$P(A \cup B) = P(A) + P(B) \text{ if } A \cap B = \varnothing. \tag{16}$$

The probability measure of two (or more) is *additive*. This property, defined by axiom 3, is one of the key features in Probability Theory. Once a probability $P(A)$ is set, the complement $P(A^c)$ is fixed. In other words, the event $\theta \in \Theta$ can either be *true* or *false*. If for each possible outcome $\theta \in \Theta$, the probability $p(\theta)$ is given, then a function $p : \Theta \rightarrow [0,1]$ is called *probability density*. The function $p(\theta) = P(\{\theta\}), \forall\theta \in \Theta$ satisfies the following conditions:

$$0 \leq p \leq 1, \tag{17}$$

$$\sum_{\theta \in \Theta} p(\theta) = 1. \tag{18}$$

Following the additivity axiom, the probability of A can be determined by its individual probabilities of each of its elementary events:

$$P(A) = \sum_{\theta \in A} p(\theta). \tag{19}$$

Again, we follow the example of genuine or counterfeit banknotes. The proposition A corresponds to the possibility the banknote is genuine, whereas the proposition B reflects the proposition that the banknote is counterfeit; then $\Theta = \{A, B\}$ is the set of events. The probabilities P(A) and P(B) represent the chances in genuineness and counterfeit of the banknote. As, in this case, $B = A^c$, Eq. (16) changes to $P(A \cup A^c) = P(A) + P(A^c)$ and $A \cap A^c = \varnothing, A \cup A^c = 1 \Rightarrow P(A) + P(A^c) = 1$. If the expert assumes the genuineness of the banknote, one sets P(A) near to 1 and $P(A^c) = 1 - P(A)$. In case one believes in the opposite, one sets $P(A^c)$ to a high value and $P(A) = 1 - P(A^c)$. If the expert is uncertain about his decision, the only thing one can do, is to set $P(A) = 0.5$, that is, $P(A^c) = 0.5$ (total ignorance). The above mentioned property is essential in Probability Theory.
In the subjective view, the Bayesian view, knowledge and uncertainty is represented based on probability measures $P(\cdot)$, where the measure describes a *likelihood* of the occurrence of an event A. In this context, $P(\cdot)$ is often referred to as *a-priori probability* which is determined on previous knowledge based on adequate data *and* taken for granted. The probability

of an event A is interpreted in a sense of *degree of certainty* or *degree of belief*. Knowledge about the occurrence of an event is either modelled by $P(A) = 1$ in case of a certain event or $P(A) = 0$ for an impossible event. Once an event A is given by a certain probability $P(A)$, the complement $P(A^c)$ is naturally given. Therefore, the only possibility to model uncertainty in a set Θ is, as there is no evidence for a certain event, the probability density $p(\theta) = \frac{1}{|\Theta|}, \theta \in \Theta$. The cardinality of the set is $|\Theta|$. The modelling of uncertainty is a weakness of Probability Theory. As wider the set Θ is, as more improbable is a single event θ_i, but this is counter-intuitive, because a certain event will occur to $p(\theta_i) = 100\,\%$ and not to $p(\theta_i) = \frac{1}{|\Theta|}$.
The basis of fusion in PT is well-grounded on the *Bayes Rule* (Bayes, 1763). Two events $A \in \Theta_1$ and $B \in \Theta_2$ are assumed to be conjoint, $A \cap B \neq \varnothing$. That is, that the joint probability can be determined by the joint elements θ in the joint set $\Theta = \Theta_1 \times \Theta_2$, defined by the cross product, and is denoted by $P(A \cap B)$.
The *conditional probability* $P(A|B)$ is defined as the probability of A being true under the condition of the event B. It is defined as:

$$P(A|B) = \frac{P(A \cap B)}{P(B)}, \; P(B) > 0, \; P(A) > 0. \tag{20}$$

The conditional probability is called *a-posteriori probability* of the event A. That is, that the probability is not only based on the previous knowledge, but also takes into account the evidence of the event B.

$$P(B|A) = \frac{P(B \cap A)}{P(A)} = \frac{P(A \cap B)}{P(A)} = \frac{P(A|B) \cdot P(B)}{P(A)}. \tag{21}$$

In Eq. (21) $P(A)$ is usually defined as the *total probability*, which is defined as

$$P(A) = \sum_{i=1}^{n} P(A|B_i) \cdot P(B_i). \tag{22}$$

It has to be pointed out, that $B_1, ..., B_n$ have to be *pairwise exclusive* and *mutually exhaustive*. Furthermore, $P(B_i) > 0 \, \forall i$. Now, we show basically how information is fused in the sense of Bayesian inference. Let F_i be a measured machine readable feature on a banknote (e. g. a certain coding) which has to be detected by an optical sensor S_i as a "true"—S_i—or "false"—S_i^c—information. Furthermore, it can be assumed, that the accuracy of the sensors $P(S_i|F_i)$ may be 0.9, that is, the probability of a correct measurement (the sensor delivers a correct result under the condition of a correct code) is 90 %. Then, the probability of the detection of the feature F_i under the condition of the correct sensor information is:

$$P(F_i|S_i) = \left(\frac{P(S_i|F_i)}{P(S_i)}\right) \cdot P(F_i). \tag{23}$$

With Eq. (20) and Eq. (22) the sensor probability $P(S_i) = P(S_i \cap F_i) + P(S_i \cap F_i^c)$ can readily be formulated:

$$P(F_i|S_i) = \frac{P(S_i|F_i) \cdot P(F_i)}{P(S_i \cap F_i) + P(S_i \cap F_i^c)} = \frac{P(S_i|F_i) \cdot P(F_i)}{P(S_i|F_i)P(F_i) + P(S_i|F_i^c)P(F_i^c)}. \tag{24}$$

The a-priori probability $P(F_i)$ is given under the assumption, that it is very likely that the correct feature is used in the measurement session (e. g. 99.5 %), that is, $P(F_i^c) = 0.005$.

Furthermore, as $P(S_i|F_i) = 0.9$ is assumed, the probability $P(S_i|F_i^c) = 0.1$, that is, the detection of a feature which is not present. It has to be pointed out, that all four combinations, based on the equivalent a-priori probabilities, have to be determined: $P(F_i|S_i)$, $P(F_i|S_i^c)$ (false positive), $P(F_i^c|S_i)$ (false negative), and $P(F_i^c|S_i^c)$ to get the full picture.
In case of multiple measurements for one feature we get $P(F_1|S_1, S_2)$, which can be interpreted, e. g., as two temporarily *independent* measurements $P(S_1, S_2) = P(S_1) \cdot P(S_2)$. In this case we obtain:

$$P(F_1|S_1, S_2) = \frac{P(S_1, S_2|F_1) \cdot P(F_1)}{P(S_1, S_2)} = \frac{P(S_1|F_1)P(S_2|F_1) \cdot P(F_1)}{P(S_1)P(S_2)} = \frac{P(S_2|F_1)P(S_1|F_1) \cdot P(F_1)}{P(S_2)P(S_1)}. \tag{25}$$

With Eq. (23) $P(F_1|S_1, S_2)$ can be simplified to

$$P(F_1|S_1, S_2) = \left(\frac{P(S_2|F_1)}{P(S_2)}\right) \cdot P(F_1|S_1). \tag{26}$$

The conditional probability $P(F_1|S_1)$ from the first measurement is combined with the appropriate probabilities of the second measurement to generate a fused result. The first conditional probability serves as the new a-priori probability for the second measurement. This is how Bayesian inference combines old information with new information to update the belief states.
Probability and its frequentist representations concerns random effects. The events in Probability Theory are either assumed to be true or false. The probability of an event is connected directly to the event and to its complement. The Bayesian view leads to modelling of uncertainties. However, Probability Theory has its weakness in the area of epistemic uncertainty because of its additivity axiom. A-priori probabilities are the decisive part in PT. Furthermore, getting these a-priori probabilities is not easy to achieve. Probability Theory is very well adapted to problems based on independent random variables, but in fusion processes this independence is seldom to occur. The limitation of Boolean logic in PT leads to Fuzzy Theory.

3.3 Fuzzy Set Theory

As we have noticed, DST and PT are able to describe epistemic and aleatoric uncertainty. However, both theories are based on crisp sets of a frame of discernment. In many measurement or sensor related concepts we have to handle both, noise and systematic effects. The latter ones are assumed to be *systematic errors* which can possibly be compensated, as they are treated like a known bias or offset. Unfortunately, in complex systems it is nearly impossible to obtain all systematic measurement data. Therefore, in many instances we have to handle these errors as systematic uncertainties. The normal situation is as follows: The systematic input is recognised, but the exact information (data, values) is unknown. However, in most cases it is possible to locate a data value in a closed measurement interval—an *interval of confidence*. A systematic contribution to a measurement always takes the same value, even if unknown, within a certain interval. That is, not each value has the same probability to occur. A certain unknown value will occur to 100 %, others will definitely not occur. However, values in the interval have, in absence of further evidence, the possibility to occur. The above mentioned idea is in contrast to Probability Theory.
This fact leads to Fuzzy Set Theory (Fuzzy Theory, FT). The concept of fuzzy sets was first introduced by Zadeh in 1965 (Zadeh, 1965) in order to represent incomplete knowledge.

However, in 1937 the mathematician and philosopher Max Black developed the concept of *vagueness* and fuzzy *membership* (Black, 1937).
The traditional set approach is based on *crisp sets*. That is, a crisp variable can only take the value 1 or can be interpreted to be *true*, if the variable belongs to the set, or 0 (*false*), if it does not belong to the set. Fuzzy sets, in contrast, are not based on a Boolean value representation, but on a *membership function*, which describes the membership of a variable to its appending fuzzy set in a closed interval [0, 1]. It represents a multi-valued logic. Fuzzy sets can thus be described as a generalisation of crisp sets. This is not an antagonism in respect of PT, but rather an addendum to it, since both theories address a different kind of uncertainty: *randomness* and *vagueness* (fuzziness).
Let Θ be a frame of discernment, then a fuzzy set A of Θ is defined by a membership function $\mu_A : \Theta \to [0,1]$. The membership function $\mu_A(\theta) \in \mathbb{R}$ is the degree of membership for a fixed element θ in the fuzzy set A, that is, the degree of belonging $x = \theta$ and $x \in A$, the degree of certainty of the proposition $\theta \in A$. The *normalised* membership function of a fuzzy set A satisfies the following axioms; cf. Eq. (27) - Eq. (29):

- Axiom 1:
$$\exists x_0 \in A : \mu_A(x_0) = 1. \tag{27}$$
- Axiom 2:
$$0 \leq \mu_A(x) \leq 1. \tag{28}$$
- Axiom 3:
$$\exists x, x_0 \in A \forall x \neq x_0 : \mu_A(x) \leq 1. \tag{29}$$

If A is a fuzzy set defined by its membership function $\mu_A(x)$ and B is another fuzzy set defined by $\mu_B(x)$, then $\mu_A(x) \leq \mu_B(x)$, iff A is a subset of B:

$$A \subseteq B \Rightarrow \mu_A(x) \leq \mu_B(x). \tag{30}$$

The membership function $\mu_A(x)$ of a fuzzy variable x can also be described in terms of α-cuts—different vertical levels $\alpha \in [0,1]$. Each α-cut is defined as

$$A_\alpha = \{x \in A \mid \mu_A(x) \geq \alpha\}. \tag{31}$$

The representation of a fuzzy variable in terms of its α-cuts describes a group of nested sets (consonant sets) A_α in terms of *intervals of confidence* with its associated *levels of certainty* α (cf. Fig. 3). The level of certainty describes how certain a measurement value is—this is the case iff a crisp value is taken into account. The level of confidence is mapped to a set and represents the available knowledge based on an unknown value in the range of the set A_α.
Under the assumption of a finite support of A, we can assume that a membership function is fully represented by its α-cuts. Furthermore, the α-cuts of each fuzzy variable are closed intervals of real numbers $\mathbb{R}$.
The more the level of certainty increases, the more the interval of confidence decreases. This fact is in line with the human knowledge representation: If the level of certainty increases, the *level of confidence* must decrease. The level of certainty indicates how sure a measurement is, while the level of confidence specifies the probability that a certain value falls in its interval. In other words, as higher α is, as lower the level of confidence must be. The level of confidence is associated to $(1 - \alpha)$. The above mentioned fact will be a subject-matter in Possibility Theory.
In the framework of fuzzy sets, knowledge is represented by membership functions, which map a measurement value in a sense of a proposition to $x = \theta \in A$ to $\mu_A(x)$, that is, the belief

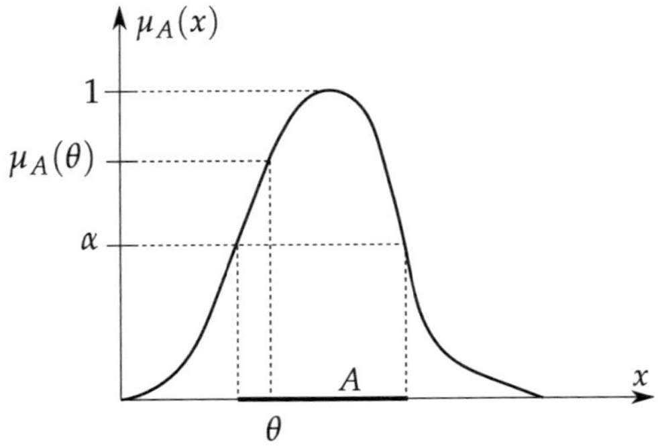

Fig. 3. Membership function and α-cut.

in the proposition is expressed by $\mu_A(x)$. This fact corresponds to the concept of vagueness as a value θ may belong to more than one membership function with different degrees of membership (e. g, $\mu_B(\theta) < \mu_A(\theta)$ and $A \cap B \neq \varnothing$), cf. Fig. 4).

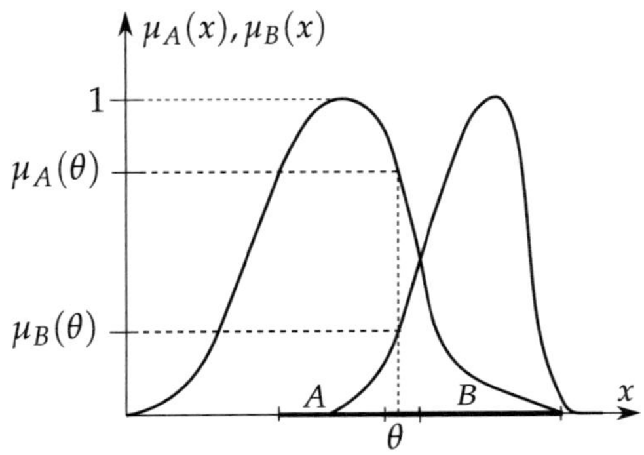

Fig. 4. Membership functions $\mu_A(x)$ and $\mu_B(x)$ with conjoint subset $A \cap B \neq \varnothing$.

Here, set A belongs to $\mu_A(x)$ and set B belongs to $\mu_B(x)$. The value $x = \theta$ is a member of both sets—a value which belongs to the subset $A \cap B$. As the degree of membership $\mu_A(\theta)$ is higher than the degree of membership $\mu_B(\theta)$, the value θ belongs more *certain* to set A than to set B. Fuzzy Set Theory can handle vagueness and *non-specificity* as it is an extension of classical set theory.

Total uncertainty (total ignorance) is expressed in terms of a rectangular shaped membership function with $\mu_A(\theta_i) = 1 \,\forall\, \theta_i \in A$.

Aggregation of information is based on elementary fuzzy set operations. The set operators aim for combining information in order to generate a single fuzzy variable. The basic operations are the equivalents to the above mentioned operations in classical set theory. Hence, we are able to define the complement, intersection and union of fuzzy variables. Following the base fuzzy set operations defined by Zadeh (Zadeh, 1965) the complement of a normalised membership value of the fuzzy set A is defined as:

$$\mu_{A^c}(x) = 1 - \mu_A(x). \tag{32}$$

The complement $\mu_{A^c}(x)$ describes the degree to which x does not belong to the fuzzy set A. The *fuzzy union* $\mu_{A \cup B}(x)$, based on a s-norm, $s : [0,1] \times [0,1] \to [0,1]$, is defined as:

$$\mu_{A \cup B}(x) = \max(\mu_A(x), \mu_B(x)). \tag{33}$$

Again, contrary to the probability theory, fuzzy union is *non-additive*. The *fuzzy intersection* $\mu_{A\cap B}(x)$, based on a t-norm, $t : [0,1] \times [0,1] \rightarrow [0,1]$, is defined as:

$$\mu_{A\cap B}(x) = \min(\mu_A(x), \mu_B(x)). \tag{34}$$

The Eq. (33) and Eq. (34) are based on so-called triangular norms, *t-norms* and *t-conorms* (or *s-norms*). As many operators exist, we refer to the literature (Klir & Yuan, 1995).
The union and intersection can be interpreted as *aggregation operations*. In general, aggregation operations combine several fuzzy sets or membership functions resulting in a single membership function. The aggregation operator, $h : [0,1]^n \rightarrow [0,1]$, generates the output function by

$$\mu_{agg}(x) = h(\mu_A(x), \mu_B(x), ...). \tag{35}$$

The aggregation operator h is a bounded, monotonic, increasing, symmetric, and idempotent operator (Klir & Yuan, 1995). Beside the fuzzy union and intersection, which can be defined in different ways, the so called *averaging operators* play an important role in aggregation. A mapping $M : [0,1] \times [0,1] \rightarrow [0,1]$ is an averaging operation iff it is continuous and satisfies the properties Symmetry: $M(x_0, x_1) = M(x_1, x_0)$; Idempotency: $M(x_0, x_0) = x_0$; Monotonicity: $M(x_0, x_1) \leq M(x_2, x_3)$ if $x_0 \leq x_2$ and $x_1 \leq x_3$; Boundaries: $M(0,0) = 0$ and $M(1,1) = 1$; $x_i \in [0,1]$ (Dujmović & Larsen, 2007). Furthermore, the following inequality is valid for all averaging operators

$$\min(\mu_A(x), \mu_B(x)) \leq M(\mu_A(x), \mu_B(x)) \leq \max(\mu_A(x), \mu_B(x)). \tag{36}$$

By applying different averaging operators it is possible to construct a more "and-like" (min) or a more "or-like" (max) aggregation. Often the following operators are used: Harmonic mean, Geometric mean, and Arithmetic mean which are certain forms of the *generalised mean*. Another important basic operator covers the full range of Eq. (36). This class of operators is called *ordered weighted averaging operators* (OWA) (Dujmović, 1974), (Yager, 1988), and (Dujmović & Larsen, 2007). A *weighting vector* $\vec{w} = (w_0, w_1, ..., w_{N-1})$ with $w_i \in [0,1]$ with the constraint

$$\sum_{i=0}^{N-1} w_i = 1. \tag{37}$$

The OWA operator h_{OWA} is defined by first ordering the membership functions $\mu_i(x)$ in *non-increasing order*, denoted with index (i), for a given value $x = \theta$, based on their individual fuzzy sets $A, B, ...$, and then applying

$$\mu_{agg}(\vec{w}, \theta) = \sum_{i=0}^{N-1} w_i \cdot \mu_{(i)}(\theta). \tag{38}$$

OWA aggregation can be freely positioned between the borders of Eq. (36). Dujmović (Dujmović, 1974) introduced a measure (degree of disjunction) that indicates how much an operator is near to the max-operation. Yager (Yager, 1988) called this operator *orness*. The operator models any degree of orness between 1 (pure OR) and 0 (pure AND) by adjustment of its corresponding parameters, the OWA weights after Eq. (37).

$$\text{orness}(h_{OWA}(\vec{w})) = \frac{1}{N-1} \sum_{i=0}^{N-1} (N-i) \cdot w_i. \tag{39}$$

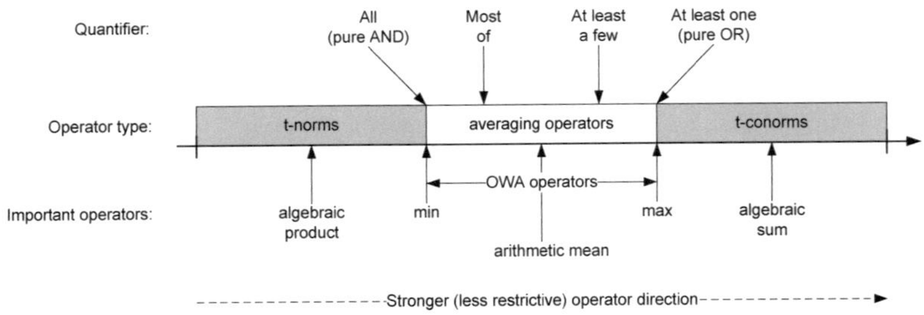

Fig. 5. Different rules of aggregation, with their associated operators (Larsen, 1999).

The *andness* degree is defined as: andness$(h_{OWA}) = 1 - \text{orness}(h_{OWA})$.
Fuzzy Set Theory operates with vague and non-specific information, based on a human-centred way of thinking. Usually, knowledge is represented by membership functions and their aggregation. In contrary to PT the propositions are multi-valued. The construction of membership function can be established by different concepts; e. g. cf. (Bocklisch & Priber, 1986), (Klir & Yuan, 1995), (Bossé et al., 2007), (Lohweg & Mönks, 2010a).

3.4 Possibility Theory

Possibility Theory (PosT) was introduced in 1978 by Zadeh (Zadeh, 1978). As imprecision and vagueness are rather possibilistic than probabilistic (frequentist's view) in its description (*... it is possible that we will have snow in summer, however, it is highly improbable...*), it is reasonable to describe the meaning of information, especially the meaning of *incomplete* information within a possibilistic framework. Possibility Theory reflects Probability Theory insofar, as it is also based on set-functions. Instead of one function, PosT is based on two functions, which are called *possibility measure* and *necessity measure*. We will concentrate on numerical possibility theory with its special branches based on consonant (nested) sets (Shafer, 1976) and on membership functions (Zadeh, 1978).
It has to be pointed out that PosT has been exhaustively studied by Dubois and Prade; e. g. (Dubois & Prade, 1978; 1983; 2000; 2003). Possibility Theory can be interpreted as a framework for handling incomplete information and aggregate information coming from multiple sources (sensors, experts, databases, etc.). If $\Theta = \{\theta_1, \theta_2, ..., \theta_n\}$ denotes the so called *finite frame of discernment*, $\theta_i, i \in \mathbb{N}$, then a *possibility measure* Π is a mapping $\Pi : 2^{\Theta} \rightarrow [0,1]$, which follows the axioms (cf. Eq. (40) - Eq. (42)):

- Axiom 1:
$$\Pi(\varnothing) = 0. \tag{40}$$
- Axiom 2:
$$\Pi(\Theta) = 1. \tag{41}$$
- Axiom 3:
$$\Pi(A \cup B) = \max(\Pi(A), \Pi(B)). \tag{42}$$

The measure $\Pi(A)$ denotes the possibility that an element θ belongs to the *crisp* set A. As it can be noted from Eq. (42) the union of two sets is *non-additive*. The *necessity measure* $N(A)$ is the dual set-function of $\Pi(A)$:
$$N(A) = 1 - \Pi(A^c). \tag{43}$$
The intersection of two sets (Eq. (44)) is determined as follows:
$$N(A \cap B) = \min(N(A), N(B)). \tag{44}$$

Interpreting Eq. (43), the necessity measure describes the fact that $N(A)$ is entirely true, iff the "complement of A" is impossible. The following implications hold for possibility and necessity measures:

$$N(A) > 0 \Rightarrow \Pi(A) = 1,$$
$$\Pi(A) < 1 \Rightarrow N(A) = 0.$$

A fuzzy variable can be interpreted as a representation of an element of a fuzzy set with a graded membership within its set. The membership function can also be interpreted differently: Let $\mu(\theta)$ be a *possibility level* for which $\theta \in A$ represents an existing, but unknown element (incomplete information) within in the crisp set A. We describe an unknown element (number, measurement value, etc.) with a certain level of certainty $(\mu(x) \mid x = \theta)$. In this case the normalised fuzzy set is called *possibility distribution* $\pi_x(\theta)$, where x is a fuzzy variable of Θ, and θ or θ_i are appointed values of Θ that x can take. A possibility distribution is a mapping $\pi : \Theta \rightarrow [0,1]$ with at least one $\pi(\theta_i) = 1$, $\theta_i \in \Theta$.
Possibility and necessity measures are represented by the associated possibility distribution $(\theta \in A)$, that is,

$$\Pi(A) = \max(\pi_x(\theta)), \tag{45}$$
$$N(A) = \min(1 - \pi_x(\theta)). \tag{46}$$

If Θ is infinite, then $\Pi(A) = \sup_{\theta \in A}(\pi_x(A))$ and $N(A) = \inf_{\theta \in A}(1 - \pi_x(A))$.
Now, we will give a short insight in the interconnection between PosT and DST in the case that the focal elements within particular sets are nested, that is, the sets are ordered in a way that each is contained in the next one. Belief and Plausibility functions are termed to be *consonant* (Shafer, 1976). Let Θ be the frame of discernment with its subsets $A_i = \{\theta_0, \theta_1, ..., \theta_{i-1}\}$ for $i = 0, 1, ..., N-1$. The subsets A_i match the relationship:

$$\Theta \equiv A_0 \subset A_1 \subset A_2 \subset ... \subset A_{N-1} \tag{47}$$

In the case of nested sets Belief and Plausibility satisfy the following relationship with $A, B \in \mathcal{P}(\Theta)$, cf. Eq. (42), Eq. (44):

$$\mathrm{Pl}(A \cup B) = \max(\mathrm{Pl}(A), \mathrm{Pl}(B)), \; \mathrm{Bel}(A \cap B) = \min(\mathrm{Bel}(A), \mathrm{Bel}(B)). \tag{48}$$

When Plausibility and Belief functions satisfy Eq. (48), they are called Possibility and Necessity functions. Possibility Theory is defined axiomatically via Eq. (40) - Eq. (42) as an independent theory, though, as a second branch it can be defined by the possibility distribution as a special form of the membership function (Zadeh, 1978) .

$$\pi_x(\theta) = \mu_x(\theta). \tag{49}$$

However, it has to be pointed out that the membership function may not be readily interpreted as a possibility distribution function, because the meaning of both functions is completely different. Whereas the membership function describes a fuzzy variable $\theta \in A$, the possibility distribution function characterises a representation of what is known (expert, sensor measurement, etc.) about the value of some quantity θ ranging on $A \subseteq \Theta$ (not necessarily a random quantity). The function $\pi_x(\theta)$ reflects the more or less plausible values of the unknown quantity roughly. We tend to another notation proposed by Dubois and Prade to circumvent any ambiguity:

$$\pi_x(\theta \mid A) = \mu_x(A \mid \theta). \tag{50}$$

The possibility that $x = \theta$, knowing that $\theta \in A$ is $\pi_x(\theta \mid A)$, whereas $\mu_x(A \mid \theta)$ is the level of possibility for $\theta \in A$. An interesting interrelationship exists between α-cuts (cf. Eq. (31)), basic belief assignments m, possibility distributions $\pi_x(\theta \mid A)$, and membership functions $\mu_x(A \mid \theta)$ iff the sets are nested (cf. Eq. (47)) . Let α_i be a certain value of $\mu_x(A_{\alpha_i} \mid \theta)$, then A_{α_i} denotes it corresponding α-cut. The associated possibility function is then defined by (Klir & Yuan, 1995):

$$\pi_x(\theta) = \sum_{\substack{i=0, \\ \theta \in A_{\alpha_i}}}^{N-1} m(A_{\alpha_i}), \tag{51}$$

under the condition, that

$$m(A_{\alpha_i}) = \alpha_i - \alpha_{i+1}$$

and $i = 0, 1, ..., N-1$, and $\alpha_N = 0$. Furthermore, keeping in mind, the higher α_i, the lower the possibility is that an element θ will fall in its associated α-cut, that is, $\Pi(A^c_{\alpha_i}) = \alpha_i$. Therefore, with Eq. (43),

$$N(A_{\alpha_i}) = 1 - \alpha_i$$

— *the level of confidence*. For details refer to Salicone (Salicone, 2007).

Possibility Theory is well suited for sensor and information fusion. Let $\pi_A(x)$ and $\pi_B(x)$ be two possibility distributions from two different signal sources, then, proposed by Dubois and Prade (Dubois & Prade, 1994), the sources are fused by:

$$\pi_{AB}(x) = \max\Big[\frac{t(\pi_A(x), \pi_B(x))}{1 - k_c(\pi_A(x), \pi_B(x))}, \min[1 - k_c(\pi_A(x), \pi_B(x)), s(\pi_A(x), \pi_B(x))]\Big]. \tag{52}$$

The functions $t(\cdot)$ and $s(\cdot)$ are t-norms and s-norms, where $K(\cdot)$ is the conflicting factor known from Dempster's rule of combination (cf. Eq. (13)). The fusion concept takes into account the reliability of different sources. If both sources are reliable an "and-like" fusion is applied; otherwise, an "or-like" fusion is executed.

Possibility Theory provides a promising framework for sensor and information fusion based on incomplete or party unknown data. It is able to handle uncertainty, ignorance, and vagueness based on crisp sets instead of fuzzy sets. However, being a distinct theory, its branches interconnect Dempster-Shafer Theory, Fuzzy Set Theory, and Probability Theory. Insofar PosT can be interpreted as an unifying Theory.

4. Human behaviour based sensor fusion

Human beings have remarkable potential for reasoning and rational decision-making, even when confronted with such uncertain data. Therefore, it is highly desirable to design technical systems in a way which is inspired by human behaviour. Zadeh called this human level machine intelligence (HLMI), cf. (Zadeh, 2008). Although HLMI has been intensively investigated under the umbrella of artificial intelligence (AI) and its subdivision multi-source data fusion for the past fifty years, it is still a hot research topic (Zadeh, 2008).

4.1 Two-layer Conflict Solving Fusion

Dempster-Shafer based fusion creates counter-intuitive results in high conflict situations. If two sensors, that are measuring the same effect, deliver different information, then a conflict situation occurs. This inherent defect pointed out by Zadeh (Zadeh, 1986), brings criticism to the DST. Therefore, many other alternatives were researched. For instance, we refer to Campos' rule (Campos, 2006) and the work of Dezert and Smarandache (DSmT) (Smarandache & Dezert, 2006). No matter whether the original DST rule or other ad-hoc rules are applied, none of them have been regarded as a superior method compared to others. Therefore, a Two-Layer Conflict Solving (TLCS) data fusion approach is suggested, which includes two layers to combine pieces of evidence. The conflict is solved in some degree during combination—therefore, it is named as conflict solving. The first layer resolves the conflict in some extent, and the second continues to solve it and achieves more stable results. Psychologically, as clearly stated in (Lipshitz et al., 2001) ,*"Decision making has been traditionally studied at three levels: individual, group and organizational."*, also cf. (Sunita, 1999). This shows that decision is made at three layers, in which conflict is unavoidable to be considered and solved: The individual level is the basic element that holds conflict; group level has a larger range which includes conflict while organisational level is the largest. The latter can be instantiated by connecting several layer 2 outputs via TLCS to generate a higher hierarchical level—the organisational one. In such a way, humans believe that conflict can be solved in an straightforward way, although it is impossible to totally eliminate its negative impacts. The approach is also applicable if several groups of sensors are considered in a larger system. Fig. 6 depicts the scheme of TLCS. In Fig. 6, layer 1 is regarded as working at the individual

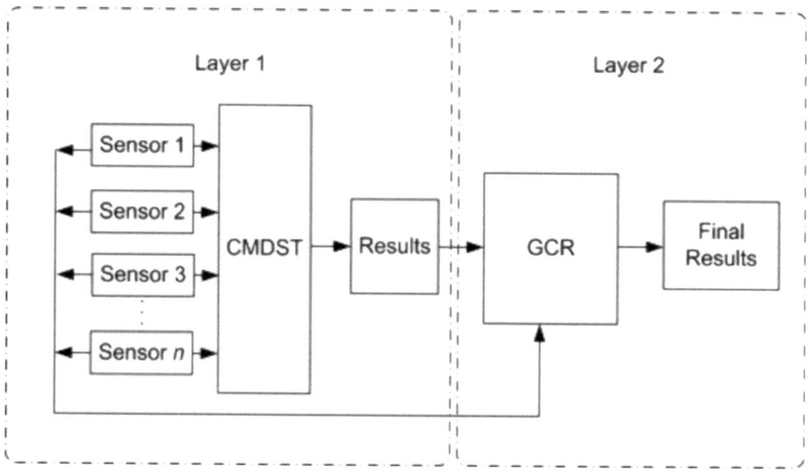

Fig. 6. Two-Layer Conflict Solving System.

level because the Conflict-Modified DST (CMDST) is an approach which combines every two sensors' data so that a conflict is considered and solved between two individuals. After receiving the results from the previous layer, layer 2 collects all sensors' original information and fuses it with the combined results from the CMDST. The conflict is further resolved at a group level.

4.1.1 Aggregation rules

Especially due to the inherent defect residing in the DST (Zadeh, 1986), (Campos, 2006), many other authors have proposed their own data fusion approaches which serve as ad-hoc alternatives. Just to name only a few, e. g., Murphy's rule (Murphy, 2000)

$$m_M = \frac{m_1(A) + m_2(A)}{2}, \tag{53}$$

is a trade-off rule, which takes the arithmetic average value of two masses $m_1(A)$ and $m_2(A)$. Yager's rule (Murphy, 2000) regards that the universal set Θ should include the mass from the conflicting parts, so that the universal set (set with all propositions) is always introduced in Yager's rule:

$$m_Y(C) = \sum_{A,B \in 2^\Theta, A \cap B = C} m_1(A)m_2(B), \tag{54}$$

$$m_Y(\Theta) = m_1(\Theta)m_2(\Theta) + \sum_{A,B \in 2^\Theta, A \cap B = \emptyset} m_1(A)m_2(B). \tag{55}$$

Campos' rule

$$m_C(A_i) = \frac{DST}{1 + \log\left(\frac{1}{1 - \sum_{A_1 \cap .. \cap A_n = \emptyset} \prod_{i=1}^{n} m_i(A_i)}\right)}, \tag{56}$$

is explained in (Campos, 2006). Dezert-Smarandache Theory (DSmT) is rather a comprehensive theory; we refer to (Smarandache & Dezert, 2006).

4.1.2 Conflict-modified DST

Based on the idea of DST, CMDST calculates the conflict in a different manner as shown in the formula:

$$k_{cm} = \sum_{\substack{A_1 \cap A_2 = \emptyset, A_1 \cap A_3 \emptyset, ..., \\ A_i \cap A_k = \emptyset, ..., An-1 \cap A_n = \emptyset}} \prod_{i=k}^{n} m_i(A_k). \tag{57}$$

Within this definition, conflicts are calculated between every two sensors instead of all the sensors together (which is used in DST). This difference can be seen from the condition of summation in Eq. (13), $A_1 \cap ... \cap A_n = \emptyset$ and $A_1 \cap A_2 = \emptyset, A_1 \cap A_3 = \emptyset, ..., A_1 \cap A_n = \emptyset, ..., A_{n-1} \cap A_n = \emptyset$. Due to the specified way of determining conflicts in Eq. (57), this kind of conflict will very likely be larger than one, whereas in DST the denominator is $1 - k_c$. Therefore, the denominator in DST must also be modified. First of all, the denominator is modified as:

$$\binom{n}{2} - k_{cm} = \binom{n}{2} - \sum_{\substack{A_1 \cap A_2 = \emptyset, A_1 \cap A_3 = \emptyset, ..., \\ A1 \cap A_n = \emptyset, ..., A_{n-1} \cap A_n = \emptyset}} \prod_{i=k}^{n} m_i(A_k). \tag{58}$$

The reason for choosing $\binom{n}{2} = B_c(n)$ (binomial coefficient) is that there are $B_c(n)$ possible combinations for calculating conflicts (n is the number of sensors). Thus, K_{cm} (K in DST) is:

$$K_{cm} = \frac{1}{B_c(n) - \sum_{\substack{A_1 \cap A_2 = \emptyset, A_1 \cap A_3 = \emptyset, ..., \\ A_i \cap A_k = \emptyset, ..., An-1 \cap A_n = \emptyset}} \prod_{i=k}^{n} m_i(A_k)} \tag{59}$$

with $\frac{1}{B_c} \leq K_{cm} \leq \infty$. The CMDST is formed as:

$$\text{CMDST}(A) = \bigoplus_{i=1}^{n} m_i(A) = K_{cm} \sum_{\substack{A_1 \cap A_2 = A, \ldots \\ A_1 \cap A_3 = A, \ldots \\ A_{n-1} \cap A_n = A}} \prod_{i=1}^{n} m_i(A). \tag{60}$$

4.1.3 Group Conflict Redistribution

As pointed out in (Sunita, 1999), decision making is also studied in a group level. Hence, Group Conflict Redistribution (GCR, layer 2 in Fig. 6) acts as group conflict solving strategy, solving a conflict in a larger extent compared to individual level (CMDST). Distinguishing from layer 1 (CMDST), GCR combines sensors' propositions in a group manner, which means all sensors will participate in this procedure (Li & Lohweg, 2008):

$$m(a) = \frac{\sum_{A_1 \cap .. \cap A_n = A} m_i(A) + (B_c(n) + |\log(B_c(n) - k_{cm})|) \cdot \text{CMDST}(A)}{n + B_c(n) + |\log(B_c(n) - k_{cm})|}. \tag{61}$$

The variable n defines the number of sensors and $\log(x)$ is logarithm to the base 10. The denominator includes the number of sensors and how many combinations $B_c(n)$ among sensors are possible as well as conflict evaluation term $|\log(B_c(n) - k_{cm})|$. In the numerator part, sensors' original propositions $\sum_{A_1 \cap .. \cap A_n = A} m_i(A)$ are calculated with corresponding CMDST results, which are obtained from layer 1. Finally, the sum of final fused results remains '1'. Numerical examples on TLCS which describe the aggregation effect can be found in (Li & Lohweg, 2008).

4.1.4 Balanced TLCS

The intention to optimise the TLCS concept is threefold. Balanced TLCS (*BalTLCS*) intends to cover three topics: First, the general TLCS approach should be utilised again. Second, the result for two information sources (sensors, cognitive knowledge) should generate intuitive results. Third, the conflict redistribution scheme for data fusion should be adapted and simplified in a way that the fusion process i) should be an additive function of all inputs and ii) could be extended to multiple sources and focal elements (masses). The approach is based on the normalised version of CMDST. The normalisation is based on the Eq. (57) - Eq. (60). The normalised conflicting factor ${}^{N}K_{cm}$ is defined as follows:

$${}^{N}K_{cm} = \frac{B_c(n)}{B_c(n) - k_{cm} + \epsilon} < \infty \tag{62}$$

where $\forall \epsilon > 0, \forall n > 1, n \in \mathbb{N}, k_{cm} = 0 \, \exists L \in \mathbb{R} : L = (1 - {}^{N}K_{cm}) < \epsilon$. Therefore, as $0 \leq k_{cm} \leq B_c(n)$, the normalised conflict factor ranges between

$$\frac{B_c(n)}{B_c(n) + \epsilon} \leq {}^{N}K_{cm} \leq B_c(n) \cdot \epsilon^{-1} < \infty. \tag{63}$$

The non-conflicting part is determined and coupled with the conflicting part according to Eq. (60) as

$${}^{N}\text{CMDST}(A) = \bigoplus_{i=1}^{n} m_i(A) = \frac{{}^{N}K_{cm}}{B_c(n)} \sum_{\substack{A_1 \cap A_2 = A, \ldots \\ A_1 \cap A_3 = A, \ldots \\ A_{n-1} \cap A_n = A}} \prod_{i=1}^{n} m_i(A). \tag{64}$$

The *BalGCR* (Balanced Group Conflict Redistribution) approach combines the sensors' propositions in a group manner which implies that in this case all sensors will participate additively in this procedure.
The intention is to utilise the inverse of the normalised conflict factor as a control parameter. If no conflict has occurred, mainly the ^{N}CMDST fusion result should contribute to the overall result. If the conflict is high, then all different information sources have to be taken into account. None of the information sources is allowed to dominate the other ones, and none of them is allowed to play a part in the overall result with more than $\frac{1}{n}$.
Furthermore, in a strong conflict case it is not intended to shift the information content (hypotheses) to the universal set Θ, defining ambiguity or ignorance, because a conflict has to be solved in any case. As the set must be complete regarding the sources, all information sources (hypotheses) which can appear must be covered in a set. The only situation where hypotheses can be transferred to the universal set Θ occurs when a source delivers no reliable data. Formally *BalGCR* fusion consists of two parts. The first one describes the non-conflicting part:

$$m_{nc} = {}^{N}K_{cm}^{-1} \cdot \mathrm{CMDST}(A) = \frac{1}{B_c(n)} \sum_{\substack{A_1 \cap A_2 = A, \dots \\ A_1 \cap A_3 = A, \dots \\ A_{n-1} \cap A_n = A}} \prod_{i=1}^{n} m_i(A). \tag{65}$$

The second part characterises the conflicting part. Eq. (65) tends to zero in heavy conflicts ($k_{cm} \rightarrow B_c(n)$). Therefore, it is proposed here that all masses supporting a certain hypothesis have to be averaged to fulfil the above mentioned statements. Furthermore, the average value has to be controlled, e. g., by the normalised conflicting factor or coefficient. It is proposed to use

$$m_c(A) = \frac{k_{cm}}{B_c(n)} \cdot \frac{1}{n} \sum_{i=1}^{n} m_i(A) \tag{66}$$

for the conflicting part. It can be recognised that, in the case of maximum conflict, the average value of all sensory hypotheses is determined. In the case of minimum conflict $m_c(A) \rightarrow 0$ both parts (Eq. (65) and Eq. (66)) are additively connected.

$$m(A) = m_c(A) + m_{nc}(A) = \frac{k_{cm}}{B_c(n)} \cdot \frac{1}{n} \sum_{i=1}^{n} m_i(A) + \frac{1}{B_c(n)} \sum_{\substack{A_1 \cap A_2 = A, \dots \\ A_1 \cap A_3 = A, \dots \\ A_{n-1} \cap A_n = A}} \prod_{i=1}^{n} m_i(A). \tag{67}$$

This equation describes the data fusion approach by the balanced TLCS concept. All masses are fused regarding their pro-and-con-hypotheses. Finally, the sum of final fused results remains always '1', if all sensors deliver acceptable results in the sense of an adequately functioning sensor. As the fusion process is based on averaging, the final result is relatively insensitive against noise effects. The fused mass of the universal set is determined for n sensors as follows:

$$m(\Theta) = 1 - [m(A) + m(B) + m(C) + \dots] . \tag{68}$$

The masses take into account all information from the hypotheses. For further details and results we refer to (Li & Lohweg, 2008) and (Lohweg & Mönks, 2010b).

4.2 Majority-Observation-Guided Possibilistic Fusion

One promising mathematical theory that appears to be capable of (partially) modelling a human-like decision process is *Possibility Theory*. Although Possibility Theory was introduced in 1978 by Zadeh (Zadeh, 1978) and has been primarily advanced by Dubois and Yager over the past few decades (Dubois & Prade, 2000; Yager, 2004), it is still in its infancy. This section is focused on possibilistic multi-sensor systems for industrial automation. With reference to the Sect. 4.1, *Dempster-Shafer Theory*-based fusion approaches were recently applied to this problem statement and reached satisfactory results (Lohweg & Mönks, 2010b). On the contrary, possibilistic approaches are rarely applied to those industrial-automation processes that can be classified under the problem theme of context-based anticipatoric machine conditioning. As Possibility Theory appears to be crucial for modelling the human behaviour when confronted with vague data (Zadeh, 2008), this section presents a novel possibilistic multi-source data fusion approach, the so-called *Majority-Observation-Guided Possibilistic Fusion Rule (MOGPFR)*. The MOGPFR is capable of handling the imprecision of sensors. Moreover, it has the ability of merging imprecise, incomplete or inconsistent data in a human-like group-decision process.

The rest of this section is organised as follows: Section 4.2.1 provides the basics needed for the understanding of the framework. Sections 4.2.2– 4.2.6 introduce the novel possibilistic fusion rule.

4.2.1 Fusion operators

As pointed out in Sect. 3.4, conjunctive aggregation of possibility distributions is debatable when not all sources are completely reliable. In this context, when all sources are completely unreliable, the disjunctive rule is most appropriate. However, when confronted with a group of reliable and unreliable sources the class of averaging operators, which lies between the conjunctive and the disjunctive rule (cf. Fig. 5), might be beneficial. OWA operators consider all fuzzy sets equally important. However, in multi-criteria decision problems, such as data fusion, different criteria, represented by different fuzzy sets, might have different importance. Hence, the effect on the overall outcome of satisfying a criterion should decrease as the importance of satisfying the criterion decreases. To meet this desire, the concept of OWA aggregation is expanded by attaching a so-called *importance weight*, w_i, to each criteria, a_i to the importance weighting vector $\vec{v}$ of a n-dimensional attribute A. This leads to the class of *Importance Weighted Ordered Weighted Averaging (IWOWA)* operators as defined below:

$$\lambda_{\text{IWOWA}}(\vec{v},\vec{w},\vec{a}) = \sum_{i=1}^{n} w_i b_{(i)}, \tag{69}$$

where i is a permutation on $\{1,...,i,...,n\}$, such that $b_{(1)} \geq ... \geq b_{(i)} \geq ... \geq b_{(n)}$ and $b_i = h^{\alpha}_{\text{OWA}}(v_i, a_i)$ (Larsen, 1999). Larsen has shown (Larsen, 1999) that the class of IWOWA operators is order equivalent to the *Weighted Arithmetic Mean (WAM)* (Yager, 1994). Order-equivalence is sufficient when the operator is applied to provide preference ordering (Larsen, 2002). However, in situations where the aggregated value is used for other purposes, such as multi-source data fusion, a full equivalence to the WAM is needed. This property can be obtained by normalizing Eq. (69) in the interval of $\lambda_{\text{IWOWA}}(\vec{v},\vec{w},0)$ and $\lambda_{\text{IWOWA}}(\vec{v},\vec{w},1)$, which leads to the class of *Implicative Importance Weighted Ordered Weighted Averaging (IIWOWA)* operators as defined as follows (Larsen, 2002):

$$\lambda_{\text{IIWOWA}}(\vec{v},\vec{w},\vec{a}) = \frac{\lambda_{\text{IWOWA}}(\vec{v},\vec{w},\vec{a}) - \lambda_{\text{IWOWA}}(\vec{v},\vec{w},0)}{\lambda_{\text{IWOWA}}(\vec{v},\vec{w},1) - \lambda_{\text{IWOWA}}(\vec{v},\vec{w},0)}. \tag{70}$$

4.2.2 Possibilistic fusion frameworks

Various multi-source data-fusion frameworks based on the Possibility Theory have been proposed in the past thirty years (Yager, 1979; 1994). These frameworks were mainly influenced and developed by Dubois and Yager (Dubois et al., 2001). Therefore, this section introduces the two possibilistic fusion frameworks previously mentioned.
Dubois' adaptive fusion is grounded on the discussions on the usefulness of the *maximum specificity principle*. It is based on the supposition that $j \leq n$ sources of a n-dimensional fusion problem are reliable. First, the j reliable sources are combined conjunctively. Secondly, the intermediate result of the combination of the j reliable sources is combined disjunctively with the $(n-j)$ unreliable sources. However, in real-life, a precise number of reliable sources is sometimes unknown or is complex to evaluate. For this reason, a good alternative is an *optimistic* estimation, j_o, and a *pessimistic* estimation, j_p, of the number of reliable sources using the following derivation (Dubois et al., 2004): Let $T \subseteq S$ be a subset of sources, $S = \{S_1, ..., S_i, ..., S_n\}$ be the set of sources to be combined, and h be the consistency index

$$h(\pi_v^1(x), \pi_v^2(x)) = \sup_{x \in X} \left[\min[\pi_v^1(x), \pi_v^2(x)]\right] \tag{71}$$

of the subset T. Then, j_o and j_p are determined as stated below:

$$T_p = \sup[T|h(T) = 1], \tag{72}$$

$$T_o = \sup[T|h(T) > 0], \tag{73}$$

$$j_p = |T_p|, \tag{74}$$

$$j_o = |T_o|, \tag{75}$$

where Eq. (72) describes the biggest subset in S that satisfies $h(T) = 1$, that is, it is completely consistent; and Eq. (73) describes the biggest subset in S that satisfies $h(T) > 0$, that is, it is not completely inconsistent. With reference to Eq. (72) and Eq. (74), it is feasible that at least j_p sources are reliable, because j_p represents the number of sources with full agreement. With this in mind, the numerical adaptive combination rule can be stated as follows (cf. Eq. (52)):

$$\pi_v^{1,...,n}(x) = \max\left[\frac{\pi_v^{j_o}}{h(T_o)}, \min\left[\pi_v^{j_p}, 1 - h(T_o)\right]\right]. \tag{76}$$

Yager's fusion approach (Yager, 1979) is based on Zadeh's extension principle (Zadeh, 1978) and is defined as follows:

$$\pi_v^{1,...,n}(x) = \left\{\frac{\min\left[\pi_v^i(x_i)\right]}{F(x_1, x_2, ..., x_n)}\right\}, \tag{77}$$

where $x_i \in X$ and $F(x)$ is an averaging operator. One problem that arises in blindly using the extension principle is that the aggregated possibility distribution may not be satisfactory because it does not reflect any of the source's opinion. For example, if there are two estimates of a person's age, $p_1 = 30$ and $p_2 = 50$, then the average is 40. However, this is a value that is not very compatible with either of the estimates. The reason for this lack of compatibility is the attempt to aggregate disparate values. To solve this problem, Yager introduced an "intelligent" component, the so-called *compatibility* relationship, R, in the merging process to address conflicts in the data to be fused. In this context, the compatibility relationship is defined by expert knowledge. Fusion by considering compatibility is defined as follows (Yager, 1994):

$$\pi_v^{1,\dots,n}(x) = \left\{ \frac{\min\left[R(x_1,x_2,\dots,x_n), \min\left[\pi_v^i(x_i)\right]\right]}{F(x_1,x_2,\dots,x_n)} \right\}, \tag{78}$$

where $x_i \in X$, $R(x_1,x_2,\dots,x_n) = \min_{i,j}\left[R(x_i,x_j)\right]$, and F is an averaging operator. As observed in Eq. (78), the merging of incompatible data is indicated by a low possibility degree.

4.2.3 Fusion rule

The possibilistic frameworks introduced in Sect. 4.2.2 offer a promising setting for merging observations from different sensors. However, both approaches do not consider sensor's reliability. In case of highly unreliable or defective sensors, this might affect the fusion result negatively, which makes both approaches less suited to real-life machine conditioning applications. To overcome the drawbacks mentioned before, this section presents a novel possibilistic data-fusion approach, which is based on Yager's and Dubois' frameworks for data fusion (Dubois et al., 2001; Yager, 2004) and the *theory of collective group decisions* (Eisenführ, 2003). It is an extended close-to-practice approach, which has the ability of considering sensor reliability. Before defining the fusion rule in detail, the hypotheses based on Yager's and Dubois's concepts have been widened are discussed in the following paragraphs.

H1. In real life, group decision-making by humans is mainly driven by the concept of majority opinion. This is derived from the fact that unanimous consensus rarely occurs during group decisions. Hence, it is natural to provide a fusion approach for the ability of majority-guided aggregation.

H2. During group decision-making processes, decision-makers might have different priorities or reliabilities. It is natural that group decisions are mainly based on the voting of prioritised and reliable decision-makers.

Undoubtedly, the desired characteristics are intrinsically vague. Accordingly, different fuzzy approaches assessing majority-guided decisions are ideally suited. In this context, the class of IIWOWA operators, introduced in Sect. 4.2.1, is relevant. However, the IIWOWA operator cannot be applied directly to sensors' observations, because it is explicitly defined for inputs in the range of $[0,1]$. Therefore, the sensor observations are fuzzified to make the data measured applicable to IIWOWA operators. The fuzzification of the sensor observations is accomplished using the *Modified-Fuzzy-Pattern-Classifier* (Lohweg & Mönks, 2010a). In addition to the necessity of fuzzification because of application of an IIWOWA operator, the fuzzification of the measurement scale has an important practical benefit: the fuzzification process facilitates the usage of different sensors that measure different physical quantities on the same attribute. Considering the preceding discussion, the so-called *Majority-Opinion-Guided Possibilistic Fusion Rule (MOGPFR)* is defined as follows:

$$\pi_v^{1,\dots,n}(x) = \max_i\left[\rho^i\right] \cdot \hat{\pi}_v^{1,\dots,n}(x) + 1 - \max_i\left[\rho^i\right], \tag{79}$$

where

$$\hat{\pi}_v^{1,\dots,n}(x) = \left\{ \frac{\lambda_{\mathrm{IIWOWA}}\left(\vec{v}, \vec{w}_{\mathrm{p}}, \vec{\pi}_v\right)}{\lambda_{\mathrm{IIWOWA}}\left(\vec{v}, \vec{w}_{\mathrm{m}}, \vec{\mu}\right)} \right\}, \tag{80}$$

$$\vec{\pi}_v = \left(\pi_v^1(x_1), ..., \pi_v^i(x_i), ..., \pi_v^n(x_n)\right),$$
$$\vec{\mu} = (\mu_1(x_1), ..., \mu_i(x_i), ..., \mu_n(x_n)).$$

The variable $\mu_i(x_i)$ represents the fuzzified sensor observation x_i of the ith sensor S_i. Furthermore, $\pi_v^i(x_i)$ represents the corresponding possibility distribution. The characteristics of IIWOWA operators can be flexibly adjusted by modifying the importance weighting vector $\vec{v}$ and the weighting vectors $\vec{w}_p$ and $\vec{w}_m$ (Larsen, 1999).
The following paragraphs describe the methods for a problem-dependent evaluation of these parameters, so that the stated hypotheses are satisfied:
Consider a n-dimensional fusion problem of a set of n sensors, $S = \{S_1, ..., S_i, ..., S_n\}$, with their associated reliabilities, $\{\rho^1, ..., \rho^i, ..., \rho^n\}$. Then, the following properties, which have to be fulfilled by the MOGPFR, can be derived from H1 and H2:

Lemma 1. *If at least one sensor is completely reliable, the MOGPFR combines the possibility distributions conjunctively, whereas when all sources are unreliable, the MOGPFR combines the possibility distributions disjunctively (cf. Sect. 4.2.2).*

Lemma 2. *The fusion approach of the MOGPFR proceeds progressively from the conjunctive to the disjunctive mode, when the highest reliability among all sources decreases from one to zero (cf. Sect. 4.2.2).*

The function defined by Eq. (79) satisfies Lemma 1 and 2 iff the andness-degree of the IIWOWA operator of Eq. (80) used for fusing the possibility distributions is determined as follows:

$$\alpha_p = \min_{i \in 1,...,n} \left[\rho^i\right]. \tag{81}$$

Proof. The satisfaction of Lemma 1 and 2 follows directly from the definition of the OWA and IIWOWA operators (Larsen, 1999; Yager, 1994).

□

Lemma 3. *If only one source is fully reliable and all other sources are totally unreliable, the fusion result of the MOGPFR models the possibility distribution provided by the reliable source (cf. H2).*

The function defined by Eq. (79) satisfies Lemma 3 iff the importance weighting vector $\vec{v}$ of Eq. (80) is defined as stated below:

$$\vec{v} = \left(\rho^1, ..., \rho^i, ..., \rho^n\right). \tag{82}$$

Proof. Consider that S_1 is totally reliable and all other sources, $\{S_2, ..., S_i, ..., S_n\}$, are completely unreliable. Hence, with reference to Eq. (82), the importance weighting vector accounts to $\vec{v} = (1, 0, ..., 0)$ and Eq. (79) results in

$$\pi_v^{1,...,n}(x) = \hat{\pi}_v^{1,...,n}(x), \forall\, x \in X.$$

Consider the definition of IIWOWA operators (Larsen, 1999), the importance weighting function mutates to

$$h(0, \pi_v^i) = I \;\; \forall\, i \neq 1,$$
$$h(1, \pi_v^1) = \pi_v^1.$$

Therefore, the importance weighting function models for all sources except S_1 the identity element I and hence, $\pi_v^{1,...,n}(x) = \hat{\pi}_v^{1,...,n}(x) = \pi_v^1(x) \;\forall\, x \in X$. □

Lemma 4. *When all sources are completely unreliable, then the MOGPFR models total ignorance (cf. Sect. 4.2.2).*

The function defined by Eq. (79) satisfies Lemma 4.

Proof. Consider that all sources are totally unreliable. Hence, with reference to Eq. (82), the importance weighting vector accounts to $\vec{v} = (0, ..., 0)$ and Eq. (79) can be written as follows:

$$\pi_v^{1,...,n}(x) = 1, \forall\, x \in X.$$

Thus, Eq. (79) models total ignorance.

□

Lemma 5. *If there is a conflict between sources, it is solved by the MOGPFR so that the majority observation is prioritised (cf. H1).*

To meet the requirement postulated in Lemma 5, the majority observation of the sensors has to be evaluated. The evaluation of the majority observation is discussed in Sect. 4.2.4.

4.2.4 Data consistency

The evaluation of the majority observation, necessity to satisfy Lemma 5, arises the need of a *consistency* measurement amongst sensors' observations, which is introduced in the following: Consider a set of sensors $S = \{S_1, ..., S_i, ..., S_n\}$ with their corresponding reliabilities $\{\rho^1, ..., \rho^i, ..., \rho^n\}$. Let each sensor provide an observation through a possibility distribution $\pi_v^i, i \in \{1, ..., n\}$. Then, the largest set of sensors T_c, which forms a consensus observation is defined as follows:

$$T_c = \sup\left[T | h(T) > 0\right], \tag{83}$$

where $T \subseteq S$, $h(T)$ is the consistency index as defined in Eq. (71).

Naturally, observations of sensors which are elements of T_c are completely consistent with the consensus observation and therefore, their consistency should be equal to one. In the same way, the consistency of an observation should decrease when the distance to the consensus observation increases. Hence, the measurement of consistency rises the need for evaluating the range of the consensus observation and an appropriate definition for measuring the distance amongst observations.

To simplify the measure of consistency, the possibility distribution (observation) of each sensor is defuzzified using its *centre of gravity*, denoted as C (Klir & Yuan, 1995). The range of the consensus observation, $[c_{\min}, c_{\max}]$, can then be calculated as follows:

$$c_{\min} = \min_{S_i \in T_c}\left[C\left(\pi_v^i\right)\right], \tag{84}$$

$$c_{\max} = \max_{S_i \in T_c}\left[C\left(\pi_v^i\right)\right]. \tag{85}$$

Knowing the extent of the consensus observation, the measure of consistency can be defined by the following:

$$Co_c\left(\pi_v^i, T_c\right) = \begin{cases} \left|c_{\min} - C\left(\pi_v^i\right)\right|, & C\left(\pi_v^i\right) < c_{\min}, \\ \left|c_{\max} - C\left(\pi_v^i\right)\right|, & C\left(\pi_v^i\right) > c_{\max}, \\ 1, & \text{otherwise}. \end{cases} \tag{86}$$

Considering Eq. (86) sensors that have their *centres of gravity* within the range of the consensus observation are 100% consistent and, thus, are modelled by $Co_c(\pi_v^i, T_c) = 1$. Furthermore, when the distance of a sensor's observation to the boundaries of the consensus observation increases, the consistency of the observation with reference to the consensus observation decreases, and hence, $Co_c(\pi_v^i, T_c) = [0, 1)$.
However, in practical applications, no consensus observation may be found amongst all the sensors, e. g. $|T_c| = 1$. Then, the measurement of consistency, as defined by Eq. (86), becomes inadequate. Nevertheless, a consistency measure that evaluates the amount of conflict among the sensors is both obvious and adequate. It should be noted that reliabilities amongst sensors can vary. It is debatable whether a reliable sensor, that is in conflict with an unreliable sensor should be considered inconsistent. Accordingly, the consistency measurement should take the reliabilities of sensors into account. WAM meets the desired features stated above and thus, is used as follows:

$$Co_a(\pi_v^i, \pi_v^*) = \max\left[1 - \max_{j \neq i}\left[r_j\right], \lambda_{\text{WAM}}\left(\vec{\rho}^i, \vec{d}_i\right)\right], \tag{87}$$

where

$$\pi_v^* = \{\pi_v^j\}_{j \neq i}, \; \vec{\rho}^i = (\rho^j)_{j \neq i}, \; \vec{d}_i = (d_{i,j})_{j \neq i},$$

and

$$d_{i,j} = 1 - \left|C\left(\pi_v^i\right) - C\left(\pi_v^j\right)\right|. \tag{88}$$

With reference to Eq. (87), the consistency of a sensor's observation is the weighted average of all distances from the observation of the ith sensor π_v^i to the observations of all other sensors π_v^*. The weighting vector of the weighted arithmetic mean corresponds to the reliabilities of the sensor' observations π_v^*. Moreover, due to the max-operator in Eq. (87), the consistency of π_v^i will be equal to one, if all observations of π_v^* are completely unreliable. Both versions of evaluating the consistency of a sensor's observation are brought together in the following equation:

$$Co(\pi_v^i) = \begin{cases} Co_c(\pi_v^i, T_c), & |T_c| > 1, \\ Co_a(\pi_v^i, \pi_v^*), & \text{otherwise.} \end{cases} \tag{89}$$

After determining the consistency of each observation it is natural to consider this information during the fusion process. Thus, an observation with a low consistency should contribute little to the fusion result and vice versa. Therefore, the MOGPFR is modified, so that the consistency index of each observation is explicitly taken into account.
From Eq. (80), it follows that the determination of the weighting vector $\vec{v}$ as defined in Eq. (82) has to be modified to meet the desired features mentioned above and stated in Lemma 5. A functional description, that meets the needs postulated previously is called *credibility*, denoted as $Cr(\rho^i, Co(\pi_v^i)) = \min[\rho^i, Co(\pi_v^i)]$ (Yager & Kelman, 1996). Thus, Eq. (82) is modified as follows:

$$\vec{v} = \left(Cr(\rho^1, Co(\pi_v^1)), ..., Cr(\rho^n, Co(\pi_v^n))\right). \tag{90}$$

Next, whether Eq. (90) is in accordance with Lemma 2 is checked.

Proof. The modification of Eq. (82) does not violate Lemma 2, since if all sources $\{S_2, ...S_n\}$ except S_1 are unreliable, e. g. $\vec{v} = (1, 0, ..., 0)$ (cf. Eq. (82)) then

$$\vec{v} = \left(Cr(1, Co(\pi_v^1)), ..., Cr(0, Co(\pi_v^n))\right) = (1, 0, ..., 0).$$

With respect to the proof of Lemma 2, only π_v^1 is considered. □

Using Eq. (90) for determining the importance weighting vector $\vec{v}$ of Eq. (79) satisfies Lemma 5.

Proof. Consider a n-dimensional fusion problem of a set of n sources $S = \{S_1, ..., S_i, ..., S_n\}$ and let all sources be reliable, e.. $\vec{v} = (1, ..., 1)$. Moreover, let $T_c = \{S_1, ..., S_k\}$. With reference to Eq. (89), the importance weighting vector $\vec{v}$ becomes $\vec{v} = (1, ..., 1, a_{k+1}, ..., a_n)$, where $\{a_{k+1}, ..., a_n\} < 1$ and therefore, the group of sources building the consensus observation, that is the majority observation, is prioritised. □

4.2.5 Sensor reliability

As information provided by physical sensors can vary in quality, reliability is important for the success of multi-source data fusion. Two general concepts of obtaining source's reliability are established in literature (Rogova & Nimier, 2004): *static* and *dynamic reliability*. Static reliability indicates the probability that a source is working correctly. However, it does not consider unpredictable disturbances, such as environmental influences, mis-applications, or failures caused by accidents. Paying no heed to those ascendancies effectively lowers the performance of a fusion system. That is why here both static reliability, based on expert know-how, and consensus-driven dynamic reliability estimation are used.

Consensus-driven approaches presume that unreliable sources result in conflicting situations. In this context, it is assumed that a higher conflict results in a larger distance from the consensus observation. Accordingly, the measurement of consistency (cf. Eq. (89)) is ideally suited for the determination of the reliability of a source. If there are unreliable sources that do not cause conflicts, consensus-driven reliability estimation methods become inadequate.

The consistency of a source can vary widely between fusion steps, because of noise in real-life fusion problems. Therefore, it makes sense to smooth the reliability of a source by considering its former reliabilities. The family of *exponential moving average filters*, introduced below, allows modelling the needs postulated herein.

The exponential moving average filter can be classified under *infinite impulse response* filters and is defined as follows:

$$y[n] = \omega \cdot x[n] + (1 - \omega) \cdot y[n - 1], \tag{91}$$

where $\omega \in [0, 1]$. The filter output $y[n]$ and the filter input $x[n]$ are defined for the discrete time step n. Moreover, $y[n - 1]$ is the filter output at the previous discrete time step $[n - 1]$.
Exponential filtering of the consistency of a source allows modelling of its dynamic reliability, $\rho^i_{\mathrm{d}}[n]$, intuitively. Additionally, the static reliability, $\rho^i_{\mathrm{s}}[n]$, can be used as a convex envelope for the smoothed consistency of a source. This leads to the following definition of dynamic reliability:

$$\rho^i_{\mathrm{d}}[n] = \min\left[\rho^i_{\mathrm{s}}[n], \hat{\rho}^i_{\mathrm{d}}[n]\right], \tag{92}$$

where

$$\hat{\rho}^i_{\mathrm{d}}[n] = \omega \cdot Co(\pi^i_v) + (1 - \omega) \cdot \hat{\rho}^i_{\mathrm{d}}[n - 1], \tag{93}$$

and

$$\hat{\rho}^i_{\mathrm{d}}[n] = 1, \forall\, n < 0. \tag{94}$$

The inertia of a process monitored by a source can be taken into account by aligning ω. Thus, a high inertia should be accounted by a small ω and vice versa. To illustrate the effect of exponential filtering, Fig. 7 depicts the filtered and non-filtered reliability of an arbitrary source. It is clearly visible that the influence of process noise is considerably reduced due to exponential averaging.

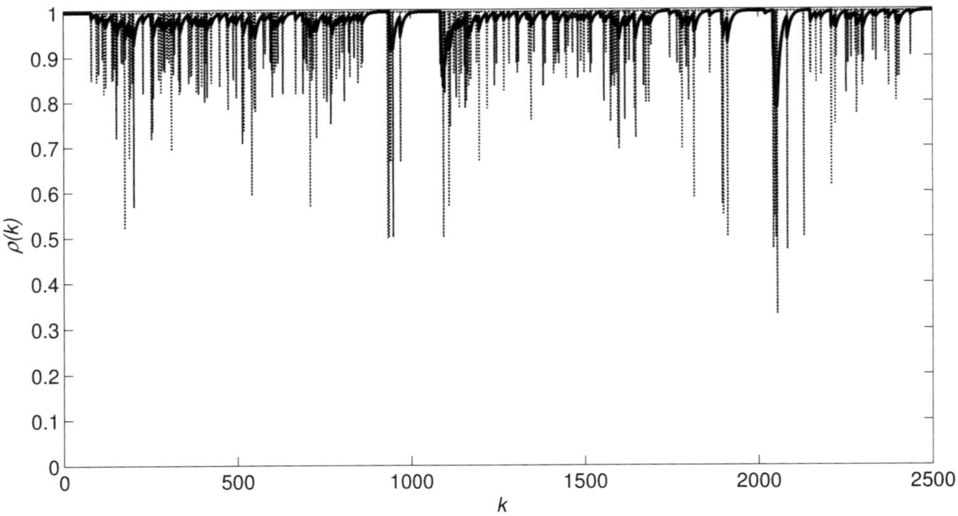

Fig. 7. The smoothed and non-smoothed reliability of an arbitrary source. The dashed line visualises the non-smoothed reliability. The thick line illustrates the smoothed reliability when an exponential moving average filter with $\omega = 0.2$ is applied.

4.2.6 Trustworthiness of the fusion result

An important problem after obtaining a fusion result is the question of its trustworthiness. Delmotte et al. (Delmotte, 2007) proposed an approach that measures the trustworthiness of a fusion result by considering internal contradictions and reliabilities of sources. Here, Delmotte's idea of determining the trustworthiness of fusion results is captured and modified so that it qualifies for the MOGPFR.

In Sect. 4.2.4, the measurement of credibility was introduced. Here, the average credibility of all sensors is accounted for the trustworthiness of the fusion result. In this context, a high overall credibility should result in a high trustworthiness and vice versa. As reliabilities among sources can vary, it is natural that the trustworthiness should mainly be influenced by highly reliable sources, whereas sources with low reliability should influence the trustworthiness of the fusion result only to a small amount. Consequently, the weighted arithmetic mean is appropriate for calculating the trustworthiness. In this context, the importance weighting vector $\vec{v}$ equals the reliabilities of the sources to be merged.

With reference to the MOGPFR, the consistency index $h(T_c)$, (cf. Eq. (71)), contains information about the trustworthiness of the fusion result too. That is, a high consistency index indicates a high agreement amongst the sources that build the consensus observation. On the contrary, a low consistency index indicates a low harmony in the consensus observation. Accordingly, a high consistency index should result in a high trust in the fusion result and vice versa.

As both the overall credibility of the sources and the consistency index are equally important to the trustworthiness of a fusion result, denoted by *Tr*, it is defined as the average of both the indexes:

$$Tr\left(\pi_v^{1,\dots,n}(x)\right) = 0.5 \cdot h(T_c) + 0.5 \cdot \lambda_{\mathrm{WAM}}\left(\vec{\rho}, \vec{c}\right), \tag{95}$$

where

$$\vec{c} = \left(Cr(\rho^1, Co(\pi_v^1)), \dots, Cr(\rho^n, Co(\pi_v^n))\right),$$

and

$$\vec{\rho} = \left(\rho^1, ..., \rho^n\right).$$

4.3 Two-Layer Possibilistic Fusion

Complex fusion tasks, such as condition monitoring of printing machines, require the fusion of different attributes, because sophisticated processes cannot be described sufficiently by one attribute. In general, an attribute describes the state of a variable that influences the condition of a process to be monitored. Furthermore, an attribute consists of several criteria that determine the attribute's condition. These criteria are monitored by a set of sensors. Thus, the fusion of highly sophisticated processes is a twofold process: i) the criteria corresponding to one particular attribute are fused; ii) the attributes are aggregated into one evidence about the condition of the machine. The fusion approach introduced in this section is called the *Two-Layer Possibilistic Fusion Rule (TLPFR)*. The TLPFR is presented in an application-oriented manner for the purpose of machine conditioning. However, other applications of the TLPFR are also conceivable.

Consider a n-dimensional fusion problem of a set of n attributes $A = \{A_1, ..., A_i, ..., A_n\}$. Furthermore, let each attribute A_i consist of m_i criteria observed by m_i sources $S_{A_i} = \{S_{A_{i,1}}, ..., S_{A_{i,j}}, ..., S_{A_{i,m_i}}\}$. Moreover, let each source $S_{A_{i,j}}$ provide its observation using a possibility distribution $\pi^j_{A_i}(x)$. Thus, the fusion of m_i criteria into one result about the condition of one particular attribute A_i can be considered as a m_i-dimensional possibilistic fusion problem. The fusion on the first layer, the so-called *criteria-fusion layer*, is accomplished using the MOGPFR. Furthermore, the compound consequence about the state of a particular attribute A_i is defuzzified using the *mean of maxima* (Klir & Yuan, 1995) and additionally, the attribute's trustworthiness (cf. Eq. (95)) is considered. Thus, for each attribute, one obtains numerical values indicating its current condition x_{A_i} and its corresponding trustworthiness $Tr\left(\pi^{1,...,m_i}_{A_i}(x)\right)$. Figure 8 illustrates the criteria-fusion layer in its lower part.

After obtaining the condition of each attribute and its related trustworthiness, the attributes' conditions are merged on the second layer. In this context, it is obvious that when one attribute of the machine indicates that the machine is in a "bad" condition, then the holistic condition of the machine should accommodate this condition up to a certain degree. This degree depends on the trustworthiness of an attribute's observation. That is to say, if the trustworthiness of an attribute is high, the holistic condition accounts for this condition to a high degree and vice versa. Therefore, the fusion method on the second layer, denoted as *attribute-fusion layer*, should pay attention to the trustworthiness of an attribute's observation and should model a high degree of pessimism. Once again, the class of IIWOWA operators qualifies for the fusion on the second layer. The attribute-fusion layer method is defined as follows:

$$x_{A_1,...,A_n} = \lambda_{\text{IIWOWA}}\left(\vec{t}, \vec{w}, \vec{\pi}\right), \tag{96}$$

where

$$\vec{\pi} = \left(\pi^{1,2,...,m_1}_{A_1}(x), ..., \pi^{1,2,...,m_n}_{A_n}(x)\right),$$

$$\vec{t} = \left(Tr(\pi^{1,2,...,m_1}_{A_1}(x)), ..., Tr(\pi^{1,2,...,m_n}_{A_n}(x))\right),$$

where $\vec{w}$ is determined with reference to the desired andness degree, using O'Hagens approach of maximum entropy (O'Hagen, 1987). The second layer of the TLPFR is depicted in the upper part of Fig. 8.

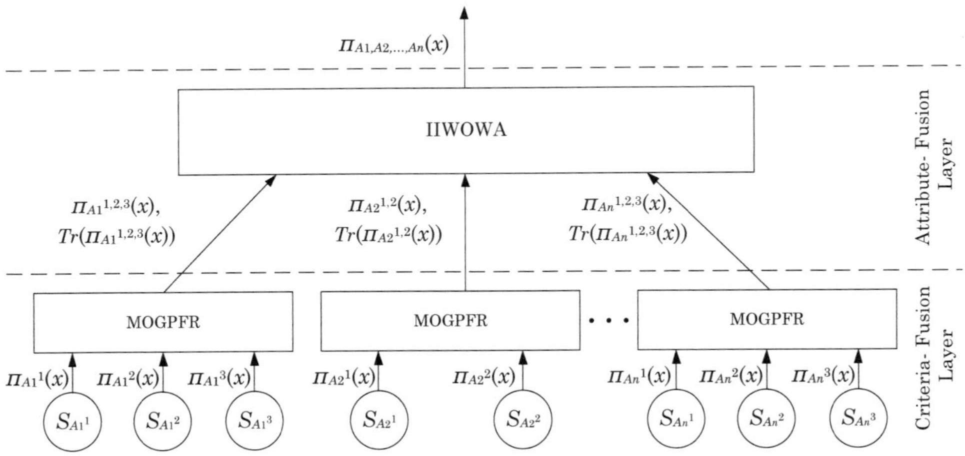

Fig. 8. The TLPFR for machine conditioning.

5. Application example

Nowadays, security prints, such as banknotes are highly sophisticated and exhibit many security features. To reach this high security level, several printing methods like line offset, letter-press printing, foil printing, and intaglio printing are applied. Especially the latter one named printing technique—intaglio printing—is of major importance for the security of banknotes. The term "intaglio" is of Italian origin and means "to engrave". The printing method of the same name uses a metal plate with engraved characters and structures. Large areas and differences in brightness are produced by different hatchings. During the printing process the engraved structures are filled with ink and pressed under huge pressure directly onto the paper. As a result, a tactile relief and fine lines are formed, unique to the intaglio printing process and almost impossible to reproduce through commercial printing methods.

5.1 Experiments

In the steel engraving machine the wiping unit is the most observed part. It is responsible for removing surplus ink around the engravings. Even small parameter manipulations cause errors on products (cf. Fig. 9). Experienced machine operators are able recognise errors before they occur and stabilise the production by changing mainly the wiping unit parameters. In

Fig. 9. Wiping error (right) in an intaglio printing process (Glock et al., 2011).

total, six sensors are mounted on the intaglio printing machine. The main focus of attention is on the wiping process of the printing machine, because here a large number of printing flaws can be caused (Lohweg & Mönks, 2010a). The following functional components are monitored:

- The motor current (MC) of the machine's main drive. It is mainly influenced by the friction between the wiping and the plate cylinder.
- The printing pressure of side 1[1] (PPS1) and the printing pressure of side 2 (PPS2).
- The pressure of the drying blade onto the wiping cylinder of side 1 (DBPS1) and side 2 (DBPS2).
- The flow rate of the wiping solution (WSFR), that is, the amount of wiping solution that is sprayed onto the wiping cylinder.

An experiment was performed by measuring all six sensor signals when the intaglio printing machine was running (Dyck et al., 2007). The test was performed in two steps: i) the machine was operated in "good" production mode; ii) the wiping pressure was decreased little-by-little until the machine printed only flawed sheets. For both experiments, a sample frequency of 7 kHz was used and the production rate kept constant at 6500 sheets per hour. During the first and the second experiment, 479 sheets (within approximately 4.4 minutes) and 796 sheets (within approximately 7.4 minutes) were printed (Dyck et al., 2007).
Sensors with equal causal relationships were arranged to the same attribute. The causality relationships were determined using expertise of the printing machine manufacturer. In this context, a specific machine attribute is related to one particular malfunction, which may cause printing errors and may give rise to machine faults. In this experiment, three machine attributes were monitored. They are stated in the following:

A1. *Drying Blade Pressure*
A misalignment of the drying blade pressure, that is, the pressure between the wiping cylinder and the drying blade, may cause so-called *wiping errors*. In particular, an excessive pressure may ruin the drying blade, whereas an undersized pressure may directly cause flaw print errors. The drying blade pressure is proportional to the motor current (MC) and of course, can be measured by the DBPS1 and DBPS2 sensors.

A2. *Cleanness of the Wiping Cylinder*
A dirty wiping cylinder may provoke another type of wiping error. Naturally, the cleaning power of the wiping unit is influenced by the amount of detergent (wiping solution) that is sprayed onto the wiping cylinder. This flow of detergent is monitored by the WSFR sensor. Moreover, dried ink that adheres onto the wiping cylinder rubs against the drying blade and increases the friction between the wiping cylinder and the drying blade. This leads to an increase of the drying blade pressure that can be detected using the DBPS1 and DBPS2 sensors.

A3. *Printing Pressure*
A misalignment of the printing pressure may bring out imperfect printed structures. Naturally, the printing pressure is monitored by the PPS1 and PPS2 sensors.

5.2 Performance evaluation of the Two-Layer Possibilistic Fusion Rule

With reference to Fig. 10, it is clearly observable that the holistic machine condition decreases gradually because the wiping pressure is decreased little-by-little. With reference to the optical inspection system (the green line in Fig. 10) from sheet number 560 onward, the printing machine produced only low-quality products, that is, wiping errors occurred. Consider a threshold of 50 % (0.5) of $\max(\pi(x)) = 1$; the TLPFR diagnosed this "bad" machine condition

[1] In the field of printing machines, the side where the machine's main drive is located is called "side 1" and the side where the machine's control centre is located is called "side 2".

approximately 160 sheets, or 1.47 minutes, prior to the occurrence of the wiping errors. The "bad" machine condition was detected by the machine's attributes A_1 and A_2 (wiping pressure and cleanness of the wiping cylinder), whilst attribute A_3 indicated a "good" machine condition. This behaviour of the machine's attributes is intuitive, because the wiping pressure was decreased little-by-little, whilst the printing pressure was kept constant.

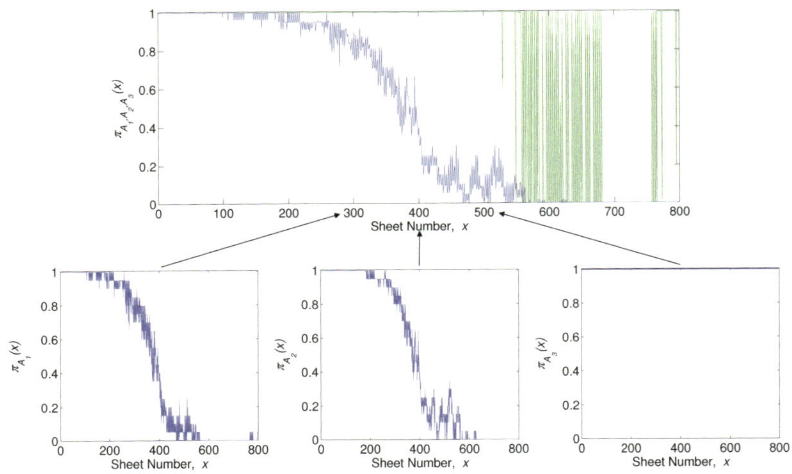

Fig. 10. The fusion results of the wiping error experiment. In the plot above, the blue line illustrates the holistic machine condition determined by the TLPFR; and the green line illustrates the membership grades of the optical inspection system. The plots underneath depict the conditions of the attributes.

The fuzzyfied values of all the sensors obtained during the wiping error experiment are depicted in Fig. 11. With reference to Fig. 11 (a), all sensors that describe attribute A_1

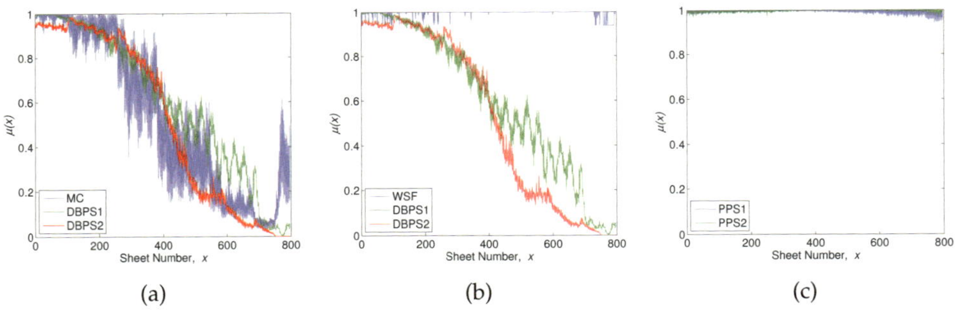

Fig. 11. The membership grades of the sensors during the wiping error experiment. Panel (a) illustrates the fuzzified sensor outputs of attribute A_1, (b) – of attribute A_2, and (c) – of attribute A_3.

detected the wiping error. Furthermore, the motor current was highly afflicted with additive noise, which caused slight conflicts. This arose with a decrease of the trustworthiness of A_1 (cf. Fig. 12 (a)) and in a discounted reliability of the motor current sensor (cf. Fig. 12 (d)).

The conflicts were mainly solved at sheet number 560, that is, when the printing machine was producing only erroneous sheets. Therefore, the reliability of the MC sensor was upgraded. Considering Fig. 11 (b), the WSFR sensor did not detect the decrease of the wiping pressure, whereas the DBPS1 and the DBPS2 sensors did detect. As a result, the reliability of the WSFR sensor was decreased, until it was considered totally unreliable from sheet number 600 onwards. The trustworthiness of A_2 was lowered because of the conflict between the WSFR sensor and both DBPS sensors. From sheet number 600 onwards, that is, when the reliability of the WSFR sensor accounted to zero, the trustworthiness of A_2 increased steadily until it accounted to one, because the observation of the WSFR sensor was no longer considered during the fusion. As mentioned above, the third attribute did not detect the loss of the wiping pressure. Thus, the membership grades of both sensors remained constant and accounted to one (cf. Fig. 11 (c)). Therefore, no conflicts occurred, and the reliability of both sensors, as well as the trustworthiness of the attribute accounted constantly to one (cf. Fig. 12 (c) and (f)). The decrease of the wiping pressure was clearly indicated in the holistic machine condition, because of the high degree of pessimism ($\alpha = 0.8$) for the fusion of the attributes, although it was not detected by A_3.

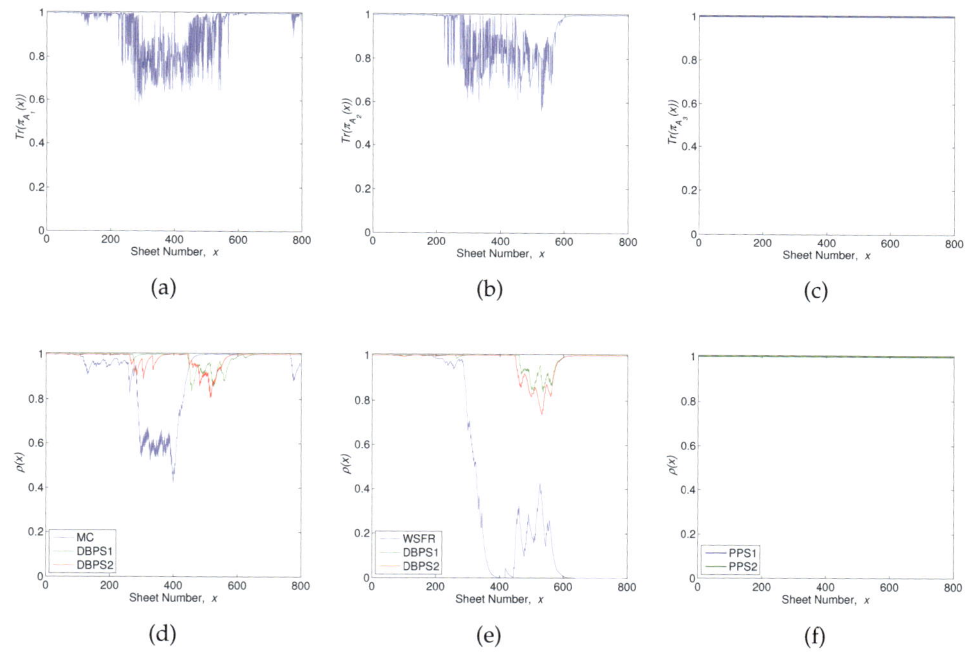

Fig. 12. The trustworthiness of the attributes conditions and the reliabilities of the sensors during the wiping error experiment. Panel (a) illustrates the trustworthiness of attribute A_1, (b) – of A_2, and (c) – of A_3. Furthermore, panel (d) illustrates the reliabilities of the sensors forming attribute A_1, (e) – of A_2, and (f) – of A_3.

Considering the discussions in this section, one can conclude that the TLPFR is insofar superior to an optical inspection system in respect of print flaws generated by machine errors. This is mainly because faulty machine conditions can be detected prior to the occurrence of printing errors. This makes it possible to readjust the printing machine to avoid print flaws.

6. Conclusion and outlook

In this chapter we have reviewed the nature of knowledge perceived by experts and the representation of incomplete, inconsistent, and vague information in technical systems. Based on various theories, namely: Probability Theory, Evidence Theory, Fuzzy Set Theory, and Possibility Theory, different concept of information fusion in the appearance of incomplete and unreliable data was presented. Furthermore, two human-centric fusion approaches were characterised. In particular, a new possibilistic framework for sensor fusion was presented. In the future we will concentrate on implementation aspects of the above mentioned framework.

7. References

Ayyub, B. M. & Klir, G. J. (2006). *Uncertainty Modeling and Analysis in Engineering and the Sciences*, Chapman and Hall/CRC, Taylor and Francis Group.

Bayes, T. P. (1763). An Essay Towards Solving a Problem in the Doctrine of Chances, *Philosophical Transactions of the Royal Society of London, London* 53: 370–418.

Black, M. (1937). Vagueness. an exercise in logical analysis, *Philosophy of Science* 4(4): 427–455.

Bocklisch, S. F. & Priber, U. (1986). A parametric fuzzy classification concept, *Proc. International Workshop on Fuzzy Sets Applications*, Akademie-Verlag,Germany: 147–156.

Bossé, E., Roy, J. & Wark, S. (2007). *Concepts, Models, and Tools for Information Fusion*, Intelligence and Information Operations Library, Artech House, Boston, Mass. URL: `http://www.gbv.de/dms/bowker/toc/9781596930810.pdf`

Campos, F. (2006). *Decision Making in Uncertain Situations: An Extension to the Mathematical Theory of Evidence*, Boca Raton, Florida USA.

Carnap, R. (1947). Probability as a guide in life, *Journal of Philosophy* 44(6): 141–148.

Delmotte, F. (2007). Detection of Defective Sources in the Setting of Possibility Theory, *Fuzzy Sets and Systems* 5(158): 555–571.

Dempster, A. P. (1967). Upper and Lower Probabilities Induced by a Multivalued Mapping, *Annals Mathematics Statistics* 28: 335–339.

Dubois, D., Esteva, F., Godo, L. & Prade, H. (2001). An Information-based Discussion of Vagueness, *The 10th IEEE International Conference on Fuzzy Systems, Toulouse* 2(3): 781–0784.

Dubois, D., Foulloy, L., Mauris, G. & Prade, H. (2004). Probability-Possibility Transformations, Triangular Fuzzy Sets, and Probabilistic Inequalities, *Reliable Computing* 10(5): 273–297.

Dubois, D. & Prade, H. (1978). Operations on Fuzzy Numbers, *International Journal of Systems Science* 9(6).

Dubois, D. & Prade, H. (1983). Unfair Coins and Necessity Measures Towards a Possibilistic Interpretation of Histograms, *Fuzzy Sets and Systems* 1(10): 15–20.

Dubois, D. & Prade, H. (1994). Possibility Theory and Data Fusion in Poorly Informed Environments, *Control Engineering Practice* 5(2): 811–823.

Dubois, D. & Prade, H. (2000). Possibility Theory in Information Fusion, *Proceedings of the Third International Conference in Information Fusion, Paris* 3(1): 6–19.

Dubois, D. & Prade, H. (2003). Possibility Theory and its Applications: A Retrospective and Prospective View, *The 12th IEEE International Conference on Fuzzy Systems, San Antonio* 5(1): 5–11.

Dujmović, J. J. (1974). Weighted Conjunctive and Disjunctive Means and their Application in Systems Evaluation, *of the University of Belgrade, EE Dept., Series Mathematics and Physics, Belgrade* 483: 147–158.

Dujmović, J. J. & Larsen, H. L. (2007). Generalized Conjunction/Disjunction, *International Journal of Approximate Reasoning* 46(2): 423–446.

Dyck, W., Türke, T. & Lohweg, V. (2007). A Fuzzy-Pattern-Classifier-Based Adaptive Learning Model for Sensor Fusion, *2007 IEEE Workshop on Machine Learing for Signal Processing, Thessaloniki* pp. 282–287.

Eisenführ, F. (2003). *Rationales Entscheiden*, 3 edn, Springer, Berlin.

Glock, S., Voth, K.,Schaede, J. G. & Lohweg, V. (2011). A Framework for Possibilistic Multi-source Data Fusion with Monitoring of Sensor Reliability, *World Conference on Soft Computing. In: San Francisco, CA, USA, May 23-26, 2011 May 2011.*

Hacking, I. (1975). *The Emergence of Probability*, Cambridge University Press, Cambridge.

Hall, D. L. & Llinas, J. (2001). *Handbook of a Multisensor Data Fusion*, The electrical engineering and applied signal processing series, CRC Press, Boca Raton, Fla.
URL: `http://www.gbv.de/dms/hbz/toc/ht013184121.pdf`

Keynes, J. M. (1921). *A Treatise on Probability*, Dover Publications, London.

Klir, G. J. & Wierman, M. J. (1998). *Uncertainty-Based Information: Elements of Generalized Information Theory (Studies in Fuzziness and Soft Computing)*, Physica-Verlag.

Klir, G. J. & Yuan, Y. (1995). *Fuzzy Set and Fuzzy Logic: Theory and Applications*, Prentice-Hall Inc., New York.

Kolmogorov, A. (1933). *Grundbegriffe der Wahrscheinlichkeitsrechnung*, 1 edn, Julius Springer, Berlin.

Larsen, H. L. (1999). Importance Weighted OWA Aggregation of Multicriteria Queries, *18th International Conference of the North American Fuzzy Information Processing Society, New York* pp. 740–744.

Larsen, H. L. (2002). Efficient Importance Weighted Aggregation Between Min and Max, *9th Conference on Information Processing an Management of Uncertainty in Knowledgebased Systems, Annecy* 1(1): 1203–1208.

Li, R. & Lohweg, V. (2008). A Novel Data Fusion Approach using Two-Layer Conflict Solving, *IAPR Workshop on Cognitive Information Processing, Santorini* pp. 132–136.

Lipshitz, R., Klein, G., Orasanu, J. & Salas, E. (2001). Taking Stock of Naturalistic Decision Making, *Journal of Behavioral Decision Making* 14: 331–352.

Lohweg, V. & Mönks, U. (2010a). *Fuzzy-Pattern-Classifier based Sensor Fusion for Machine Conditioning*, Sensor Fusion, I-TECH Education and Publishing, Vienna.

Lohweg, V. & Mönks, U. (2010b). Sensor Fusion by Two-Layer Conflict Solving, *The 2nd International Workshop on Cognitive Information Processing, Elba* 2(1): 65–79.

Murphy, C. K. (2000). Combining Belief Functions when Evidence Conflicts., *Decision Support Systems* 29(1): 1–9.

O'Hagen, M. (1987). Fuzzy Decision Aids, *Twenty-first Asiomar Conference on Signal, Systems and Computers, , Pacific Grove* 21(2): 624–628.

Popper, K. R. (1934). *The Logic of Scientific Discovery*, Routledge; Reprint edition (October 8, 1992), London.

Ramsey, F. P. (1960). *The Foundations of Mathematics and Other Logical Essays*, Paterson, N.J.,Littlefield, Adams.

Rogova, G. L. & Nimier, V. (2004). Reliability in Information Fusion: Literature Survey, *Proceedings of the Seventh International Conference on Information Fusion, Stockholm* pp. 1158–1165.

Salicone, S. (2007). *Measurement Uncertainty - An Approach via the Mathematical Theory of Evidence*, Springer, New York.

Schubert, J. (2007). An information fusion demonstrator for tactical intelligence processing in network-based defense., *Journal Information Fusion* 8(1).

Shafer, G. (1976). *A Mathematical Theory of Evidence*, 1 edn, Princeton University Press, Princeton.

Smarandache, F. & Dezert, J. (2006). *Advances and Applications on DSmT for Information Fusion: Collected Works 1 (2)*, American Research Press, Rehboth USA.

Sunita, S. A. (1999). Order Effects and Memory for Evidence in Individual versus Group Decision Making in Auditing, *Journal of Behavioral Decision Making* 12: 71–88.

Yager, R. R. (1979). Possibilistic Decisionmaking, *IEEE Transactions on Systems, Man and Cybernetics* 9(2): 388–392.

Yager, R. R. (1987). On the Dempster-Shafer Framework and New Combination Rules, *Information Sciences* 41(2): 93–137.

Yager, R. R. (1988). On Ordered Weighted Averaging Aggregation Operators in Multi-criteria Decision Making, *IEEE Transactions on Systems, Man and Cybernetics* 18: 183–190.

Yager, R. R. (1994). Aggregation Operators and Fuzzy Systems Modeling, *Fuzzy Sets and Systems* 67(2): 129–145.

Yager, R. R. (2004). A Framework for Multi-source Data Fusion, *Information Sciences* 163(1-3): 175–200.

Yager, R. R. & Kelman, A. (1996). Fusion of Fuzzy Information With Considerations for Compatibility, Partial Aggregation, and Reinforcement, *International Journal of Approximate Reasoning* 15(2): 93–122.

Yager, R. R. & Liu, L. (2008). *Classic Works of the Dempster-Shafer Theory of Belief Functions*, Springer, New York.

Zadeh, L. A. (1965). Fuzzy Sets, *Information and Control* 8.

Zadeh, L. A. (1978). Fuzzy Sets as a Basis for a Theory of Possibility, *Fuzzy Sets and Systems* 1: 3–28.

Zadeh, L. A. (1986). A simple view of the dempster-shafer theory of evidence and its implication for the rule of combination., *AI Magazine* 7(2): 85–90.

Zadeh, L. A. (2008). Toward Human Level Machine Intelligence - Is It Achievable? The Need for a Paradigm Shift, *IEEE Computational Intelligence Magazine* 3(3): 11–22.